Serological Methods

for Detection and Identification of Viral and Bacterial Plant Pathogens

---

A Laboratory Manual

R. Hampton, E. Ball, S. De Boer, editors

APS PRESS
The American Phytopathological Society
St. Paul, Minnesota USA

Front cover: left, transmission electron photomicrograph of two immunoglobulin G (IgG) molecules from anti-SMV hybridoma, S1, interacting with a virion of soybean mosaic potyvirus. Specimen mounted on 600-mesh grid without membrane support, stained with sodium silica tungstate, pH 8. Right, computer-enhanced image of the two molecules shown in the photomicrograph. (Courtesy R. L. Mernaugh, Southwest Foundation for Biomedical Research, San Antonio, TX)

Back cover: Diagrammatic structure of immunoglobulin G (IgG), R. Kennedy version, showing isotype and paratope regions of the molecule. (From "The Immune Response: Antigens, Antibodies, Antigen-Antibody Interactions" by R. L. Mernaugh, G. R. Mernaugh, and G. R. Kovacs; see page 8)

This book has been reproduced directly from computer-generated copy submitted in final form to APS Press by the editors of the volume. No editing or proofreading has been done by the Press.

Library of Congress Catalog Card Number: 90-83302
International Standard Book Number: 0-89054-115-9

Printed in the United States of America

The American Phytopathological Society
3340 Pilot Knob Road
St. Paul, Minnesota, 55121, USA

TABLE OF CONTENTS

SEROLOGICAL METHODS FOR DETECTION AND IDENTIFICATION OF VIRAL AND BACTERIAL PLANT PATHOGENS
A Laboratory Manual

Editors: R. O. Hampton, E. M. Ball, and S. H. De Boer

Table of Contents, Cont'd.

Table of Contents, cont'd.

ACKNOWLEDGMENT

In addition to sincere thanks extended for invaluable advice and support provided by Drs. E. Ball, S. DeBoer, and J. Hill throughout the development of this book, special appreciation is also expressed to:

Drs. J. B. Jones and M. R. McLaughlin for the excellence of their suggestions as external reviewers of the book draft;

Each of the thirty-four authors for prompt submission of well-prepared book sections, and for many enjoyable interactions on technicalities and innovations of serological approaches and methods;

Luree Hughes, secretary, language specialist, and friend for untiring devotion to the task of typing and re-composing the sections of this book, several times over;

and to Willa Mae Hampton, my wife, for meticulous proof reading of the original and numerously revised versions of all sections of the book, and for moral support during countless hours of seemingly endless endeavor.

R.O.H.

Serological Methods
for Detection and Identification of Viral and Bacterial Plant Pathogens

A Laboratory Manual

SEROLOGICAL METHODS FOR DETECTION AND IDENTIFICATION OF VIRAL AND BACTERIAL PLANT PATHOGENS

A Laboratory Manual

Introduction

Statement of Purpose

This manual presents methods for preparing and treating antigens, polyclonal antisera, and monoclonal antibodies, as well as state-of-the-art serological principles, methods, step-by-step procedures, and applications for detection and differentiation of viral and bacterial plant pathogens. The manual is formatted for instructing students, apprising and challenging established researchers, and rapid review/reference for all. Although we have emphasized materials and procedures, we have endeavored toward informational self-sufficiency, to minimize user dependence on other supportive literature for efficacious applications of the methods.

Some of the methods presented have been in use for several decades; others represent a recent evolution of diverse and versatile serological assays and applications. Although detection and identification are emphasized in the manual title, sections of the manual address other important applications: serotypic and taxonomic relationships, epitope and paratope characterization, analyses of in vitro nucleic acid translation products, and analyses of in vivo constituents related to pathogenesis. Traditional methods (e.g., microprecipitation, microagglutination, and agar diffusion assays) are still used for titration of serum antibodies. We urge users of the manual to equally consider the specific usefulness of both traditional and newer methods (e.g., not all antigenic determinants are equally detectable by each method). Users of the manual must be responsible for: (a) assuring correct use and standardization of methods and specific applications, (b) assuring appropriate test environments and controls, and (c) critically testing the adaptability, effectiveness, and precision of each method used.

Status report, 1974 (prior APS publication on serology) to 1990

Treatment of both viral and bacterial plant pathogens, though extremely different antigenic entities, was conceivable because serological progress with both pathogens, since 1974, is attributable to the same technological advancements. For both, monoclonal antibody technology has provided unparalleled epitope specificity, with resultant near-zero reactions to non-pathogen-related antigens. Antigen-antibody reaction-enhancement methods have provided 1,000-fold increases in the number of assays possible per ml of polyclonal antiserum. Concurrently, reaction speed and accuracy in detection and diagnosis have been greatly improved, and computerized procedures for automated quantitation have now emerged.

Rapid advances have occurred during the 1980's, but in publishing this manual, we anticipate continued adaptations, innovations, and enhancements during the 1990's and beyond. Polyclonal antibodies have marvelously provided "order out of chaos" in the 20th century and will continue to fulfill many needs far into the future. However, monoclonal antibodies (McAbs) are now being selected to react either to generically shared or type-specific epitopes; to define, enumerate, and systematize epitopes among pathogen isolates, or epitopes within pathogen groups, as well as to characterize the antibodies produced against those epitopes.

Current views of serology

A skeptical view of serology as an instrument of "pathogen systematics" in past decades probably would have been justified. For instance, the combination of serology and electron microscopy (virion morphology) conceivably might have been less elucidating of viral relationships in times of otherwise limited virological methods. We are only now beginning to understand the significance of domain conservation in viral genomes, and thus to develop explanations of 'coat protein gene' conservation relative to virion infectivity and structural integrity and, in turn, relative to taxonomic orderliness among serologically related viruses.

Bacterial taxonomy has developed independently of serological advancements. However, for both viruses and bacteria, it has become possible to define specific epitopes (e.g., N terminal amino acid sequences of virion coat proteins; surface epitopes of bacteria) in relation to pathogenicity, pathotypic variants, or pathogen characteristics. We believe that future applications of serology can provide increased understanding of all strategic pathogen nucleic acid translation products, and thereby help define structure and function of both nucleic acids and the proteins assembled as nucleotide translation products. Likewise, we expect yet greater advancements in antigen-detection sensitivity, accuracy, speed, and automation.

Incidentally, dates of literature citations in most manual sections of this manual were current to 1988-89. Ironically, authors exercising greatest facility and promptness in completing sections also had less opportunity to incorporate very recent citations.

Serological terminology

It has been our objective to communicate in a standardized terminology. However, in a few cases, contributing authors were granted strong preferences in terminology (e.g., agar double diffusion vs. Ouchterlony double diffusion; dot immunobinding assays vs. dot-blot assays; immuno-electron microscopy vs. serologically specific electron microscopy). We especially endeavored to standardize the Glossary of Terms and associated definitions, notwithstanding subtle slanting toward plant pathogens. Accordingly, we encourage users to master the vocabulary and concepts represented in both Section I and in the glossary.

Sources of supply

Sources of supply and brands of materials indicated in each section of the manual are those familiar to and proven satisfactory by contributing authors. Such names are provided for the convenience of manual users, but do not infer that other sources or brands may not be equally satisfactory for prescribed uses. We are acutely aware and regretful of the short life of listed sources of supply, due to current rapid changes of business names, affiliations, mergers, acquisitions, and addresses.

It has been our pleasure to publish a manual that appears to be the best of its kind in the world. Our motivation is to facilitate research excellence, and it is to this purpose that we gladly dedicate the manual and all efforts required for its coordination, editing, codification, and final assembly.

R. O. H.

Section I.

THE IMMUNE RESPONSE: ANTIGENS, ANTIBODIES, ANTIGEN-ANTIBODY INTERACTIONS

R. L. Mernaugh, G. R. Mernaugh, and G. R. Kovacs
Texas A & M University
College Station, TX 77843

When plant pathologists become involved in immunology, the goal, generally, is to generate an antibody probe which will specifically identify a target antigen in an assay. This goal should be facilitated by an understanding of the basics of immunology.

Within the last two decades, advances in the field of immunology have provided insight into the mechanisms by which the immune system operates to produce antigen-specific antibodies. However, the concepts and technical jargon accompanying these advances may remain obscure to the plant pathology researcher. The purpose of this article is to acquaint plant pathologists with some basic immunology relative to the immune response, antigens, antibodies and antigen-antibody interactions.

Immune Response

An **antigen** is anything that elicits an **immune response** in an animal. This response may be either **humoral** leading to antibody formation, or it may be **cellular**. In some cases, an antigen may be degraded during a cellular response in which case antibodies would not be formed (12,36).

When an animal has been exposed to an antigen, **antigen-presenting cells (APCs), T cells** and **B cells** act in concert to stimulate the production of antigen-specific antibodies (12,25). Unless these cells interact successfully the B cells fail to produce antigen-specific antibodies.

When antigen is introduced into the body, APCs, such as macrophages or B cells, interact with it, ingest it, and break down or process it (Fig. 1A). The processed antigen is then returned to the surface of the APC (Fig. 1B) where it is presented by the cell's **major histocompatibility molecules (MHC)** (24,25). Subsequently, a T cell with a receptor specific for both the processed antigen and the MHC molecule (Fig. 1B) on the APC interacts with the APC (Fig. 1C) (9). A trimolecular complex is formed between the T cell antigen receptor, the processed antigen and the MHC molecule on the APC (10). This complex triggers the APCs and T cells to produce a variety of soluble factors such as **interleukin** 1 (IL-1) (Fig. 1D) and IL-2 (Fig. 1E) which stimulate T cell proliferation and differentiation. Properly activated T cells bind to antigen presenting B cells (Fig. 2). These T cells then produce IL-4 and IL-5 which stimulates B cell multiplication and differentiation (Fig. 2D). If properly stimulated the B cell's immunoglobulin genome, which is responsible for the antibody's makeup, is spliced to yield the genetic information used to direct antibody formation. When an animal's immune system is further stimulated with the same antigen, that part of the immunoglobulin genome which is responsible for the antigen-binding region of an antibody undergoes a series of **somatic mutations**. These mutations lead to the production of antibodies with higher or lower affinity for the antigen.

Those B cells producing antibodies with higher antigen affinity will attract more antigen and will be preferentially stimulated to grow, divide and produce more of the high affinity antibody. T cell secreted IL-5 along with other signals, allow B cells to differentiate into either short-lived (3-4 days) plasma cells which secrete large amounts of antibody or long-lived memory cells (Fig. 2D). Memory B cells are primed to respond to future antigen stimulation and do not secrete but rather express antibody on their surface (12).

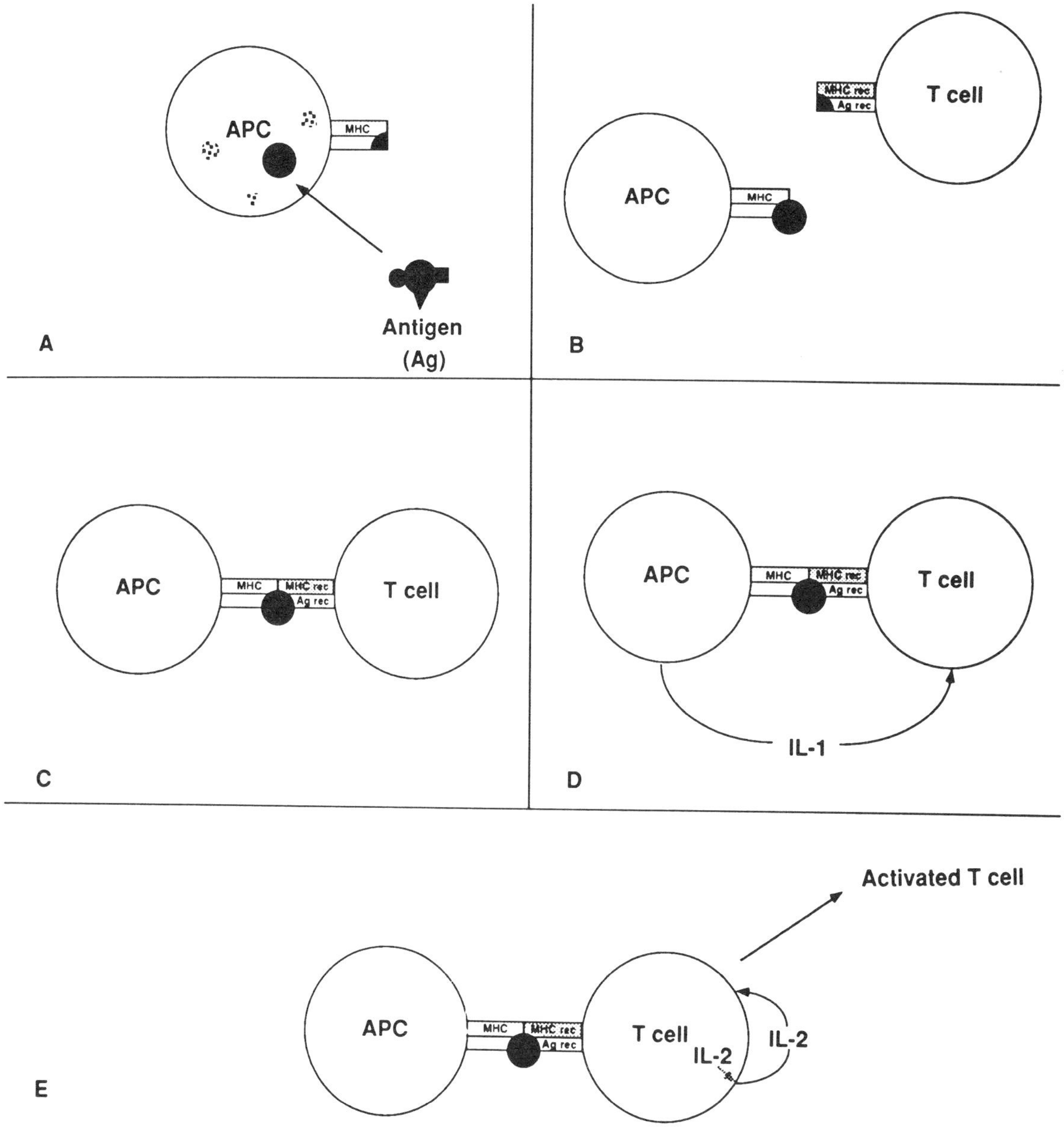

Fig. 1. Interaction between an antigen (Ag), an antigen presenting cell (APC) and a T cell. Fig. 1A. An antigen interacts with an APC and is taken up and degraded. Fig. 1B. The antigen is presented by the APC's MHC molecule. Fig. 1C. The T cell with a receptor (MHC rec) for the antigen and the APC's MHC molecule interacts with the APC and the antigen. Fig. 1D. Interleukin-1 (IL-1) produced by an APC stimulates the T cell which has interacted with the antigen and the APC's MHC molecule. Fig. 1E. IL-1 along with the antigen stimulate T cell IL-2 production which, in turn, activates the T cell.

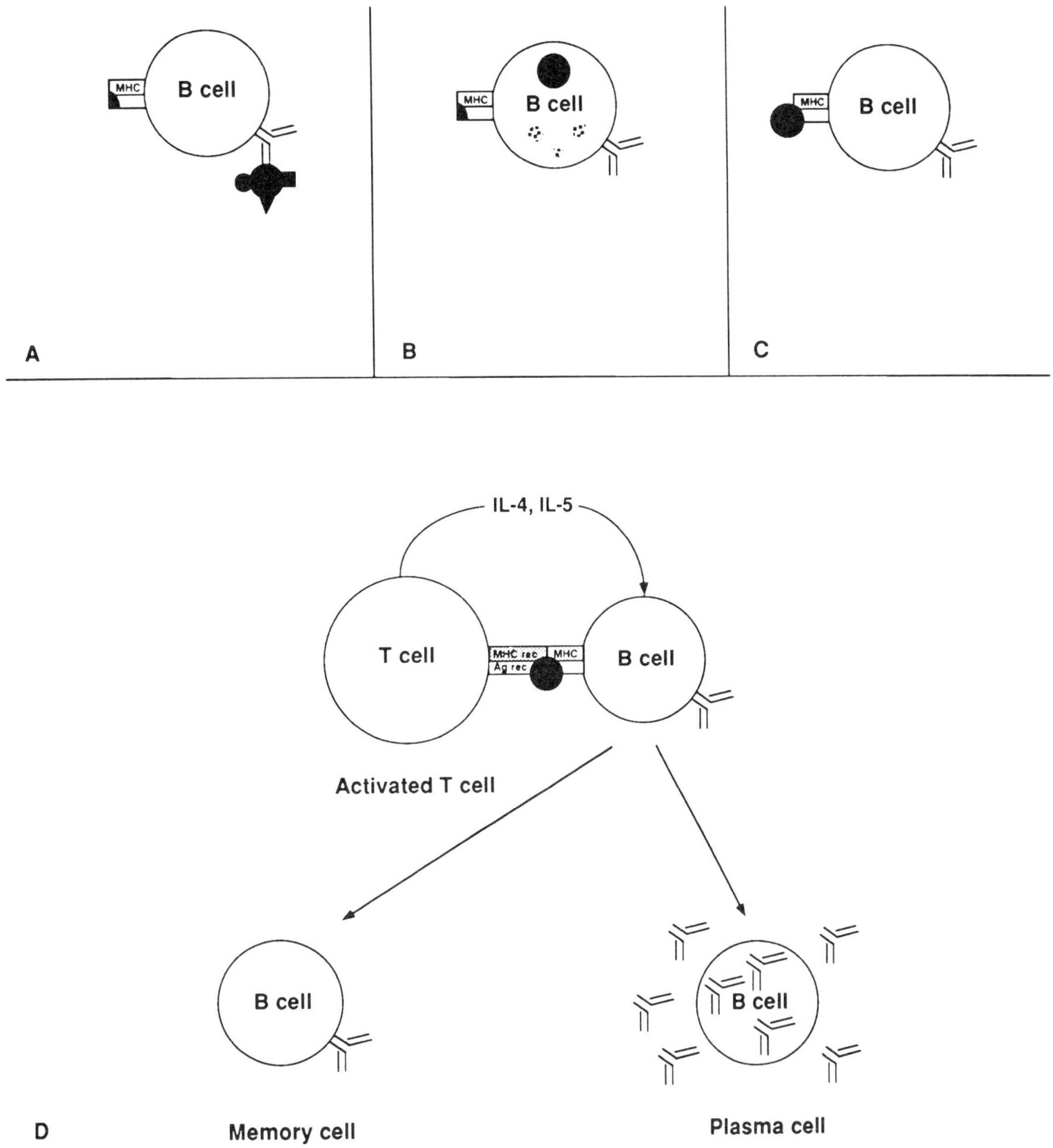

Fig. 2. Interaction between an antigen, a B cell and an activated T cell. Fig. 2A. A B cell with a surface antibody specific for the antigen, interacts with the antigen. Fig. 2B. The antigen is taken up and processed by the B cell. Fig. 2C. The antigen is presented by the B cell's MHC molecule. Fig. 2D. The activated T cell interacts with the antigen and the B cells MHC molecule and the T cell produces IL-4 and IL-5 which, in turn, stimulates B cells to differentiate into memory and plasma cells.

Antigen

A substance is antigenic when it elicits either a cellular or humoral immune response in an animal. If an animal's B cells do not express antibodies on their surface capable of interacting with a substance, then antibodies to that substance will not be made. If a substance is rapidly and sufficiently degraded before it can be processed or presented by APCs, it will not be antigenic. An unstable/degradable substance can be rendered antigenic by chemical fixation (formaldehyde) or when presented by slow release by mixing with adjuvant e.g. Freund's or alum (12).

A low-molecular-weight substance, such as a synthetic peptide (10-15 amino acids), may be efficiently processed and presented by an APC's MHC molecule. However, it may be too small to be recognized simultaneously by both the APC's MHC molecule and a receptor on a T cell. A small polypeptide can be made more antigenic by chemically fixing it to a larger molecule (e.g. keyhole limpet hemocyanin) exhibiting a ligand capable of interacting with a T cell receptor. Some researchers also add T cell recognizing synthetic peptides to the synthetic peptide to stimulate the antibody response in animals (12).

In some cases, a large molecule may fail to elicit antibody formation in an animal because the animal's MHC molecules do not present or recognize the molecule. Since MHC molecules are controlled by the genetic makeup of the animal, desired antibodies can be produced by switching to different animal strains or species possessing suitable antigen presenting MHC molecules (12).

The route of immunization can influence the final outcome of an immune response to a foreign substance. Intradermal, intrasplenic or intravenous injections present a potential antigen to APC's associated with the Langerhan's cells of the skin, dendritic cells of the spleen and monocytes of the blood, respectively. The same foreign substance presented to other areas of the body which may be free of APC's may not always elicit an antibody response (12).

A substance may exhibit one or more antigenic sites (**epitopes**) which are capable of eliciting and interacting with the antibodies produced by an animal. Generally, the smallest peptide that demonstrates antigenic reactivity represents the epitope and is about 5-7 amino acids in size (36).

Antibodies to synthetic peptides which represent fragments of native molecules have been used to describe and define various types of epitopes available to antibodies (36).

Antibodies capable of interacting with random coil forms of a peptide define a **sequential epitope**. Antibodies which interact with a peptide exhibiting a specific conformation define a **conformational epitope**. Antibodies to a conformational epitope will not react with the same peptide if it is unfolded.

An expanse of amino acid residues constituting two or more groups of amino acids linked to one another are called **continuous** or **contiguous epitopes**. Antibodies to an intact molecule which also interact with a fragment of that molecule (or corresponding synthetic peptide) define a continuous epitope.

Antibodies that recognize a group of amino acids consisting of two joining, separate peptide chains or a folded single peptide chain define **discontinuous** or **discontiguous** epitopes. Continuous or discontinuous epitopes, in essence, are conformational epitopes.

Antibodies which interact with a buried epitope, such as a denatured molecule obtained from a polyacrylamide gel band, but which do not interact with the corresponding native molecule recognize epitopes known as **cryptotopes**.

A **neotope** (new epitope) is an antigenic site on a monomeric subunit created by conformational changes brought about by the interaction of the subunit with another molecule. An antibody which interacts with an epitope on a substrate only when the substrate is bound to an enzyme would define a neotope.

Generally, the degree of antigenicity of a molecule subunit is a reflection of its location or prominence on the molecule or its physical or chemical makeup. Hydrophilic amino acid residues which tend to correspond to continuous epitopes usually appear on an antigen's exterior. Hydrophobic residues remain buried and are less accessible to the immune system unless the molecule is denatured (36).

Epitopes that are mobile and capable of flexing 1-2 Å usually interact more strongly with an antibody than rigid molecules. Hydrophilic regions of a molecule that protrude and exhibit segmental mobility correspond to continuous epitopes.

The amino and carboxy terminal ends of a molecule exhibit a high degree of mobility. These termini are usually close to one another, occupy sites on a molecule's surface and typically constitute continuous epitopes.

Sites on a molecule that exhibit little amino acid sequence conservation constitute epitope uniqueness useful for diagnostic distinctions. Apparently, these regions can tolerate changes brought about by DNA mutations since they are probably not involved in interactions responsible for the molecule's folding (36).

Antibody

Antibodies are immunoglobulin molecules produced by an animal's B lymphocytes in response to an antigenic stimulus. Each molecule has two or more antibody-binding sites termed **paratopes** (see Fig. 3 for all references to antibody structure). A paratope is capable of recognizing a single antigenic determinant known as an epitope (14).

Polyclonal antibodies, obtained from the serum of an animal inoculated with an antigen bearing many antigenic sites, consist of a population of antibodies reactive with more than one epitope. A single antibody-producing B lymphocyte, immortalized either by mutation, fusion with an established cell (e.g SP2/0-Ag14)(12) or transformation with a virus (Epstein Barr Virus) (7), will produce antibodies (**monoclonals**) reactive with a single epitope.

There are five **classes** (**IgG, IgM, IgA, IgD** and **IgE**) and eight **subclasses** (**IgG1, IgG2, IgG3, IgG4, IgM1, IgM2, IgA1** and **IgA2**) of antibody molecules. IgG, IgD, IgE and serum IgA antibody molecules (Fig. 3A) each consist of a pair of heavy and light polypeptide chains. The chains are held together by disulfide bonds or, in the case of IgA2, by noncovalent linkages. IgM antibodies (Fig. 3C) consist of five pairs of heavy and light chains while secretory IgA (Fig. 3B) consists of two to three such pairs. Pairs of heavy and light chains in IgM pentamers and IgA dimers are united by a **J** chain through disulfide bonds (Fig. 3B, 3C). In addition, secretory IgA also contains an **Sc** component (Fig. 3B) (3).

Digestion of IgG with papain yields three fragments (Fig. 3E). Two of these fragments, each with a heavy and light chain, are equal in size and are antigen-binding (**Fab** fragments). The third fragment, consisting of antibody heavy chains only, does not bind antigen and crystallizes at 4 C (**Fc** fragment). The region between the Fab and Fc fragments is called the **hinge** region. This region gives the antibody molecule flexibility needed for Fab segments to operate independently of one another. Digestion of IgG with pepsin yields two Fab fragments [$\mathbf{F(ab')_2}$] joined by disulfide bonds and an **Fc'** fragment (Fig. 3F) (3,12).

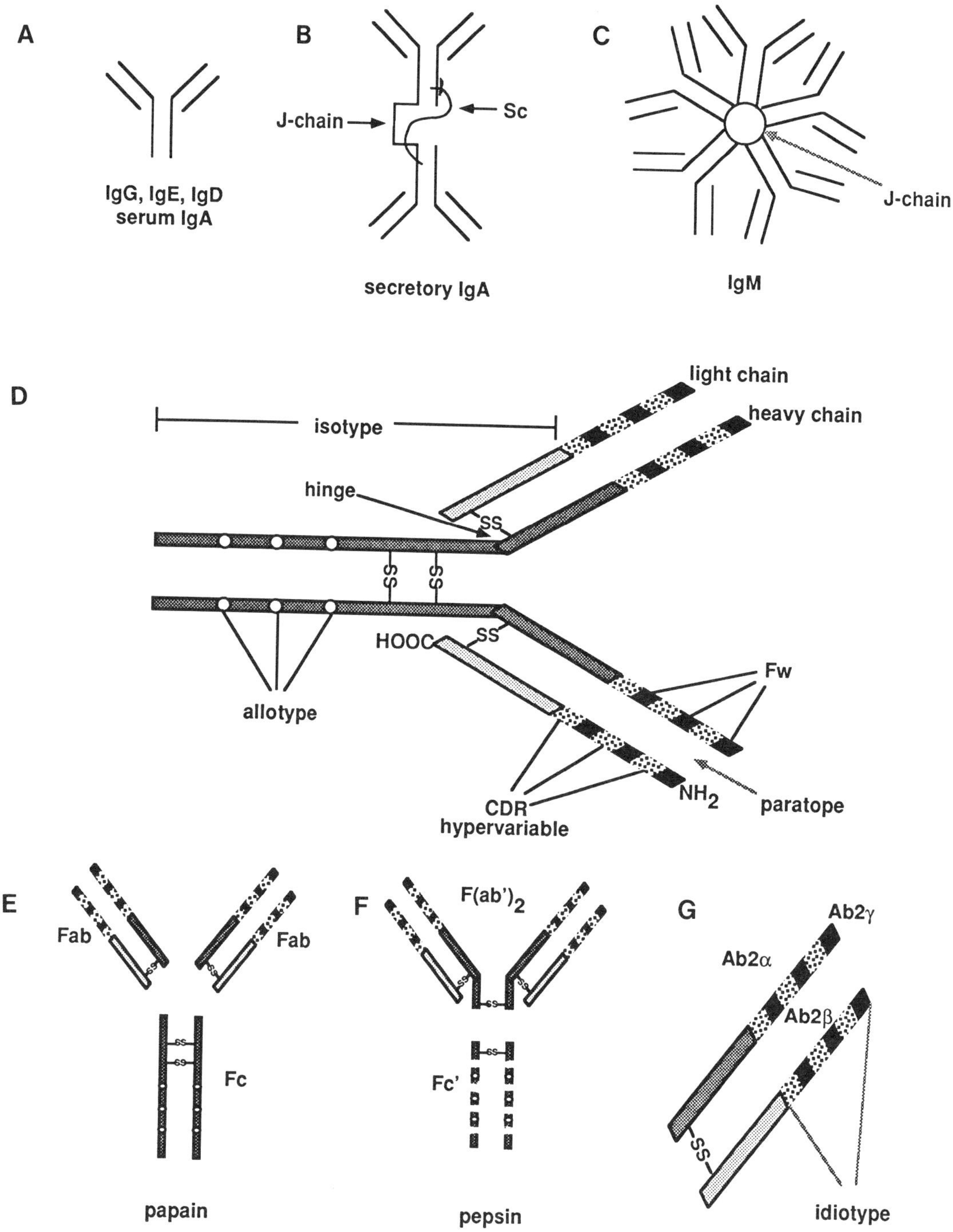

Fig. 3. Features of immunoglobulin molecules. Fig. 3A. Y-shaped configuration of an IgG, IgE, IgD and serum IgA molecule. Fig. 3B. Secretory IgA dimer with J chain and secretory component (Sc). Fig. 3C. IgM pentamer with J chain. Fig. 3D. Monomeric immunoglobulin molecule (based on version by R. Kennedy; see Ref. 20) illustrating location of the allotype, hypervariable or complementary determining region (CDR), framework (Fw), hinge, heavy and light chain, isotype and paratope regions of an antibody molecule and location of disulfide bonds. Fig. 3E. IgG molecule after hydrolysis with papain illustrating the Fab and Fc fragments. Fig. 3F. IgG molecule after hydrolysis with pepsin illustrating the $F(ab')_2$ and Fc′ regions. Fig. 3G. Idiotype portion of an antibody molecule illustrating the locations of idiotopes recognized by Ab2 alpha, beta and gamma.

The amino terminal end of the heavy and light chains of Fab fragments which make up the paratope each contain three **complementary determining regions (CDRs)** (Fig. 3D). The CDRs are oriented or anchored by **framework (Fw)** (Fig. 3D) regions to the remainder of the Fab fragment. Essentially, the six CDRs found within the antibody's paratope, and to some extent the Fw regions, are responsible for binding to or interacting with an antigen. Since the amino acid sequence for the CDRs varies among antibodies to different antigenic sites, they are also referred to as **hypervariable regions** (Fig. 3D). Regions outside of the CDR and Fw regions vary less for a class or subclass of antibody and are referred to as **constant regions** (Fig. 3D). The antibody class and subclass can be immunologically identified with **anti-isotype** (anti-class or anti-subclass) antibodies directed to unique immunoglobulin heavy chain constant regions. Antibodies directed to other antigenically distinct constant regions on an immunoglobulin molecule can also be used to determine an immunoglobulin's light chain (**lambda or kappa**) or allotypic characteristics. **Allotypic** characteristics (Fig. 3D) are genetically inherited antigenic sites. These sites are present on a class of antibodies in some individuals and absent from other individuals of the same species (34).

The first class of antibodies produced by an antigenically stimulated animal during a primary immune response are IgM antibodies. Serum IgM antibody concentration peaks approximately ten days after antigenic stimulation and then drops. If the antigen successfully elicits an appropriate T cell response, IgG antibodies will then be produced. Serum IgG antibody concentration peaks at about fourteen days. If the same animal is antigenically restimulated to induce a secondary immune response, the time frame for the appearance of IgM and IgG antibodies will be similar to that seen during the primary immune response; however, the IgG concentration will generally be much greater during the secondary response (3). It must be remembered that if the MHC molecules of an animal are incompatible with the antigen only IgM antibodies will be produced (12). Since IgA, IgD and IgE antibodies are not as prevalent as IgG or IgM, they are seldom used as diagnostic immunological reagents.

Antibodies directed to antigenically distinct regions associated with an immunoglobulin's paratope or hypervariable region define an immunoglobulin's idiotype. Each epitope within an idiotype is termed an **idiotope** (Fig. 3G) (14). Idiotypes were first described by Kunkel et al. (23) and Oudin and Michel (29), and the antibodies directed against these regions were termed **anti-idiotype** (30). These antibodies do not recognize the antigenically distinct constant regions associated with either an immunoglobulin's isotype or allotype (19). The potential and value of anti-idiotope antibodies in plant pathology are only now being recognized (27). The terms idiotypic and anti-idiotypic, respectively, are adjectives that are used to describe the nature of either Ab 1 which bears an idiotope or Ab 2 which is capable of interacting with an idiotope.

Anti-idiotope antibodies are characterized on the basis of their interactions with idiotopes on other antibodies from the same animal species or from different species; their interactions with idiotype-presenting antibodies in the presence of competing antigens; and their in vivo biological activity.

Antibodies on B cells recognize a wide variety of idiotopes on antibodies from intrastrain, interstrain and interspecies animals (5). Shared idiotopes among antibodies are referred to as either **public idiotopes (IdX)** or **cross-reactive idiotopes (CRIs)** (13,28). Those idiotopes which are not shared by immunoglobulins are termed **private idiotopes (IdI)** (6).

Interactions between an anti-idiotope antibody (Ab 2) and the idiotope on one or more antibodies (Ab 1) are used to characterize an anti-idiotope antibody. The topographical location of the recognized idiotope on the surface of an Ab 1 with respect to its paratope can then be used to further characterize the anti-idiotope antibody. Based on the location of the idiotope recognized by the anti-idiotope antibody, four or possibly five different categories of anti-idiotypic antibodies can be described.

Four of the categories of anti-idiotypic antibodies, which are given Greek letter designation, include **Ab 2 alpha, Ab 2 beta, Ab 2 epsilon**, and **Ab 2 gamma** (11,15,22,31,35). A fifth class of Ab, which may or may not be considered anti-idiotypic in nature, is termed an **autobody** (17,18,22).

Distinctions between different types of anti-idiotypic antibodies are initially based upon in vitro competition assays involving the idiotope bearing antibody, the anti-idiotope antibody and the antigen.

Anti-idiotypic antibodies designated Ab 2 alpha (Fig. 3G) recognize idiotopes on the variable regions of another antibody that are spatially removed from the paratope of that antibody. Consequently, idiotope/Ab 2 alpha interactions are not inhibited by the antigen (22,31). Generally, Ab 2 alpha anti-idiotypic antibodies administered in vivo suppress antibody production by B cells which exhibit an idiotope recognized by Ab 2 alpha (14,15). However, in one case, Ab 2 alpha administered to animals previously unexposed to a virus stimulated an antiviral antibody response (32).

Anti-idiotypic antibodies referred to as Ab 2 beta (Fig. 3G) are also known as **internal image** antibodies (14,16,21,26), **antigen-mimicking** antibodies (37) or **homobodies** (4,5,21). These anti-idiotypic antibodies interact with the paratope of another antibody, and this interaction is inhibited by the antigen. When Ab 2 beta is introduced into animals previously unexposed to the original antigen, Ab 2 beta can stimulate the production of Ab 3. This antibody will have antigen binding characteristics similar to those displayed by the original idiotope bearing antibody (Ab 1) (37).

Antibodies referred to as Ab 2 gamma (Fig. 3G) recognize idiotopes in or near the paratope of another antibody molecule. Therefore, the idiotope/Ab 2 gamma interactions are sterically hindered by the antigen (22). It is difficult to discriminate between Ab 2 beta and Ab 2 gamma anti-idiotype Abs with in vitro immunoassays. Sometimes an anti-idiotope antibody can only be categorized as Ab 2 beta when it mimics the antigen and recognizes the same biological features as the antigen, such as antigenic receptors (11).

Ab 2 epsilon represents a fourth category of anti-idiotypic antibodies which were first described by Bona and his coworkers (4). These investigators found that these anti-idiotypic antibodies interacted with an idiotope on Ab 1 and an epitope on the antigen. The Ab 1/Ab 2 epsilon interaction was also partially inhibited by the antigen. The term epibody was coined to designate those anti-idiotypic antibodies which interacted with the idiotope on Ab 1 and with the antigen used to induce Ab 1.

Antibodies which bind to the antigen and to themselves are termed **autobodies** (18). Autobody characteristics were first ascribed to an anti-T15 idiotope IgM (Ab 2). This Ab interacted with the T15 idiotope on an IgA (Ab 1) to pneumococcal antigen and also with the antigen. The antibody further demonstrated self-binding properties which were antigen inhibitable. It was later discovered that the anti-pneumococcal T15 idiotope bearing Ab 1 was also self-binding, and this self-binding was again antigen inhibitable (17). Since, in this situation, Ab 1 and Ab 2 both exhibited self-binding characteristics, autobodies may be considered a special category of antibodies.

In the literature, anti-idiotopic antibodies which mimic antigens are generally placed in the Ab 2 beta category since biological properties which distinguish them from Ab 2 gamma have been determined. In the absence of pertinent information concerning the biological properties of an anti-idiotope antibody, the antibody is described on the basis of its in vitro interactions with idiotope bearing antibodies in the presence or absence of the antigen and is not always categorized as either Ab 2 beta or Ab 2 gamma.

Antigen-Antibody Interactions

The interaction between an antigen and an antibody involves hydrogen bonds, salt bridges, electrostatic charges, and coulombic, hydrophobic and van der Waal's forces (12,33,36). This interaction is reversible; the strength of the interaction (**affinity**) can be affected by temperature, pH and solvent conditions (12). If these conditions are varied sufficiently, assays such as enzyme-linked immunosorbent assays (ELISAs) which employ the same antigen and antibody, can give divergent results.

Stable, semi-reversible immune complexes can be formed when multimeric interactions occur between antibodies and antigens which exhibit either multiple copies of a single epitope or multiple copies recognized by different polyclonal antibodies. Such multimeric interactions enable low affinity antibodies to bind antigens tightly (12).

The ability to form stable immune complexes is known as **avidity**. IgM antibodies with ten paratopes are more multimeric and have greater antigen avidity than IgG antibodies with two paratopes. A paratope of an antibody may exhibit a low degree of affinity for an antigen. However, if this weak interaction is broken, the remaining paratopes of an antibody, attached to nearby epitopes, can act to stabilize the antigen-antibody complex. A low-affinity IgM antibody has nine other paratopes by which to stabilize this interaction, whereas a low-affinity IgG antibody has only one. Consequently, IgM antibodies which interact weakly with more than one antigen can appear to be less specific than IgG antibodies, in general, simply because IgM antibodies have a greater capacity for stabilizing the antigen-antibody interaction. Therefore, a low affinity IgG antibody with the same antigenic specificity as an IgM antibody may not remain bound to an antigen and may not be detected in an immunoassay while an IgM antibody would be.

The hypervariable regions of the heavy and light chain which make up the antibody's paratope are arranged to form a pocket or groove. Typically, an antigen interacts with the CDRs found within or surrounding this groove. It is generally believed that this interaction assumes a lock (antibody) and key (antigen) configuration (12).

X-ray crystallography has been used to determine the three-dimensional configuration of the antigen-antibody interaction. Only three of these studies have been performed. In one study (8), the structure of the influenza neuraminidase antigen-antibody complex was examined. This complex exhibited characteristics of a hand shake, and as a result of the antigen-antibody interaction, the neuraminidase, and possibly its homologous antibody, underwent conformational change.

The two remaining studies (2,33) dealt with the lysozyme-antibody complex. The anti-lysozyme antibody combining site was not a typical groove. It consisted of a relatively flat surface with protuberances and depressions formed by its amino acid side chains. The site of the antigen-antibody interaction occupied approximately 690 Å and consisted of sixteen lysozyme and seventeen antibody residues in close contact. All of the CDRs and two amino acid residues outside of the CDR made contact with the antigen. A difference in one of the lysozyme amino acid residues which made contact with the antibody obviated the antigen-antibody interaction (2). Binding of the antibody to lysozyme, in one of these studies, also resulted in changes in the conformation of lysozyme (33).

Structural changes apparently do occur in the antigen and/or antibody as a result of their interaction. Although their conformational changes may at first seem minor, conformational changes induced by the interaction of two molecules can produce misleading results and erroneous conclusions. For example, it has been demonstrated that several anti-idiotypic monoclonal antibodies (Ab 2) produced against idiotype-bearing monoclonal antibodies (Ab 1) inhibited the antigen-Ab 1 interaction. Based on these results and without further study, one could conclude that some of these anti-idiotypic antibodies are Ab 2 beta, recognize the paratope of Ab 1, and could mimic the antigen. However, it was determined that some of these anti-idiotypic antibodies (Ab 2) were not directed to the paratope of Ab 1. The non-paratope directed anti-idiotypic antibodies apparently induced allosteric changes in Ab 1 which inhibited the antigen-Ab 1 interaction (1).

It is apparent from the foregoing discussion that immunology can no longer be viewed as a single matter of injecting an antigen to induce antibody formation. The immune system and the products resulting from an immune response as well as their interactions with other molecules can provide powerful tools if one's conceptual approaches are sound. A knowledge of immunological mechanisms should thus allow the researcher to generate efficient probes which could not only be used to serologically detect, identify, group and type many plant pathogens but could also be used to characterize their antigenic makeup.

Literature Cited

1. Agius, M. A., Geannopoulos, C. J., Fairclough, R. H., and Richman, D. P. 1988. Monoclonal anti-idiotypic antibodies against myasthenia-inducing anti-acetylcholine receptor monoclonal antibodies. Preponderance of nonparatope-directed antibodies affecting antigen binding. J. Immunol. 140:62-68.
2. Amit, A. G., Marinzza, R. A., Phillips, S. E. V., and Poljak, R. J. 1986. Three-dimensional structure of an antigen-antibody complex at 2.8 Å resolution. Science 233:747-753.
3. Barrett, J. T. 1988. Textbook of Immunology. C. V. Mosby Co., St. Louis, MO. 455 pp.
4. Bona, C. A., Finley, S., Waters, S., and Kunkel, H. G. 1982. Anti-immunoglobulin antibodies. III. Properties of sequential anti-idiotypic antibodies to heterologous anti-gamma globulins. Detection of reactivity of anti-idiotype antibodies with epitopes of Fc fragments (homobodies) and with epitopes and idiotopes (epibodies). J. Exp. Med. 156:986-999.
5. Bona, C. A., Kang C.-Y, Kohler, H., and Monestier, M. 1986. Epibody: The image of the network created by a single antibody. Immunol. Rev. 90:115-127.
6. Bona, C. A., and Victor-Korbin, C. 1986. Questions related to a functional idiotype network. Ann. Immunol. (Inst. Pasteur) 137C:87-90.
7. Campling, B. G., Cole, S. P. C., Atlaw, T., Roder, J. C., and Kozbor, D. 1987. The EBV hybridoma technique and its application. Pages 3-22 in: Human Hybridomas. Diagnostic and Therapeutic Applications. A. J. Strelkauskas, ed. Marcel Dekker, Inc., New York. NY. 289 pp.
8. Colman, P. M., Laver, W. G., Varghese, J. N., Baker, A. T., Tulloch, P. A., Air, G. M., and Webster, R. G. 1987. Three-dimensional structure of a complex of antibody with influenza virus neuraminidase. Nature 326:358-365.
9. Ezquerra, A., and Coligan, J. E. 1988. T cell receptors: Structure and genetics. Current Opinion in Immunol. 1:77-83.
10. Fox, B. S. 1988. Interaction of major histocompatibility complex molecules with immunogenic peptides. Current Opinion in Immunol. 1:103-106.

11. Gaulton, G. N., and Greene, M. I. 1986. Idiotypic mimicry of biological receptors. Annu. Rev. Immunol. 4:253-280.
12. Harlow, E., and Lane, D. 1988. Antibodies. A Laboratory Manual. Cold Spring Harbor Laboratory. Cold Spring Harbor, NY. 726 pp.
13. Hornbeck, R. V., and Lewis, G. K. 1985. Idiotype connectance in the same system. II. A heavy chain variable region idiotope that dominates the antibody response to the para-azobenzenearsonate groups is a minor idiotope in the response to trinitrophenyl group. J. Exp. Med. 161:53-71.
14. Jerne, N. K. 1974. Towards a network theory of the immune system. Ann. Immunol. (Inst. Pasteur) 125C:373-389.
15. Jerne, N. K. 1985. The generative grammar of the immune system. EMBO J. 4:847-852.
16. Jerne, N. K., Roland, J., and Cazenave, P.-A. 1982. Recurrent idiotypes and internal image. EMBO J. 1:243.
17. Kang, C.-Y., Cheng, H.-L., Rudikoff, S., and Kohler, H. 1987. Idiotypic self binding of a dominant germline idiotype (T15). J. Exp. Med. 165:1332-1343.
18. Kang, C.-Y., and Kohler, H. 1986. A novel chimeric antibody with circular network characteristics: Autobody. Ann. N. Y. Acad. Sci. 475:114-122.
19. Kennedy, R. C., Eichberg, J. W., Lanford, R. E., and Dreesman, G. R. 1986. Anti-idiotypic antibody vaccine for type B viral hepatitis in chimpanzees. Science 232:220-224.
20. Kennedy, R. C., Melnick, J. L., and Dreesman, G. R. 1986. Anti-idiotypes and immunity. Sci. Amer. 255:48-56.
21. Kennedy, R. C., Henkel, R. D., and Dreesman, G. R. 1985. Internal image-bearing anti-idiotype antibodies as a potential vaccine against hepatitis B virus. Pages 109-116 in: High-Technology Route to Virus Vaccines. G. R. Dreesman, J. G. Bronson, and R. C. Kennedy, eds. Am. Soc. for Microbiol., Washington, D. C. 180 pp.
22. Kieber-Emmons, T., and Kohler, H. 1986. Towards a unified theory of immunoglobulin structure-function relations. Immunol. Rev. 90:31-48.
23. Kunkel, H. G., Mannik, M., and Williams, R. C. 1963. Individual antigenic specificity of isolated antibodies. Science 140:1218-1219.
24. Long, E. O. 1988. Processing requirements for presentation of antigens to lymphocytes. Current Opinion in Immunol. 1:98-102.
25. Long, E. O., and Kindt, T. J. 1988. Antigen recognition overview. Current Opinion in Immunol. 1:71-72.
26. Lu, B.-Z., Couraud, P.-O., Schnutz, A., and Strosberg, R. D. 1983. The internal image of catecholamines: Expression and regulation of a functional network. Ann. N.Y. Acad. Sci. 418:240-247.
27. Mernaugh, R. L., Durand, D. P., and Hill, J. H. 1987. Generation of anti-idiotype hybridoma antibodies to soybean mosaic virus rabbit polyclonal antibodies. (Abstr.) Phytopathology 77:1765.
28. Metzger, D. W., Miller, A., and Sercarz, E. E. 1980. Sharing of an idiotypic marker by monoclonal antibodies specific for distinct regions of hen lysozyme. Nature 287:540-541.
29. Oudin, J., and Michel, M. Une nouvelle forme d'allotypic de globulins gamma du serum de lapin apparenment liee à la fonction et à la specificite anticorps. C. R. Acad. Sci. (Paris) 257:805-808.
30. Oudin, J., and Michel, M. 1969. Idiotypy of rabbit antibodies. I. Comparison of idiotypy of antibodies against _Salmonella typhi_ with that of antibodies against other bacteria in the same rabbits, or of antibodies against _Salmonella typhi_ in various rabbits. J. Exp. Med. 30:595-617.

31. Rees. R. A. 1987. The antibody combing site: Retrospect and prospect. Immunol. Today 8:44-45.
32. Schick, M. R., Dreesman, G. R., and Kennedy, R. C. 1987. Induction of an anti-hepatitis B surface antigen response in mice by noninternal image (Ab2a) anti-idiotypic antibodies. J. Immunol. 138:3419-3425.
33. Sheriff, S., Silverton, E. W., Padlam, E. A., Cohen, G. H., Smith-Gill, S. J., Finzel, B. C., and Davies, D. R. 1987. Three dimensional structure of an antibody-antigen complex. Proc. Natl. Acad. Sci. USA 84:8075-8079.
34. Unane, E. R., and Benacerraf, B. 1984. Textbook of Immunology. 2nd ed. Williams and Wilkins, eds. Baltimore, MD. 324 pp.
35. UytdeHaag, F. G. C. M., Bunschoten, H., Weijer, K., and Osterhaus, A. D. M. E. 1986. From Jenner to Jerne: Towards idiotype vaccines. Immunol. Rev. 90:93-113.
36. van Regenmortel, M. H. V. 1988. Molecular dissection of protein antigens and the prediction of epitopes. Pages 1-40 in: Laboratory Techniques in Biochemistry and Molecular Biology. R. H. Burdon and P. H. van Kippenberg, eds. Elsevier Science Publ. Co., Inc., New York, NY. 227 pp.
37. Zanetti, M., Sercarz, E., and Salk, J. 1987. The immunology of new generation vaccines. Immunol. Today 8:18-25.

Section II.A.1.

PREPARATION OF ANTIGENS, VIRUSES

M. K. Brakke
University of Nebraska
Lincoln, NE 68583

Plant viruses code for several proteins. Of these, the virion capsid and inclusion body proteins accumulate in the highest concentrations in infected plants. The virion capsid proteins are most frequently used as antigens. Inclusion body proteins have more potential for use as antigens than has sometimes been realized.

The main points in purification of these antigens are selection of source tissue, assays to monitor concentration and purity, procedures for purifying the virions or inclusion bodies, and, when applicable, procedures for purifying the protein subunits of virions or inclusion bodies (for reviews see 13,14,33,39).

Other protein antigens encoded by the virus have been obtained by cloning the gene and inserting it into an expression vector to produce the protein in *Escherichia coli* or other suitable system (34). Alternatively, a synthetic peptide corresponding to a portion of the virus protein whose structure has been deduced from the nucleic acid sequence may be used as the antigen (1,5,43).

Assays

Viral Assays, Quantitative and Qualitative Considerations

A good assay is crucial to experiments for optimizing purification procedures, particularly for virions that are difficult to purify. Accuracy, specificity, and sensitivity of the assay must be determined in requisite standardization tests (15,39).

Infectivity assays reflect the specific infectivity (i.e. quality) of the virions, their concentration, and the effect of inhibitors or potentiators that might be present in the inoculum. Results are always relative. Systemic and local lesion assays are equally sensitive, but the latter is more efficient and accurate if properly done and interpreted.

Assays based on ultraviolet absorbance require some degree of virion purification because plant proteins, nucleic acids, and pigments also absorb ultraviolet light. The degree of purification and the recovery of virions are both important. In one version, a plant extract clarified by partial denaturation (see below) and low speed centrifugation is centrifuged through a sucrose density gradient to further purify the virions, which are then measured by photometric scanning of the gradient (10). The absorbance of plant components that sediment with the virions can usually be subtracted from virion absorbance by a base line correction. The method is accurate and reliable for spherical virions. However, values for rod-shaped virions may be low, as a portion of these typically sediments as polymers to the pellet of the gradient column.

Ultraviolet absorbance assays give absolute amounts of virions if calibrated by dry weight measurements (16). In absence of dry weight determinations, the specific absorbance can be estimated from the shape of the virion. Rod-shaped virions have absorbances at 260 nm of 2.5 to 3.0 for 1 mg/ml solutions and 1 cm light path. For most spherical virions, the value varies from 5 to 8.

Serological assays of several types may be used to estimate virion concentration by measuring serological reactions, or by measuring virions, virion proteins, or virion nucleic acids recovered from antibody-virion precipitates.

Nucleic acid or subunit protein may be assayed instead of virions. For example, Veerisetty and Brakke (57) and Goldberg and Brakke (30) assayed alfalfa latent and maize chlorotic mottle viruses, respectively, by density gradient centrifugation of viral RNA. This method often gives lower values than assays of virions because degraded RNA is excluded. An advantage is that essentially the same procedure applies to most ssRNA viruses, except those having RNA that co-sediments with ribosomal RNA, because RNA properties are nearly constant from one virus to another. Assays can also be based on nucleic acid hybridization (e.g. dot blots) which does not require prior separation of virus from host nucleic acid (37).

Another virion component assay is the measurement of the capsid protein by SDS-gel electrophoresis (sodium dodecyl sulphate-polyacrylamide gel electrophoresis, SDS-PAGE). This is an excellent assay for rod-shaped virions because capsid protein is recovered from viral aggregates. This assay might be used to reinvestigate the purification of barley stripe mosaic virus (BSMV). For this purpose Triton X-100 would be added to the extract (see procedure below) and the virions would be partially purified by a preliminary high speed centrifugation, without prior low-speed centrifugation. In recent direct comparisons, such an assay detected twice as many virions in an extract as did the density gradient centrifugation assay (19).

Standard curves, i.e. a plot of the test response given by a series of standard virus concentrations, should be included each time assays are done (except for ultraviolet absorption assays). If comparisons between test and standard reactions are based on visual or eyeball comparisons, the accuracy is usually two-fold or less, whether the assay is serological, dot blot, gel electrophoretic or other. For greater accuracy, the reactions should be measured instrumentally.

Most types of assays work well with stable virions present in high concentrations. However, only a few types may be suitable for unstable viruses or those present in low concentration.

Purification of Barley Stripe Mosaic Virus

The following purification procedure for BSMV is given to illustrate typical conditions (modified from Ref. 18).

1. Inoculate barley plants in the two-leaf stage by leaf rubbing with an extract (containing suspended abrasive) of infected plants.
2. Harvest leaves about two days after symptoms appear. Systemic symptoms will usually appear on the youngest leaves in 4 to 5 da at 22 to 25 C.
3. Weigh leaves, cut into two-cm segments, and grind for one min in a Waring blender with 3 ml of pH 7.0, 0.01 M phosphate buffer per gm of leaves.
4. Filter homogenate through cheesecloth to remove fibers.
5. Centrifuge for 20 min at 10,000 rpm (avg. RCF 8000 x g) in a Sorvall SS34

rotor. Discard the pellet.

6. Add 1/20th volume of 20% Triton X-100 to the supernatant fraction.
7. Layer the supernatant fraction over 4 ml of 20% sucrose dissolved in buffer in 30 ml centrifuge tubes. Centrifuge for one hr at 28,000 rpm in the Beckman Type 30 or 50.2 Ti rotor (avg. RCF 68,000 x g) at 5 C.
8. Suspend the pellets in 0.05 M borate buffer, pH 8.0, containing 1% Triton X-100, 4 ml per tube.
9. Centrifuge for 5 min at 5000 rpm (avg. RCF 2,000 x g) in the Sorvall SS34 rotor. Discard the pellet.
10. Layer two ml of the supernatant fraction on each sucrose gradient in 1 x 3.5 inch centrifuge tubes. The gradients are prepared by layering 9, 9, 9, and 5 ml of solutions of 400, 300, 200, and 100 mg of sucrose per ml of 0.01 M, pH 7.0 phosphate buffer, respectively. The layers of sucrose solutions should stand 24 hr at 4 C before use so diffusion can form a smooth gradient. Centrifuge for 2 hr at 26,000 rpm (avg. RCF 89,000 x g) in the SW 28 Beckman rotor at 5 C.
11. Remove solution from virus zone either with a syringe and needle or with a photometric scanning and gradient fractionating machine. The zone may be identified either by light scattering or ultraviolet absorbance.
12. Dilute the fraction containing virus to decrease the viscosity and centrifuge for one hour at 46,000 rpm (avg. RCF 140,000 x g) in the Beckman Type 50 TI or equivalent rotor.
13. Dissolve pellet in 0.5 ml per tube of 0.05 M sodium orthoborate, pH 8.0.

Comments and Applications

Step 1. Selection of Host Plant and Growing Conditions

Source plants should yield extracts that are low in mucilages, tannins and acids. Plants of the Solanaceae, Leguminoseae, Cucurbitaceae, Chenopodiaceae, and Gramineae are usually suitable. Extracts of fruit trees, strawberries, and sorrel (Rumex acetosa), for example, have enough acids and tannins to precipitate proteins. Extracts of sweet potatoes have mucilages that interfere with purification attempts and special attention to extraction buffer is needed (41). If possible, viruses of such plants should be transferred to more suitable hosts before purification.

The concentration of virus varies with cultivar. An easily available cultivar that supports massive virus synthesis should be selected.

Temperature is the most critical growing condition. It influences movement of BSMV, and thus the percentage of cells infected, and, to a lesser extent, the concentration of virus in each infected cell. Barley plants inoculated with BSMV require one wk to show systemic symptoms at 20 C and four wk at 15 C (45). Once the symptoms appear, the concentration of virions at the two temperatures is about the same.

It may be desirable to keep plants at a warm temperature until the virus becomes systemic and then move them to an optimal temperature for viral synthesis.

In general, light and nutrients that favor plant growth favor high viral synthesis.

Step 2. Selection of Tissue for Harvest

The concentration of virions varies with type of tissue, age of plant and tissue, and stage of disease development. Leaves are the most commonly used tissue from which to purify most virions, but other tissues may be better in particular cases (23). Systemically infected leaves usually have a higher concentration of virions than those with local lesions. The highest concentration of BSMV is in the first systemically infected leaf, when symptoms first appear (10). BSMV replicates in growing leaves, not after the leaf has matured. As leaves senesce, the concentration of virions per leaf decreases slightly. Specific infectivity may also decrease. Many mosaic viruses follow a similar pattern, but the decrease in virion concentration as the leaf senesces may be more marked than for BSMV (39).

To a first approximation, the concentration of virions in inoculated leaves is proportional to the number of infection sites (54). Therefore, leaves should be inoculated with a high concentration of virus if the inoculated leaves are to be used for purification. If the virus induces local lesions, leaves probably should be harvested when lesions first appear, but the best time can be determined by assays.

Step 3. Extraction

Small amounts of leaves are conveniently ground in a mortar and pestle. A power grinder, such as a Waring blender, is used for larger amounts. Virions are stable in a blender for periods of a few minutes if temperatures above 30 C are avoided. The grinder or blender may need to be cooled if the virions are unstable. Phloem-limited viruses, such as luteoviruses, require extensive mechanical grinding or enzyme treatment (46,53). Leaves may be pulverized in liquid nitrogen to enhance release of phloem-localized viruses and preserve viruses that are unusually unstable.

The buffer selected should stabilize virions and minimize aggregation. This usually means a buffer between pH 6 and 8, 0.1 M concentration or less, with reducing agents and chelating agents as needed (3,44). Polyvinylpyrrolidone, gelatin, and alkaloids may be added to bind or precipitate tannins (20). Insoluble polyvinylpyrrolidone (1 gm/gm tissue) may bind mucilages as well as tannins (50). Buffers of pH 5 to 6 are useful if the virions are soluble in them, because many plant proteins are not. Actually, 0.5 M sodium orthoborate, pH 9.0, would have extracted significantly more BSMV than the phosphate buffer recommended above, but subsequent purification would have been more difficult because of increased amounts of chloroplast fragments. High salt, high pH buffers generally appear to release virions from inclusions or complexes with cell wall fiber or other plant components. However, salts up to 0.1 M concentration increased the adsorption of TMV to cell wall fibers as compared to distilled water (55).

Many hydrogen ion buffers also bind or chelate polyvalent cations such as Ca^{++} and Mg^{++}. Many virions, in turn, are stabilized by Ca^{++} and Mg^{++} ions. Good et al. (31) listed buffers that do not chelate metal ions. Commonly used buffers are phosphate, borate, citrate, and Tris [tris(hydroxymethyl) amino methane].

Steps 4-6. Selective Denaturation and Clarification

Extracts of leaves contain numerous proteins, ribosomes, polysomes, fragments of membranes, broken and intact organelles, and cell wall fibers, the content varying with age of leaf. Extracts of young leaves have more ribosomes and a greater variety of proteins than those of older leaves. The stability and enzymatic activity of the proteins also vary with leaf age. Most of the proteins do not sediment with virions during high speed centrifugation but remain in the supernatant fraction. Materials that sediment with the virions may include: fragments of membranes, chloroplasts, mitochondria, and other organelles; ribosomes and polysomes; cytoskeleton fibers; RUBISCO (ribulose bis-phosphate carboxylase/oxygenase); phytoferritin; and unidentified materials, some of which seem to be polysaccharide. Almost all virus purification procedures include steps intended to either solubilize, or, more frequently, denature and precipitate these materials without affecting the virions (13).

Cell wall fibers should be removed by filtration because they do not pack well in pellets. Celite (diatomaceous earth), cellulose powder, and decolorizing charcoal, which have been added as filtering aids, appear to result in removal of some of the chloroplasts. I do not know of any studies on loss of virions due to use of these materials. The time and speed of the low speed centrifugation is not critical. The faster and longer it is, the more impurities are removed, but the more virions that are lost. Where necessary, purification efficiency can be improved by using a sucrose-solution cushion (Step 7) or reextracting the low speed pellet to remove virions.

Chloroplast and membrane fragments are the most visible of the plant components, the former because of their green color and the latter because they scatter light. Many purification procedures are quite effective in removing these visible components, but less effective in removing less visible components, such as RUBISCO and ribosomes. In the above procedure, Triton X-100 solubilizes membrane and chloroplast fragments (8). Non-enveloped virions are generally stable to most nonionic detergents such as Triton X-100 (8,42).

Treating extracts with equal volumes of chloroform (48), chloroform:n-butanol at 1:1 (51), or 8.5% n-butanol (56) are common methods to remove the visible plant components. A wide variety of organic solvents is suitable. Ribosomes and proteins are usually not completely removed. Chloroform is a very effective deproteinizing agent in presence of 1 to 2 M NaCl, but is nonselective, deproteinizing virions as well as other nucleoproteins (19). Most viruses resist chloroform in 0.1 M or lower buffers, but so do some plant proteins and ribosomes. An alternative procedure calls for lyophilizing the leaves, grinding, extracting the powder with organic solvents, and then suspending the powder in buffer (58).

Freezing extracts is more effective than freezing leaves, but both denature (i.e. decrease solubility) plant proteins. Plant protein denaturation increases with duration of freezing, but so may the destruction of virions. Incubation of extracts at 25 to 40 C denatures some plant proteins, probably results in enzymatic destruction of others, and destroys ribosomes if magnesium is chelated. Incubation is more effective for extracts of young than old barley leaves, probably because of a qualitative difference in proteins and enzymes with leaf age.

Selective adsorption may be used instead of selective denaturation. Plant components have been adsorbed to bentonite, but the amount of bentonite is critical because the virions may also be adsorbed (25). Calcium phosphate is a very effective adsorbent of plant components. It also adsorbs BSMV, but not at phosphate buffer concentration $\geq$ 0.09 M (9). The protein adsorbing properties of calcium phosphate depends on its past history, i.e. method of preparation and storage, which must be carefully controlled for consistent results (26). Plant antigens can also be selectively precipitated by incubation with specific antibodies (29).

Digestion of plant components with added enzymes has been little used except for luteoviruses. The approach could be effective in removing putative polysaccharides.

Step 7. High Speed Centrifugation

Most purification procedures depend on centrifugation to separate virions from plant components of different size and density. A few procedures depend on molecular sieving or adsorption.

High speed centrifugation is an effective concentration procedure, but not an efficient separation procedure. Up to 4 alternate low and high speed cycles have been used (differential centrifugation). Addition of a cushion of about 1 cm depth of 20% sucrose solution at the bottom of the tube improves separation of virions from slower sedimenting contaminants. Separation in the high speed step would be enhanced by stopping centrifugation just when all the virions had pelleted. However, this time is not easy to determine experimentally or to calculate because it depends on the concentration of virions and of total solutes, especially other macromolecules. The tendency to centrifuge a little longer than necessary, "just for good measure", is also a disadvantage. For instance, if a virion whose sedimentation coefficient ($s_{\omega,20}$) is 108s were centrifuged twice as long as necessary, 1/3 of the contaminating RUBISCO ($s_{\omega,20}$ = 18s) would be incorporated in the pellet.

Conditions given in Step 7 suffice to pellet all BSMV virions as determined by trial and error (Brakke, unpublished results). The centrifugation time needed is inversely proportional to the sedimentation coefficient. $S_{\omega,20}$ values for virions vary from 80 s for brome mosaic virus to about 900 s for rhabdoviruses (7). Most small spherical virions range from 100 to 130 s, potyviruses approximate 160 s, and TMV, 190 s. The three virions of BSMV sediment between 170 and 200 s (18).

Steps 8 and 9. Resuspension and Clarification

The optimal buffer for extracting virions from tissue is not necessarily optimal for resuspending pellets. The solubility and stability of virions, and of proteins in general, change with degree of purity. Borate buffer is recommended for resuspending BSMV pellets primarily because the virions seem to dissolve in it with less aggregation than in neutral phosphate buffer. High speed pellets usually have to be dispersed mechanically and allowed to stand for $\geq$ 30 min for the virions to dissolve and diffuse from clumps of insoluble material. Triton X-100 is included to disperse remaining membrane fragments. The low speed centrifugation should be kept short to avoid loss of rod-shaped virions which aggregate easily. Faster and longer "low speed" centrifugation could be used for spherical virions.

Step 10. Density Gradient Centrifugation

Sedimentation through sucrose gradients is an efficient procedure for separating particles on basis of size, shape and buoyant density (6). It is not a virion-concentration procedure. It can also be used to assay virions, measure purity, and estimate sedimentation coefficients (12). Sensitivity of photometric scanning limits the application of density gradient centrifugation to about $\geq$ 1 μg of virions. The upper limit of applicable virion concentration is 1 to 3 mg of virions per ml (11). The sucrose should be dissolved in a buffer in which the virions are stable and soluble. Centrifugation for 1 to 5 hr at about 25,000 rpm in gradients of 100 to 400 mg of sucrose per ml suffices for most virions. Centrifugation through a second sucrose gradient under identical conditions usually provides little additional purification. A second centrifugation through a gradient of higher sucrose concentration, e.g. 300 to 600 mg per ml, is useful but requires 5 to 10 times longer centrifugation.

Other materials can be used instead of sucrose. Glycerol and sugars other than sucrose give results similar to those for sucrose. Sedimentation through salt gradients can separate particles by sedimentation coefficients, as in sucrose gradients, or by buoyant density only. A sedimentation coefficient separation through a salt gradient requires less than half the centrifugation time required for a sucrose gradient. Seldom attempted, it merits greater attention.

Salt gradients are usually used for separations based on buoyant density, i.e. isopycnic density gradient centrifugation. Salts commonly used are CsCl and Cs_2SO_4, but any salt in which the virions are stable and which is soluble enough to give solutions of sufficient density can be used. The buoyant density of virions in cesium salt gradients depends on the proportion of nucleic acid in the virion, capsid hydration, and the number of cesium ions bound by virions. Thus, the cesium salt of the virion is separated in gradients of cesium salts. Ribosomes are unstable in CsCl and, therefore, should not be present in virion preparations purified through CsCl gradients, nor should proteins and lipids. However, polysaccharides with variable numbers of acid groups (which bind Cs^+) might have a range of buoyant densities overlapping those of virions and, therefore, could occur as contaminants. Most virions have buoyant densities in CsCl of 1.30 to 1.40. For best results, centrifugation should be continued until virions approach equilibrium. Formulas are available to calculate the time (12). In most cases, overnight centrifugation (15 hr) at 20,000 rpm is sufficient, with lesser times being required at higher speeds, in short preformed gradients, or in vertical tube rotors. A convenient experimental test to find if equilibrium has been reached is to mix sample and salt solution thoroughly in one pair of tubes, but to have a preformed or step gradient with the sample on the surface in another pair of tubes. Equilibrium is achieved when virions in both pairs of tubes occupy the same position after centrifugation.

Step 11. Removal of Virion Zone

Solution from the virion zone can be removed more accurately and quantitatively by pumping the gradient through a flow cell for measuring ultraviolet absorbance than with a needle pierced through the side of the tube. However, withdrawal by use of a needle is much faster.

Step 12 and 13. Concentration and Resolution

Increased speed and/or duration of centrifugation is required to pellet the virions in this stage, as compared to Step 7, because of the viscosity of the sucrose solution and the low concentration of the virions. The supernatant fraction should be removed from the pellet as soon as the centrifuge has stopped because these pellets often dissolve quickly. Small quantities of virions can be passed through an appropriate Millipore filter as a substitute for final centrifugation.

Virions may be significantly less stable after purification than in plant sap. Selection of a suitable buffer may need to be reevaluated with purified virions.

Alternatives to High Speed Centrifugation

Separation of virions based on size by molecular filtration through agar beads was investigated by Steere and Ackers (52). The method worked well with the stable viruses they investigated. Controlled pore glass has been used for a similar method (4). Virions can also be separated by non-denaturing electrophoresis through agarose gels (49). Size is an important variable for non-denaturing electrophoresis, but shape, surface charge, and tendency to adsorb to the gel are also important. Only small amounts are usually separated by this method.

Virions can be, but seldom are, purified by chromatography (2,59). Antibodies bound to a solid support provide a highly specific chromatographic medium (27).

Precipitation with ammonium sulfate or similar salts has been extensively used for tobacco mosaic virus, but many other viruses are not stable to high concentrations of ammonium sulfate (39). Proteins co-precipitate as mixtures in presence of salts, which limits usefulness of this approach for purification, but not for concentration.

Polyethylene glycol (PEG) precipitation is frequently used instead of the first high speed centrifugation (32). Low concentrations of PEG with low concentrations of NaCl precipitate rod-shaped virions, e.g. 6% PEG 6000 plus 0.24 M NaCl precipitates BSMV (35). Higher concentration of both PEG and NaCl are needed to precipitate spherical virions. The PEG- virion precipitates can be pelleted by centrifugation for 15-20 min at 10,000-15,000 x g and the virions redissolved by stirring in buffer. Some virions redissolve with difficulty.

Purification of Capsid Protein

Most purification methods are suitable for stable virions occurring in plants at high concentrations, such as TMV. It is much more difficult to purify virions that are present in low concentration, that are unstable, or that aggregate easily.

The degree of purification attained with each additional step may be matched by virion loss due to aggregation or instability. An alternate approach for antiserum production is partial purification of the virions followed by SDS gel electrophoresis to purify the capsid protein, which is then injected as antigen. The virions may be purified through the equivalent of Step 7 above and the pellet dissolved by heating in SDS-containing sample buffer preliminary to electrophoresis. For quantity preparations, a slab gel is run with one sample well (i.e., a gel-wide loading slot) (17). The proteins may be stained with Coomassie Blue, but the concentration of acetic acid should be kept low (1%) and the staining time short to prevent hydrolysis of the protein. Destaining need not be complete. Remaining Coomassie Blue is ethanol soluble and remains in the ethanol when the protein is precipitated after extraction from the gel. The protein may be eluted from the gel in a number of ways, but should be eluted at room temperature because SDS-protein complexes are insoluble below about 15 C. SDS (0.1%) may be required in the extraction buffer to keep the denatured capsid protein soluble.

Use of agarose as the supporting gel for SDS electrophoresis of proteins eliminates the need to elute the protein from the gel because the agarose is non-toxic and can be injected into the animal as a homogenate containing purified protein (47). A major advantage of SDS-gel electrophoretic purification is that the same procedure applies for all proteins, thus obviating modifications for each virus.

Purification of Inclusion Bodies

The same principles are important in the purification of inclusion bodies as in the purification of virions. Differences arise from the larger sizes of inclusion bodies, which, therefore, require slower, shorter centrifugations for sedimentation, relative to virions. Likewise, inclusion bodies are heterogenous in size, limiting the usefulness of sucrose density gradient centrifugation for their purification. Hiebert et al. (33) have thoroughly reviewed the purification procedures reported for inclusion bodies.

Harvesting tissue at the right stage of viral infection may be even more important for purifying inclusion bodies than for virions. Purification success may be monitored by examining samples by light microscopy after appropriate staining, electron microscopy of thin sections of embedded pellets, or SDS-PAGE if a protein of the inclusion body has been identified. All inclusion bodies seem to resist dissociation by Triton X-100, which is frequently included in the buffers to disperse chloroplasts. Some inclusion bodies survive treatment with chloroform and/or CCl_4.

Cylindrical inclusion bodies of potyviruses are often extracted with high salt (0.5 M) buffers (33), but those of wheat streak mosaic virus required a low salt buffer and 10% glycerol for successful extraction (17). Separation of the inclusion bodies from other components has always involved differential centrifugation. Cylindrical inclusions can be purified by repeated cycles of 500 to 1,000 x g (avg. RCF) for 10 min and about 10,000 - 15,000 x g (avg. RCF) for 10 to 20 min. Use of sucrose cushions is beneficial. Various types of isopycnic gradients have been used, including step gradients of 50 to 80% sucrose. Resuspension buffers usually differ from grinding buffers and must be carefully selected to dissolve the inclusion body while preserving its structure.

Nuclear inclusions of potyviruses, amorphous inclusions, and "viroplasms" sediment faster than cylindrical inclusions of potyviruses. Purification schemes for these faster sedimenting inclusions generally omit low speed centrifugation and rely upon Triton X-100 to disperse membranes and organelle fragments. Procedures call for repeated pelleting of the inclusions by centrifugation for 5-10 min at 1,000-4,000 x g, max RCF (22,24,33).

Maize stripe virus-infected plants produce large amounts of a non-capsid protein that accumulates as crystalline aggregates. The protein is soluble at pH 7, but crystallizes or aggregates at pH 3, which provides the basis for its purification (28).

Usually the purified inclusion bodies are not used directly as antigens. Instead, the constituent proteins are further purified by preparative SDS gel electrophoresis before immunization (33).

Viral Antigens from Expression Vectors

Availability of cloned cDNAs of plant virus genomes in principle allows production of any viral-encoded proteins in expression vectors by standard methods of molecular biology (21,37,38,40). This approach has been used by Horikoshi et al. (34) to produce antibodies to RNA replicase of brome mosaic virus. Another application involved production of antibody to a chimeric fusion protein of cucumber mosaic virus protein 3A (36).

Literature Cited

1. Albrecht, H., Geldreich, A., De Murcia, J. M., Kirchherr, D., Mesnard, J.-M., and Lebeurier, G. 1988. Cauliflower mosaic virus gene 1 product detected in a cell-wall-enriched fraction. Virology 163:503-508.
2. Atabekov, I. G. 1961. Ion exchange adsorption of plant viruses. Voprosy Virusologii 6:669-672
3. Barnett, O. W., and Fulton, R. W. 1971. Differential response of Prunus necrotic ringspot and tulare apple mosaic viruses to stabilizing agents. Virology 46:613-619.
4. Barton, R. J. 1977. An examination of permeation chromatography on columns of controlled pore glass for routine purification of plant viruses. J. Gen. Virol. 35:77-87.
5. Berna, A., Briand, J. -P., Stussi-Garaud, C., and Godefroy-Colburn, T. 1986. Kinetics of accumulation of the three non-structural proteins of alfalfa mosaic virus in tobacco plants. J. Gen. Virol. 67:1135-1147.
6. Brakke, M. K. 1953. Zonal separations by density-gradient centrifugation. Anal. Biochem. and Biophys. 45:275-290.
7. Brakke, M. K. 1958. Estimation of sedimentation constants of viruses by density-gradient centrifugation. Virology 6:96-114.
8. Brakke, M. K. 1959. Dispersion of aggregated barley stripe mosaic virus by detergents. Virology 9:506-521.
9. Brakke, M. K. 1962. Stability of purified barley stripe mosaic virus. Virology 17:131-142.
10. Brakke, M. K. 1963. Photometric scanning of centrifuged density gradient columns. Anal. Biochem. 5:271-283.
11. Brakke, M. K. 1964. Nonideal sedimentation and the capacity of sucrose gradient columns for virus in density-gradient centrifugation. Arch. Biochem. Biophys. 107:388-403.
12. Brakke, M. K. 1967. Density-gradient centrifugation. Pages 93-118 in: Methods in Virology, Vol. 2. K. Maramorosch and H. Koprowski, eds. Academic Press Inc., New York. 682 pp.
13. Brakke, M. K. 1967. Miscellaneous problems in virus purification. Pages 119-136 in: Methods in Virology, Vol. 2. K. Maramorosch and H. Koprowski, eds. Academic Press Inc., New York. 682 pp.
14. Brakke, M. K. 1969. Isolation and purification of vector-borne plant viruses. Pages 527-555 in: Viruses, Vectors and Vegetation. K. Maramorosch, ed. Interscience Publishers (John Wiley), New York. 666 pp.
15. Brakke, M. K. 1970. Systemic infections for the assay of plant viruses. Annu. Rev. Phytopathol. 8:61-84.
16. Brakke, M. K. 1979. Ultraviolet absorption spectra and difference spectra of barley stripe mosaic and tobacco mosaic viruses in buffer and sodium dodecyl sulfate. Virology 98:76-87.
17. Brakke, M. K., Ball, E. M., Hsu, Y. H., and Langenberg, W., L. 1987 Wheat streak mosaic virus cylindrical inclusion body protein. J. Gen. Virol. 68:281-287.
18. Brakke, M. K., and Palomar, M. K. 1976. Separation of components of barley stripe mosaic virions by density-gradient centrifugation. Virology 71:255-261.
19. Brakke, M. K., White, J. L., Samson, R. G., and Joshi, J. 1988. Chlorophyll, chloroplast ribosomal RNA, and DNA are reduced by barley stripe mosaic virus systemic infection. Phytopathology 78:570-574.
20. Brunt, A. A., and Kenten, R. H. 1963. The use of protein in the extraction of cocoa swollen-shoot virus from cocoa leaves. Virology 19:388-392.

21. Carroll, S. B., and Laughon, A. 1987. Production and purification of polyclonal antibodies to the foreign segment of Beta-galactosidase fusion proteins. Pages 89-112 in : DNA Cloning: a Practical Approach, Vol. 3. D. M. Glover, ed. IRL Press Ltd., Oxford. 254 pp.
22. Chang, C.-A., Hiebert, E., and Purcifull, D. E. 1988. Purification, characterization, and immunological analysis of nuclear inclusions induced by bean yellow mosaic and clover yellow vein potyviruses. Phytopathology 78:1266-1275.
23. D'Arcy, C. J., Hewings, A. D., Burnett, P. A., and Jedlinski, H. 1983. Comparative purification of two Luteoviruses. Phytopathology 73:755-759.
24. De Meija, M. V. G., Hiebert, E., and Purcifull, D. 1985. Isolation and partial characterization of the amorphous cytoplasmic inclusions associated with infections caused by two potyviruses. Virology 142:24-33.
25. Dunn, D. B., and Hitchborn, J. H. 1965. The use of bentonite in the purification of plant viruses. Virology 25:171-192.
26. Fulton, R. W. 1959. Purification of sour cherry necrotic ringspot and prune dwarf viruses. Virology 9:522-535.
27. Galvez, G. E. 1966. Specific adsorption of plant viruses by antibodies coupled to a solid matrix. Virology 28:171-187.
28. Gingery, R. E., Nault, L. R., and Bradfute, O. E. 1981. Maize stripe virus: Characteristics of a member of a new virus class. Virology 112:99-108.
29. Gold, H. A. 1960. Antihost serum improves plant virus purification. Phytopathology 51:561-565.
30. Goldberg, K. -B., and Brakke, M. K. 1987. Concentration of maize chlorotic mottle virus increased in mixed infections with maize dwarf mosaic virus, strain B. Phytopathology 77:162-167.
31. Good, N. E., Winget, G. D., Winter, W., Connolly, T. N., Izawa, S., and Singh, R. M. M. 1966. Hydrogen ion buffers for biological research. Biochemistry 5:467-477.
32. Hebert, T. T. 1963. Precipitation of plant viruses by polyethylene glycol. Phytopathology 53:362.
33. Hiebert, E., Purcifull, D. E., and Christie, R. G. 1984. Purification and immunological analyses of plant virus inclusion bodies. Pages 225-279 in: Methods in Virology, Vol. 8. K. Maramorosch and H. Koprowski, eds. Academic Press, New York. 396 pp.
34. Horikoshi, M., Mise, K., Furusawa, I., and Shishiyama, J. 1988. Immunological analysis of brome mosaic virus replicase. J. Gen. Virol. 69:3081-3087.
35. Lane, L. 1974. The components of barley stripe mosaic and related viruses. Virology 58:323-333.
36. MacKenzie, D. J., and Tremaine, J. H. 1988. Ultrastructural location of non-structural protein 3A of cucumber mosaic virus in infected tissue using monoclonal antibodies to a cloned chimeric fusion protein. J. Gen. Virol. 69:2387-2395.
37. Maniatis, T., Fritsch, E. F., and Sambrook, J. 1982. Molecular Cloning: A Laboratory Manual. Cold Spring Harbor Laboratory, Cold Spring Harbor, N.Y. 545 pp.
38. Marston, F. A. O. 1987. The purification of eukaryotic polpetides expressed in _Escherichia coli_. Pages 59-88 in: DNA Cloning: a Practical Approach, Vol. 3. D. M. Glover, ed. IRL Press Ltd., Oxford. 1154 pp.
39. Matthews, R. E. F. 1981. Plant Virology. 2nd ed. Academic Press, New York. 897 pp.

40. Mole, S. E., and Lane, D. P. 1987. Production of monoclonal antibodies against fusion proteins produced in Escherichia coli. Pages 113-140 in: DNA Cloning: a Practical Approach, Vol. 3. D. M. Glover, ed. IRL Press, Ltd., Oxford. 1154 pp.
41. Moyer, J. W., and Kennedy, G. G. 1978. Purification and properties of sweet potato feathery mottle virus. Phytopathology 68:998-1004.
42. Nozu, Y., and Yamaura, R. 1971. Use of Triton - X100 for isolation of tobacco mosaic virus from some plants other than tobacco. Virology 43:514-515.
43. Ooshika, I., Watanabe, Y., Meshi, T., Okada, Y., Igano, K., Inouye, K., and Yoshida, N. 1984. Identification of the 30k protein of TMV by immunoprecipitation with antibodies directed against a synthetic peptide. Virology 32:71-78.
44. Pierpont, W. S. 1966. The enzymatic oxidation of chlorogenic acid and some reactions of the quinone produced. Biochem. J. 98:567-580.
45. Rao, A. S., and Brakke, M. K. 1970. Dark treatment of wheat inoculated with soil-borne wheat mosaic and barley stripe mosaic viruses. Phytopathology 60:714-716.
46. Rochow, W. F., and Brakke, M. K. 1964. Purification of barley yellow dwarf virus. Virology 24:310-322.
47. Sakakibara, R., Tominaga, N., Sakai, A., and Ishiguro, M. 1987. Electrophoretic separation of proteins on agarose gel-application for direct immunization with a gel piece containing an antigen. Anal. Biochem. 162:150-155.
48. Schneider, I. R. 1953. Solution of tobacco mosaic virus in the aqueous phase of a chloroform-water emulsion and application of this phenomenon in the virus assay. Science 117:30-31.
49. Serwer, P. 1986. Use of gel electrophoresis to characterize multimolecular aggregates. Pages 116-132 in: Methods in Enzymology, Vol. 130. C. H. W. Hirs and S. N. Timasheff, eds. Academic Press, Orlando. 587 pp.
50. Shepard, J. F., and Grogan, G. 1967. Partial purification, properties, and serology of western celery mosaic virus. Phytopathology 57:1104-1110.
51. Steere, R. L. 1956. Purification and properties of tobacco ringspot virus. Phytopathology 46:60-69.
52. Steere, R. L., and Ackers, G. K. 1962. Purification and separation of tobacco mosaic virus and southern bean mosaic virus by agar-gel filtration. Nature 194:114-116.
53. Takanami, Y., and Kubo, S. 1979. Enzyme-assisted purification of two phloem-limited plant viruses: tobacco necrotic dwarf and potato leafroll. J. Gen. Virol. 44:153-159.
54. Taniguchi, T. 1963. Similarity in the accumulation of tobacco mosaic virus in systemic and local necrotic infection. Virology 19:237-240.
55. Taniguchi, T. 1966. Effect of salts on the adsorption of tobacco mosaic virus to tobacco leaf cell debris. Virology 28:131-134.
56. Tomlinson, J. A., Shepherd, R. J., and Walker, J. C. 1958. Purification and serology of cucumber mosaic virus. Nature 182:1616.
57. Veerisetty, V., and Brakke, M. K. 1978. Purification of some legume carlaviruses. Phytopathology 68:59-64.
58. Veken, J. H. van der. 1960. Some applications of freeze-drying in virological research. Tijdschr. Plantenziekten 66:1-11.
59. Venekamp, J. H., Mosch, W. H. M., and Taborsky, V. 1973. Purification of potato virus X, white clover mosaic virus, tobacco mosaic virus and ribosomes by column chromatography. J. Chromat. 75:235-246.

Section II.A.2.

PREPARATION OF ANTIGENS, BACTERIA

S. H. De Boer
Agriculture Canada Research Station
Vancouver, B. C., Canada

N. W. Schaad
Harris Moran Seed Co.
San Juan Bautista, CA 95045

Principle

Bacterial antigens may be extracellular, intracellular, or structural, and their biochemical nature may be protein, glycoprotein, polysaccharide, lipopolysaccharide, or lipid. The usefulness of an antigen for serological detection depends on its properties and the serological test being used. For example, structural (or somatic) antigens normally work best for immunofluorescence and direct agglutination tests; soluble antigens are better in precipitin tests, ELISA, and immunodiffusion. Important properties of the antigen are: immunogenicity (ability to elicit antibody response), location in the bacterium, copy number, solubility, and stability.

Serological tests based on undefined antigens can be useful for diagnosis and detection work (20). However, purified antigens permit optimization of serological tests, accurate determination of specificity, and precise delineation of strain relationships.

Antigen preparations are used for both immunization of animals to elicit antibody response and for determining reactivity with antibody. The antigen preparation used for immunization is sometimes referred to as immunogen (or inject antigen), and that used for reaction with antibody as test antigen. Since usefulness of an antigen preparation as immunogen or as test antigen may differ, they are discussed separately.

Immunogens

The use of crude antigen preparations to produce polyclonal antisera yields antisera reactive with many determinants, whereas crude antigen preparations may be used to produce monoclonal antibodies specific to a single antigenic determinant (5,9). Thus, use of purified antigen preparations is more important for production of polyclonal than monoclonal antibodies. Nevertheless, purified antigens are useful for production of monoclonal antibodies to known components (8).

Immunogen preparations may consist of whole cells, crude cell extracts, or purified bacterial components. Some procedures for their preparation are as follows:

Whole cells

Untreated - Bacterial cells grown on solid or in broth media can be used directly as antigen for immunization (12). It is recommended that cells be washed twice in 0.01 M phosphate buffered saline, pH 7.2 (PBS) by centrifugation and resuspension. Antibodies raised to whole cells tend to react with lipopolysaccharides (for Gram negative bacteria only), exposed proteins including flagellin, and extracellular polysaccharides.

Heat-treated - Heat treatment can be done either by treating washed cells in PBS in a boiling water bath at 100 C for 2 hr or by autoclaving for 2 hr at 121 C (12). Heat treatment destroys most proteins, thus permitting some degree of selection for carbohydrate antigens.

Formalin-fixed - To a suspension of bacteria in saline (0.85%) add an equal volume of 0.6% formalin in saline (12). Incubate at room temperature for 1-2 da and collect bacteria by centrifugation and resuspend in 0.3% formalized saline. Fixation stabilizes certain antigens such as flagella.

Glutaraldehyde-fixed - Wash bacterial cells three times in PBS and dialyze against 2% glutaraldehyde for 3 hr at room temperature (1). Then dialyze against PBS for 24 hr with several changes of buffer to remove residual glutaraldehyde. In addition to stabilizing cell wall antigens, glutaraldehyde fixation may decrease toxicity of some bacterial species to the host animal.

Crude Antigen Extracts

Glycoprotein - Shake 48-hr-old cells in sterile distilled water for 30 min (11). Remove cells by centrifugation and neutralize the supernatant fraction to pH 7.0. Add an equal volume of saturated ammonium sulfate to the supernatant fraction and incubate at 4 C overnight. Pellet the precipitate by centrifugation at 20,000 g at 4 C for 15 min, dissolve in sterile distilled water, and dialyze to remove the ammonium sulfate. This preparation will contain a mixture of proteins and polysaccharides from the surface of the cells.

Membrane protein complex - Suspend pelleted bacterial cells in 20 ml of 0.2 M LiCl/g wet weight of cells (14,26). Agitate the suspension at 45 C for 2-3 hr with or without the presence of glass beads (3 mm diameter). Remove cells and other debris by centrifugation at 12,000 g for 30 min and centrifuge the supernatant fraction at 100,000 g for 2 hr. Resuspend the high speed pellet in distilled water. This crude membrane protein preparation also contains cell wall polysaccharides. The antigenic moiety responsible for the specific reaction of membrane protein extracts of *Xanthomonas campestris* has been identified as a heat stable, inner membrane peptide (23).

Purified Bacterial Components

Lipopolysaccharide - Suspend washed cells in distilled water at 10 ml/g fresh weight of cells (6). Heat to 70 C and add an equal volume of warmed phenol. Maintain the suspension at 70 C for 15 min with agitation and then cool in ice. Centrifuge at 10,000 g for 15 min at 4 C and retain aqueous layer. Re-extract the pellet in the same way and pool aqueous fractions. Dialyze against distilled water for 24 hr to remove all traces of phenol. Centrifuge 2-4 times at 105,000 g for 4 hr until the supernatant has an absorbance at 28 nm of less than 0.01. Between centrifugations resuspend the lipopolysaccharide pellet in water.

Ribosome - Suspend 5 g wet weight of cells harvested from a log phase broth culture in 100 ml 0.01 Tris (Hydroxymethyl) aminomethane buffer, pH 7.4, containing 50 mM KCl and 5 mM $MgCl_2$ at 5 C (19). Add DNase to 2 mg/ml and disrupt the cells in a cold French pressure cell at 363 Kg/mm^2. Remove unbroken cells and cell debris by centrifugation. Precipitate most nonribosomal proteins by addition of ammonium sulfate to 20% and remove by centrifugation (2). Then precipitate ribosomes by increasing ammonium sulfate concentration to 40%. Harvest the ribosome precipitate by low speed centrifugation and resuspend in TSM buffer, 4 C, (10 mM Tris base, 3 mM succinic acid, 10 mM $MgCl_2$, 6 mM 2-mercaptoethanol, pH 8.0). Dialyze against TSM overnight and adjust to 0.6 M ammonium sulfate. Pellet ribosomes by centrifugation at 177,000 g, 2 hr at 2 C, and resuspend in TSM buffer, clarify by low-speed centrifugation if necessary. Ribosomal preparations may be contaminated with lipopolysaccharides and when used as immunogen may yield antibodies to these contaminants as well (13).

Extracellular polysaccharides - Harvest cells grown on solid medium in PBS and vortex at maximum speed for 1 min. Remove cells by centrifugation at 12,000 g for 1 hr. Precipitate the extracellular polysaccharide by the addition of four volumes of cold (-20 C) acetone. Incubate at 4 C overnight and centrifuge at 12,000 g for 30 min. Resuspend pellet in distilled water. Purity of such preparations will vary with the bacterial species.

Other - Purification of many other antigens such as secreted enzymes (18, 24), membrane proteins (16,22), flagella (17), pili (3), teichoic (21) and lipoteichoic acids (25), etc. is possible, but only limited use has been made of these immunogens for serology in phytobacteriological research.

Test Antigens

The choice of test antigen depends on the objective of the test and the type of test being used. For identification of isolates and comparison of strains, pure cultures of bacteria are desirable. Concentration of antigen preparation should also be standardized. For serological diagnosis of bacterial plant diseases and detection of plant pathogenic bacteria in plant or soil samples, crude preparations can often be used directly without purification of antigen.

Pure culture - For immunofluorescence, untreated whole cells are best, whereas for tests such as immunodiffusion which need a soluble antigen, sonicated or treated cells may be required. Sonicate washed cells for 2 min in 30-sec bursts in a 2 ml saline suspension, with cooling (15). To increase solubility of the lipopolysaccharide antigen in whole cell suspensions, treat suspensions with 50 μl/ml liquified phenol (7). Purified antigens from cultured cells used as immunogen can also be used as test antigen (19).

Plant samples - For many serological tests, tissue homogenates can be used directly. Diseased tissue usually contains intact pathogenic cells, bacterial cell products, and disrupted cells and, therefore, will react in serological tests that require whole cells or soluble antigens.

Preparation of tissue homogenates for specific serological tests may differ somewhat to obtain optimum reactions. For immunofluorescence, prepare homogenates and dilutions in distilled water; buffer salts may interfere with fixation of cells to glass microscope slides. Prepare samples for ELISA in PBS containing 2% polyvinyl pyrrolidone, 0.2% skim milk (i.e. 2% BLOTTO) and 0.05% Tween 20, to minimize non-specific reactions (10). For immunodiffusion, sample treatment may be necessary to prevent non-specific precipitin band formation due to plant components (4).

Literature Cited

1. Allan, E., and Kelman, A. 1977. Immunofluorescent stain procedures for detection and identification of _Erwinia carotovora_ var. _atroseptica_. Phytopathology 67:1305-1312.
2. Bouzar, H., Moore, L. W. and Schaad, N. W. 1986. Serological relationship between ribosomal subunits from strains of Agrobacterium and Rhizobium. Phytopathology 76:1265-1269.
3. Clegg, S., and Gerlach, G. F. 1987. Enterobacterial fimbriae. J. Bacteriol. 169:934-938.
4. De Boer, S. H. 1983. Evaluation of an agar immunodiffusion procedure for confirming bacterial ring rot diagnoses. Am. Potato J. 60:661-669.
5. De Boer, S. H. 1987. Use of monoclonal antibodies to identify and detect plant pathogenic bacteria. Can. J. Plant Pathol. 9:182-187.
6. De Boer, S. H., Bradshaw-Rouse, J. J., Sequeira, L., and McNaughton, M. E. 1985. Sugar composition and serological specificity of _Erwinia carotovora_ lipopolysaccharides. Can. J. Microbiol. 31:583-586.
7. De Boer, S. H., Copeman, R. J., and Vruggink, H. 1979. Serogroups of _Erwinia carotovora_ potato strains determined with diffusible somatic antigens. Phytopathology 69:316-319.
8. De Boer, S. H., and McNaughton, M. E. 1987. Monoclonal antibodies to the lipopolysaccharide of _Erwinia carotovora_ subsp. _atroseptica_ serogroup I. Phytopathology 77:828-832.
9. De Boer, S. H., and Wieczorek, A. 1984. Production of monoclonal antibodies to _Corynebacterium sepedonicum_. Phytopathology 74:1431-1434.
10. De Boer, S. H., Wieczorek, A., and Kummer, A. 1988. An ELISA test for bacterial ring rot of potato with a new monoclonal antibody. Plant Dis. 72:874-878.
11. Digat, B., and Cambra, M. 1976. Specificity of antigens in _Pseudomonas solanacearum_ E. F. Sm. and application of serology for studying bacterial wilt. Pages 38-57 in: Proceedings of the First International Planning Conference and Workshop on the Ecology and Control of Bacterial Wilt caused by _Pseudomonas solanacearum_, L. Sequeira and A. Kelman, eds. Raleigh, North Carolina. 166 pp.
12. Edwards, P. R., and Ewing, W. H. 1972. Identification of Enterobacteriaceae. 3rd Ed. Burgess Publishing Co., Minneapolis, MN. 258 pp.
13. Eisenstein, T. K. 1975. Evidence for O antigens as the antigenic determinants in "ribosomal" vaccines prepared from _Salmonella_. Infect. Immun. 12:364-377.
14. Frasch, C. E., and Gotschlich, E. C. 1974. An outer membrane protein of _Neisseria menigitidis_ group B responsible for serotype specificity. J. Exp. Med. 140:87-104.
15. Gibbins, L. N. 1974. Variation in the occurrence of extracellular diffusible antigens in temperature-induced variants of _Erwinia herbicola_ Y46, and observations of their relationships with _Erwinia amylovora_. Can. J. Microbiol. 20:643-648.

16. Hofstra, H., and Dankert, J. 1980. Major outer membrane proteins: common antigens in Enterobacteriaceae species. J. Gen. Microbiol. 119:123-131.
17. Koffler, H., and Smith, R. W. 1972. The nature of flagellar antigens. Pages 31-66 in: Cellular Antigens; Lectures and Summaries. A. Nowotny, ed. Springer-Verlag, Inc., New York. 329 pp.
18. Mazzucchi, U., Alberghina, A., and Garibaldi, A. 1974. Comparative immunological study of pectic lyases produced by soft rot coliform bacteria. Phytopath. Medit. 13:27-35.
19. Schaad, N. W. 1974. Comparative immunology of ribosomes and disc gel electrophoresis of ribosomal proteins from Erwiniae, Pectobacteria, and other members of the family Enterobacteriaceae. Int. J. Syst. Bacteriol. 24:42-53.
20. Schaad, N. W. 1979. Serological identification of plant pathogenic bacteria. Annu. Rev. Phytopathol. 17:123-147.
21. Sharpe, M. E., Brock, J. H., Knox, K. W., and Wicken, A. J. 1973. Glycerol teichoic acid as a common antigenic factor in Lactobacilli and some other gram-positive organisms. J. Gen. Microbiol. 74:119-126.
22. Smyth, C. J. 1985. Immunology of outer membrane proteins of gram-negative bacteria. Pages 177-201 in: Immunology of the Bacterial Cell Envelope. D. E. S. Stewart-Tull and M. Davies, eds. John Wiley & Sons, Ltd. 391 pp.
23. Thaveechai, N., and Schaad, N. W. 1986. Immunological characterization of a subspecies - specific antigenic determinant of a membrane protein extract of *Xanthomonas campestris* pv. *campestris*. Phytopathology 76:148-153.
24. Ward, L. J., and De Boer, S. H. 1989. Characterization of a monoclonal antibody against active pectate lyase from *Erwinia carotovora*. Can. J. Microbiol. 35:651-655.
25. Wicken, A. J., and Knox, A. J. 1975. Lipoteichoic acids: a new class of bacterial antigen. Science 187:1161-1167.
26. Yakrus, M., and Schaad, N. W. 1979. Serological relationships among strains of *Erwinia chrysanthemi*. Phytopathology 69:517-522.

Section II.B.1.

POLYCLONAL ANTIBODIES

E. M. Ball
University of Nebraska, Lincoln, NE 68583
R. O. Hampton
USDA-ARS, Oregon State University, Corvallis, OR 97331
S. H. De Boer
Agriculture Canada Research Station, Vancouver, B.C. V6T 1X2
N. W. Schaad
Harris Moran Seed Co., San Juan Bautista, CA 95045

Plant pathologists today can choose between polyclonal or monoclonal antibodies for sero-diagnostic studies of pathogens. The major advantages of the latter are: 1) the potential of specificity for a single epitope, 2) reaction spectra which may be selectively determined by screening hybridoma cell lines for the desired specificity, and 3) perpetuity of hybridoma cell lines through culturing and frozen storage. Polyclonal antibodies are easier to obtain and have higher specificity when made with highly purified antigen. Both polyclonal and monoclonal antibodies may cross-react with heterologous antigen or may be so specific that some heterologous antigen from strains of the pathogen are not recognized. The choice of whether monoclonal or polyclonal antibodies are used must be based upon the type and purpose of assays. Because polyclonal antisera consist of heterologous populations of antibodies with variable specificities, different segments of the antibody population may be reactive when using different serological tests. By contrast, monoclonal antibodies are homogeneous in type and specificity and tend to react uniformly among tests.

A. Choice of Animals

Any hemothermic animal can serve as a source of immune serum. Rabbits, chickens, guinea pigs, rats, and mice are commonly used for production of immune serum. However, rats, mice, and guinea pigs yield relatively small volumes of serum. Chickens have high body temperature, and it is reported that chicken serum requires high osmolarity for antigen-antibody reactions; such conditions, however, may be detrimental to the integrity of some antigens. Larger volumes of serum may be obtained from larger animals such as sheep, goats, and horses, but more antigen is required for immunization. Rabbits are used most often to produce immune serum because they can be housed and cared for with little effort, are relatively easy to handle and inexpensive, and yield adequate volumes of high-titered serum in return for relatively small amounts of antigen used for immunization. For precipitating antibodies, such as needed for agar diffusion tests, chickens, rabbits, and horses are the best serum sources and guinea pigs, rats, and mice are unsatisfactory sources. Rabbit and chicken antisera produce precipitin reactions over a wide range of concentration, whereas sheep and goat antisera precipitate antigen over a narrow range. To avoid problems of variation in response between animals, at least two animals of the same species should be immunized with the same antigen in any serological study. More than two animals would be better, but are usually not feasible.

B. **Immunization**

The amount of immunogen required depends on the characteristics of the antigen, the means by which it is prepared, and the method used for injection. Longer immunization schedules requiring more antigen normally result in a greater variety and quantity of antibody than shorter schedules, but may produce antibody of lower specificity.

Generally, a combination of injection methods is used to present immunogen to the antibody producing sites in the animals. Both intramuscular and intravenous injections are usually combined in a series lasting 3-4 weeks. Intravenous (IV) injections result in the most rapid distribution of antigen to many sites, but antigen may be degraded quickly in the blood stream by the body defense systems. Therefore, more than 8 IV injections may be required to obtain serum of useable antibody concentrations.

B.1. Use of adjuvants. Antigens in buffer solutions are probably best prepared for intramuscular (IM) injections by emulsifying 1:1 (v:v) with Freund's adjuvant[a] (Difco, Sigma, Calbiochem, etc.). The adjuvant consists of paraffin oil and an emulsifier, mannide monooleate. Complete Freund's adjuvant contains heat-killed *Mycobacterium tuberculosis*, *M. butyricum* or a similar acid-fast bacterium. The incomplete form contains no bacterium. The purpose of adjuvant is to act as a depot for antigen within muscle tissue, allowing for slow release, thereby stimulating more antibody production than would occur if the same amount of antigen were injected as an aqueous solution. Complete adjuvant may be more effective for enhancing the immune response than incomplete adjuvant. In an IM injection series, complete adjuvant usually is administered in the first injection and the incomplete adjuvant given thereafter. This sequence lessens the chance of developing abscesses at the site of the injected adjuvant. Development of a complete, stable antigen-adjuvant emulsion before injection is important. Essentially, there are two ways to do this. A glass syringe fitted with an 18 or 20 gauge needle, from which the tip has been removed, is used to rapidly withdraw and expel the antigen-adjuvant mix in a small tube or bottle. Use of plastic, instead of glass, containers maximizes recovery of antigen after emulsifying. Tandem syringes also may be joined by tight-fitting tubing to rapidly force the mix back and forth between the syringes. Alternatively, mechanical overhead stirrers can be fitted with wire "beaters" (e.g., large-size paper clips partially straightened) to emulsify antigen-adjuvant preparations directly in the syringe barrel to be used for injection. In this case, the Luer-tip can be sealed with parafilm until ready for injection. An electric drill fitted with a Phillips screwdriver may also be used and commercial, motor-driven instruments are available. By either method, the product for IM injection should be a thick, creamy emulsion which does not separate on standing. To test for complete emulsification, a drop of the mixture may be placed on the still clean surface of distilled water in a dish or beaker. If the droplet

---

a The use of Freund's adjuvant in laboratory animals must be approved by the local institutional Animal Care Committee in the United States.

remains intact and does not form a "slick" over the water surface, the emulsion is complete.

Other adjuvant systems are available that are very effective in stimulating production of high titer serum (see titration, Section III.D.) but have been used infrequently with plant pathogens. The Ribi Adjuvant System (RAS), from Ribi Immunochemical Research, incorporates adjuvant and antigen into an oil phase instead of the antigen-aqueous phase of Freund's adjuvant. The RAS contains 2% oil compared to 50% in Freund's, thereby reducing the viscosity of the emulsion and permitting more rapid release of antigen into the animal tissue. In the RAS, trehalose dimycolate (TDM) replaces the Mycobacterium of Freund's, and monophosphoryl lipid A (MPL), a detoxified endotoxin, is also included to enhance antibody response. Cell wall skeleton can also be used as an enhancer. Ribi reports that mice immunized with 50 μg of ovalbumin alone produced ELISA titers of 1:100 in 18 days of immunization, whereas 2-50 μg doses of antigen in RAS (with TDM and MPL) gave titers of 1:102,400 after 14 days. Also, mice immunized with 100 μg of bacterial antigen produced titers of 1:1600 after 21 days of immunization and with 2-100 μg doses of antigen with MPL gave titers of 1:12,800 at that time. Preliminary studies involving immunization with RAS and an intradermal dose of capsid protein isolated from polyacrylamide gels loaded with potato virus M, potato virus S, or potato leafroll virus showed that the immune sera were effective in blocking the viral reaction in double antibody sandwich (DAS)-ELISA and in indirect-ELISA tests (E.M.B.). Superior high titer antisera have been produced to cucumber mosaic virus, bean yellow mosaic virus, and pea seedborne mosaic virus using the Ribi adjuvant with an initial 20-site intradermal injection followed by an intramuscular injection of virus in Freund's adjuvant (R.O.H.).

Muramyl peptides, wax fractions of purified cell walls of <u>Mycobacterium</u>, have been used as adjuvants, and N-acetylmuramyl-L-alanyl-D-isoglutamine is described as the simplest active analog. Dimethyldioctadecyl ammonium bromide (DDA), a lipoidal quaternary ammonium compound also has been reported as an effective adjuvant (from Calbiochem Biochemical/Immunochemical catalog, 1987, p. 248-249).

Several companies are increasing efforts to find better, purer, and more chemically defined adjuvants. A bacterium from genus <u>Amycolate</u> has shown adjuvant activity without producing side effects. Purified bacterial ribosome aggregates elicit a strong immune response, and then degrade within the body, which eliminates the side effect caused by depot formation. Also, a glycoside extracted from tree bark (<u>Quillaja saponaria</u>) can boost vaccine activity, can be used effectively in microgram quantities, and does not produce side effects. Some synthetic compounds that are immunostimulatory include surfactants, polynucleotides, and complex carbohydrates. Combinations of different types of adjuvants may possibly be more effective stimulators of antibody production than any of these when given alone (Biotechnology 6:773-777).

B.2. <u>Injection methods</u>

B.2.a. <u>Intravenous</u>. For the <u>rabbit</u>, antigen is administered in about 0.5 ml of physiological saline (0.85% NaCl) by a small syringe fitted with a sterile 27-gauge needle. The marginal ear vein is dilated by swabbing the ear with 70% ethanol and rubbing the ear vigorously or by touching the tip of the ear with a swab moistened with xylene. An assistant can put pressure immediately at the base of the ear while the operator bends the ear forward with one hand and inserts the needle into the vein with the other, pointing toward the base of the ear (Fig. 1). Pressure on the vein is released, and the liquid is injected slowly causing momentary vein-clearing ahead of the needle. Upon withdrawing the needle, moderate pressure is applied to the wound site, using a cotton swab, until bleeding stops. For chickens, intravenous injection of the wing vein is preferred (ventral access), with the needle inserted toward the body. The tail vein of mice is used for intravenous injections.

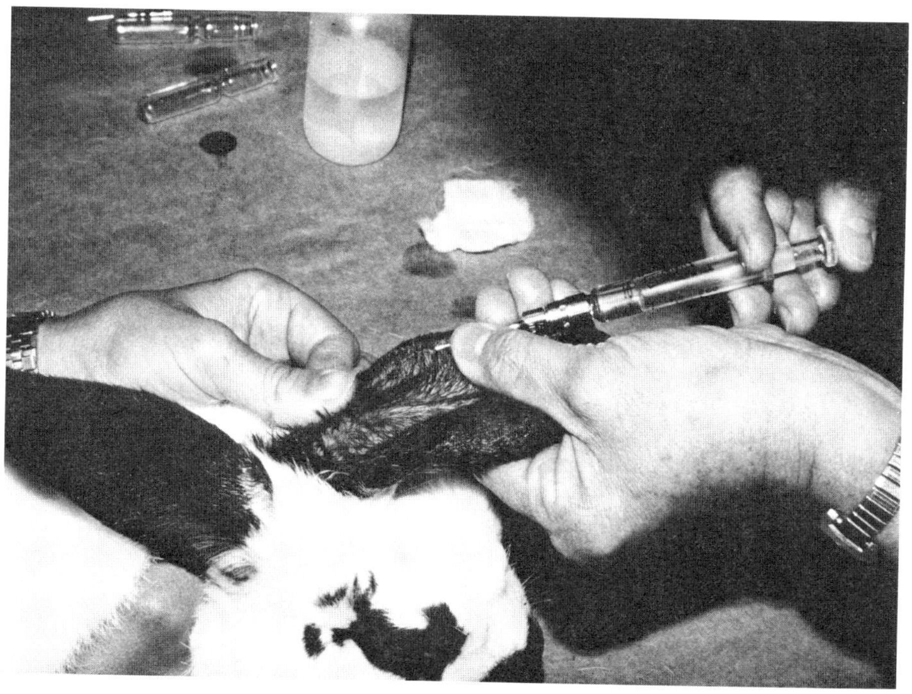

Fig. 1. Intravenous injection of a rabbit. The ear is swabbed with 70% ethanol and rubbed vigorously to dilate the marginal ear vein. Pressure is applied at the base of the ear by the technician, to block blood flow and enlarge the vein. The tip of a 27-gauge needle is slipped into the vein, and the antigenic solution is injected slowly after the pressure is released. This photograph also illustrates the central ear artery located just beneath the thumb of the technician, left. The artery is dilated due to the alcohol-rub and can be used to obtain samples of blood during immunization (see Serum acquisition, rabbits).

B.2.b. Intramuscular. For the rabbit, both the calf and thigh muscles receive multiple injections of about 0.2-0.4 ml of the antigen-adjuvant mix, several smaller sites being more effective than a single depot of antigen. Ethanol is swabbed on the leg and fur is separated to access skin area. The leg is held in the palm of the left hand, with fourth and fifth fingers curled around the heel and the first and second fingers applying slight pressure against the muscle to make it taut, thereby facilitating insertion of a sterile 20-22-gauge needle (Fig. 2). The adjuvant material is injected into the muscle, but not close to the bone. The same technique applies to the leg of a chicken. In addition, the breast muscles of the chicken can receive multiple-site injections of antigen. For subsequent injections, alternate leg or breast muscle can be used at weekly intervals.

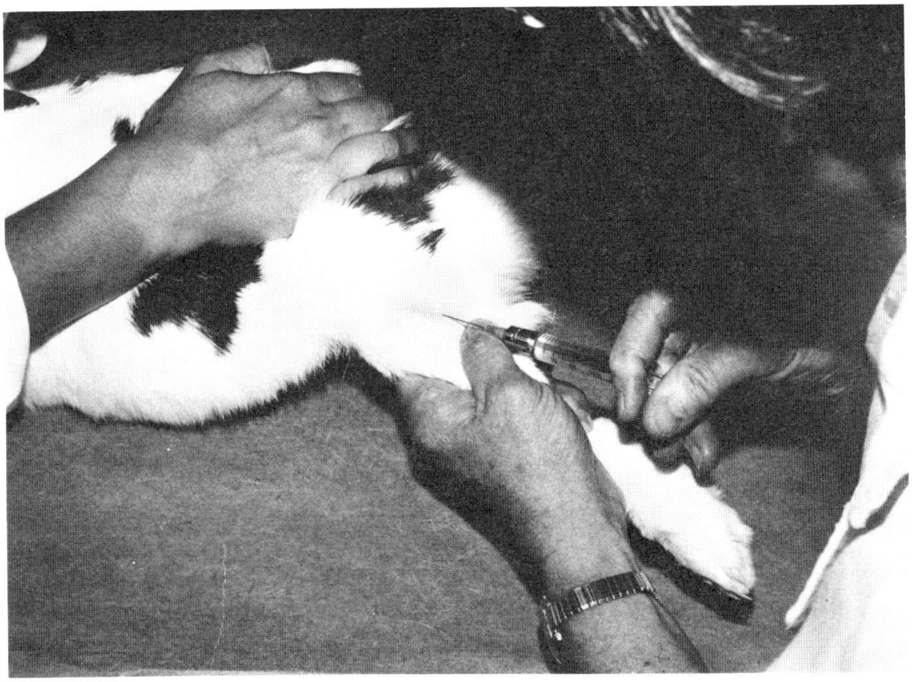

Fig. 2. Intramuscular injection of the hind leg of rabbit. The leg is extended, but not in an uncomfortable position. The technician locks the 4th and 5th finger of the left hand just above the "heel". The index and 3rd fingers extend underneath the thigh and/or calf muscles to make them taut. Ethanol is used to wet the fur permitting easy access to the skin. Antigen in adjuvant is injected in 0.2 - 0.4 ml volumes in several sites into the muscle tissue.

B.2.c. Intraperitoneal. For the rabbit, two people are needed. One person should grasp the hind legs between the second and third fingers of the left hand, with the right hand supporting the spinal column. The second person should grasp the left front foot with the left hand and the right front foot with the right hand. The two persons working in unison should flip the rabbit over on its back. The person holding the front feet should then release the feet to the second person. The person releasing the front feet is then to prepare the rabbit for injection. The area from caudal to center of abdomen is wiped with ethanol. Tap the animal over the abdomen so the muscles are tightened. Grasp the skin over the abdomen, pull up slightly and insert the needle into the peritoneal cavity. Be sure that the insertion is not just below the skin but into the peritoneal cavity. Penetration is indicated by reduced resistance to the needle. The antigen (0.5 ml) is forcibly injected immediately upon entry to help move intestines away from the injection site.

Intraperitoneal injection is the preferred method for immunizing mice with an adjuvant mix. The mouse is held by the scruff of the neck using thumb and forefinger of the left hand, and the tail and hind leg is held with the little finger (Fig. 3). Using a 22- or 24-gauge needle, about 0.2 ml of the emulsion is injected into the peritoneal cavity, being careful to avoid nicking the intestines.

Fig. 3. Intraperitoneal injection of a mouse. The mouse is caught by the tail using the right hand and allowed to stretch out on cage top. It is grasped by the thumb and index finger of the left hand at the "scruff" of the neck, the third (and fourth, if possible) finger is curled under the back to make the abdominal muscles slightly taut for easier insertion of the needle. The fourth and fifth finger holds the tail and left hind leg. Antigen is injected through a 22-gauge needle, with the tip penetrating the skin to the peritoneum but not penetrating the intestines.

B.2.d. Subcutaneous, rabbit

(1) Leg muscle area: After wiping the area with ethanol, pull the skin up to 0.5 to 1.0 cm using one hand. With the other hand, insert a 22-gauge needle parallel and underneath the skin. Inject 0.5 to 1.0 ml of antigen-adjuvant or antigen in saline. Withdraw the needle and massage the area injected. A small lump should be present.

(2) Nuchal area: Similarly, antigen in saline, or more often in adjuvant, is injected into the neck area of the rabbit by pulling up the loose skin and inserting a 22-gauge needle between the skin and muscle tissue. After withdrawing the needle, massage the injection area.

B.2.e. Intradermal, rabbit. Rabbits can be immunized very effectively by this method when only small amounts of antigen are available; for example, purified phloem-restricted viruses, antigens isolated from agarose or polyacrylamide gels, or non-capsid viral proteins. Hair is removed on either side of the backbone and over the shoulder area. One ml of antigen-adjuvant is injected through a 26-gauge needle into the skin, not under it, forming small white "blips" of 0.05 ml or less in two rows along each side. Thereby some 20 sites are made available to stimulate antibody production (31).

B.2.f. Foot, rabbit

(1) Footpads: Clean the pad of the foot with ethanol. Slowly insert a 26 gauge needle through the center of the sole until resistance ceases. Slowly inject a maximum of 0.025 ml of antigen-adjuvant/pad.
(2) Toepads: This method, like intradermal, is aimed toward getting antigen to lymph nodes as quickly as possible. Approximately 0.01 ml of an antigen-adjuvant is injected into each toepad of a rabbit.[b] Usually subsequent injections are given in IM or IV sites.

B.3. Injection schedule. General procedures for viral immunization that result in adequate antibody production in rabbits or chickens involve two IM injections, and one IV injection of 1-2 mg each given at weekly intervals (see also Table 1). Peak titer usually occurs 5 to 7 weeks after the first exposure. The schedule can be repeated depending upon the titers obtained from frequent testing of the sera throughout the immunization period. However, booster doses of antigen are effective only after a significant drop from the maximum antibody titer is demonstrated. Plant viruses vary considerably in their ability to elicit antibodies, but the injection of excessive amounts of antigen rarely improves antibody titer. Rather, alternation of injection methods or antigen stabilization is usually more advantageous.

---

b The use of Freund's adjuvant and toe- or footpad injections must be approved by the local institutional Animal Care Committee in the United States.

Table 1. Published Procedures for Immunizing Rabbits with Plant Pathogens

Immunogen	Route Number	Amount mg, ml	Reference
Virus			
Intact			
Broad bean wilt	1-IV, 4-IM*	46.2 mg	28
Blueberry shoestring	3-IM	3.6 mg	24
Kalanchoe latent	8-IP	2.0 mg	17
Lettuce infectious yellows	4-IM	1.3 mg	10
Maize dwarf mosaic	3-IM, 1-IV	4.0 mg	20
Red clover necrotic mosaic	2-IM, 1-IV	3.0 mg	27
Sweet clover necrotic mosaic	3-IM, 2-IV	10.0 mg	19
Turnip mosaic	4-IM	8.0 mg	29
Capsid proteins			
Alfalfa mosaic	3-IM	0.6 mg	1
Maize stripe	3-IM	0.3 mg	13
Papaya ringspot	2-IM	4.0 mg	26
Noncapsid proteins			
Barley stripe mosaic	2-IM, 1-ID, 1-SC	0.6 mg	2
Maize stripe - nucleoprotein	4-LN	4.0 mg	14
Maize stripe, rice hoja blanca	3-IM	1.0 mg	12
Papaya ringspot, pepper mottle - amorphous inclusion	3(IM-TP)	3.0 mg	8
Papaya ringspot, watermelon-1 - cylindrical inclusion	3(IM-TP)	3.0 mg	36
Sonchus yellow net-nucleocapsid gene	4-SC	4.0 mg	37
Tobacco vein mottling - components	3-IM	2.0 mg	18
Wheat streak mosaic - cylindrical inclusion	6-IM	3.1 mg	3
Bacterial Cells			
Corynebacterium sepedonicum	4-IM, 2-IV	3.0 ml	6
C. sepedonicum	1-SC, 3-IV	4.5 ml	5
Erwinia carotovora pv. *atroseptica*	3-IM, 2-IV	3.5 ml	7
Xanthomonas campestris pv. *phaseoli*	6-IV	3.0 ml	23
X. campestris pv. *campestris*	4-IM	5.0 ml	31
X. campestris pv. *holcicola*	7-10 SC	3.5-5.0 ml	22
Bacterial Components			
X. c. pv. *campestris*			
TCA extract	4 IP	7.5 mg	31
ribosomal extract	4 IP	12.0 mg	
membrane protein	3 IP	1.2 mg	32
PAGE-derived polypeptides	3 IP	0.7 mg	33
Clavibacter michiganense pv. *insidiosum* extra-cellular polysaccharide	1 IM, 2-IP	141-370 μg	34
Erwinia chrysanthemi membrane protein	1M, 1P	1.7 mg	35
Rhizobium and *Agrobacterium* ribosomal subunits	4-IM	12.0 mg	4

Table I, contd.

Immunogen	Route Number	Amount mg, ml	Reference
Fungi			
Ceratocystis ulmi, toxin	3 IP	1.6 mg	25
Epichloe typhini, mycelium	3 SC	3.0 ml	21
Eutypa armeniace, cell walls	5 SC	22.0 mg	16
Gaeumannomyces graminis f. sp. tritici	1 ID, 3 IM	2.1 mg	11
Ophiostoma ulmi toxin (Ceratocystis)	2 IM	4.0 mg	30
Phaeolus schweinitzii, mycelium	2 IM	1.0 mg	9
Phomopsis longicola	4 SC	132.0 mg	15

*IM = intramuscular, IV = intravenous, IP = intraperitoneal
ID = intradermal, SC = subcutaneous, TP = toepad
LN = lymph node

Many different methods of injection have been successful in producing antisera to bacteria. In most cases, antisera have been made against cells fixed with formalin or glutaraldehyde. Heat treatment may be used to destroy protein antigens and thus permit some degree of selection for carbohydrate antigens. Subcutaneous injections of Erwinia chrysanthemi cells at 3 x 10^{11} cfu/ml in adjuvant, given at weekly intervals, produced serum titers in agglutination tests of 1:2,560 after 7-10 injections (11). Intravenous injections (0.5 ml), given at weekly intervals, of Xanthomonas campestris pv. phaseoli cells at 3 x 10^{8} cfu/ml resulted in titers of 1:5,120 after 6 weeks (23). Likewise, 7 IM injections of 0.5 ml of E. carotovora pv. carotovora at 1 x 10^{9} cells/ml produced an antiserum of acceptable titer (26). In the investigation of strains of Pseudomonas syringae pv. tomato, rabbits were immunized with autoclaved cells, cell wall, somatic (0) antigens, flagellar (H) antigens, and soluble cellular and extracellular antigens. "Short term" sera were made using IV injections of 10^{11} cells/ml given every 3 days for 3 weeks and rabbits exsanguinated 7 days later; "long term" sera were made using an IM injection of 10^{11} cells/ml in adjuvant with a second dose given 4.5 months later and rabbits exsanguinated 1.5 months afterwards. Sera from long term immunization gave higher titers to somatic and flagellar antigens and reduced cross-reactivity among strains compared to short term sera (12). Graded doses in adjuvant of 0.1, 0.2, and 0.4 mg of membrane protein extract from X. campestris pv. campestris in adjuvant produced antisera that were effective in organizing serological groupings of 35 strains (30). Graded doses of 1.5, 2.5, 3.5, and 4.5 mg in adjuvant given IM at 10 day intervals were successful in producing an antiserum to ribosomes of Agrobacterium and Rhizobium (4).

Examples of variability in dosage and amount with viruses, bacteria, fungi, and their components are presented in Table 1.

C. **Serum Acquisition and Treatment**

All animals used to produce antiserum should be fasted (food withheld 12-18 hr) prior to obtaining blood samples to eliminate the accumulation of lipids in the serum.

C.1. Rabbits.

C.1.a. Blood collection. Blood can be obtained from rabbits by several pathways: cardiac,[c] arterial or veinal. The following procedure can be used for cardiac puncture. With the animal being restrained on its back, a 50 mm long, 20 gauge needle with a 10-20 ml syringe or 'vacutainer' is inserted on the left side of sternal tab, directing the point toward the chest-center. Once through the diaphragm, suction is applied as the needle is moved slowly forward. Blood is collected from the heart in 15-30 seconds and the needle is withdrawn quickly. The animal should remain still for 1-2 min then be returned to its cage without any struggling. The entire procedure is completed in a few minutes, 10-20 ml of blood withdrawn, and there is only a slight risk of doing damage to the heart.

Little risk to the life of the rabbit is involved in obtaining blood from the ear. A small swipe of xylene on the tip of the ear causes enough irritation to dilate the central ear artery. After swabbing the area with alcohol, a 20 or 22 gauge needle is gently slipped under the skin and into the engorged artery. Blood flows quickly through the needle into a beaker; 20 ml or more of blood can be collected in less than a minute. An alternative procedure involves a nick on the marginal ear vein, done with a razor blade positioned in line, not across, the vein. After collecting the blood in a tube or beaker, some difficulty may be encountered in stopping the flow of blood. Ice to cool the area or vaseline to seal the cut may be helpful. Bleeding from the central ear artery is also possible, and the nick by blade or needle usually seals quickly when slight pressure is applied to the nick.

To improve bleeding, the ear can be massaged by stroking with thumb and forefinger toward the ear tip. Massaging is very important for rabbits which bleed poorly and should be practiced before bleeding.

In addition, blood can be withdrawn from the ear by vacuum. Connect a commercial (Bellco Glass) vacuum blood collecting apparatus to a regulated vacuum line. Connect a good collecting vessel, such as a centrifuge tube. Cut the animal's ear vein as described and immediately place the collecting apparatus over the ear and against the head. Gradually apply vacuum until the apparatus has sealed against the rabbit's head. Avoid use of vacuum > 15 in Hg. After collecting the desired amount of blood, reduce and release the vacuum. Apply pressure to the ear to stop bleeding and allow clotting.

b. Serum processing. The blood is allowed to clot an hour at room temperature. The clot is released from the tube or beaker by rimming with a sterile applicator stick or glass rod. Heating the blood, after clotting, at 37 C for 30 minutes helps to shrink the clot and increase

[c] Anesthesia is required for cardiac bleeding and approval must be obtained from the local institutional Animal Care Committee in the United States.

the yield of serum. After overnight refrigeration, serum is decanted or aspirated from the clot and centrifuged at 2,000 rpm to remove cell debris. The serum is filtered through a 0.2 μm millipore or comparable filter, and brought to 0.025% sodium azide. Aliquots of serum, mixed with equal volumes of glycerol, are stored at -20 C, or lyophilized. Repeated freezing and thawing of raw serum in the absence of glycerol should be avoided.

c. Preparation of immunoglobulin (IgG)

(1) Partial purification by ammonium sulfate precipitation. The IgG fraction can be partially purified by selective precipitation with high concentrations of salts such as ammonium sulfate. Two procedures that have been widely used are as follows. The first procedure yields a higher degree of purity while the second procedure is simpler and faster; the second procedure is the one usually used prior to further purification by ion exchange, affinity chromatography, etc.

Procedure 1 (18)

(a) Add per ml of antiserum, 1 ml of demineralized water and 1.3 ml cold, saturated ammonium sulfate, pH 7.0 - 7.2, dropwise with stirring (final ammonium sulfate concentration 40%).
(b) Mix for 30 min and refrigerate overnight.
(c) Centrifuge at 8000 x g for 10 min and collect precipitate.
(d) Resuspend precipitate gently in 1 ml of 0.01 M phosphate buffer, pH 7.2, containing 0.15 M NaCl (PBS).
(e) Add 1 ml water and 1.02 ml saturated ammonium sulfate per ml of resuspended precipitate (final sulfate concentration 33%).
(f) Mix for 30 min and centrifuge as above.
(g) Resuspend pellet in PBS at one-half the original volume of antiserum.
(h) Dialyze preparation against several changes of PBS over 24-36 hr.
(i) Filter through 0.2 μm filter, add sodium azide to 0.025%, and store long term at -20 C or short term at 4 C.

Procedure 2

(a) To 1 ml antiserum add 9 ml distilled water.
(b) Add 10 ml saturated ammonium sulfate solution.
(c) Leave at room temperature for 30-60 min.
(d) Centrifuge for 5 min at 8000 x g and collect precipitate.
(e) Resuspend pellet in 2 ml half strength (1/2x) PBS.
(f) Dialyze 3 times against 500 ml of 1/2x PBS

(2) Purification of IgG by ion exchange chromatography. This procedure is relatively simple to perform (5,27) and has been used extensively to purify IgG prior to conjugation with an enzyme for use in enzyme-linked immunosorbent assays (6). Purification on an ion exchange cellulose, as described below, requires crude purification by ammonium sulfate precipitation (Procedure 2 above) prior to loading onto the column.

(a) Pre-condition DE23 cellulose (Whatman) by acid and alkali treatment according to manufacturer's instructions.
(b) Equilibrate DE23 cellulose by stirring in 1/2x PBS at 6 ml/g of wet ion exchanger. Leave 10 min and decant. Repeat procedure until pH of cellulose is equal to buffer.
(c) Add 7-10 ml equilibrated DE23 cellulose in 1/2x PBS at room temperature to a 12 x 120 mm column.
(d) Wash column several times with 1/2x PBS, connect column eluant tube to UV monitor, and adjust absorbance at 280 nm to zero.
(e) Add 2 ml ammonium sulfate-fractionated IgG to column.
(f) Wash column with 1/2x PBS and monitor effluent at 280 nm.
(g) Collect first peak from column.
(h) Dilute effluent peak with 1/2x PBS to an absorbance of 1.4 (= 1 mg/ml). Store at -20 C in Eppendorf centrifuge tubes.

NOTE: IgG does not bind to the ion exchanger under the conditions described but rather passes through the column. The unwanted serum components bind to the column. A 14 x 80 mm column can thus be used most efficiently.

Useful ion exchange products other than DE23 include DEAE-Sephadex A-50 (Pharmacia), DEAE-Sephacel (Pharmacia), Cellex D (Bio-Rad), DEAE- and CM-Affi-Gel Blue (Bio-Rad). These products bind IgG in low salt buffers and can be eluted with a salt gradient. IgG is removed from these ion exchangers at about 0.25 M NaCl. (See Antibody Purification using DEAE, P. 301, ref. 14).

(3) Purification of IgG by Protein A affinity chromatography. Protein A binds specifically with the Fc portion of IgG (caution: not all species) and can therefore be used for affinity chromatography[d] (22,33,35). A convenient form of Protein A, bound to a chromatographic bead, is Protein A-Sepharose CL-4B (Pharmacia). The procedure with this product is as follows:

(a) Saturate 1.5 g Protein A-Sepharose in 0.01 M phosphate buffer, pH 7.2, for 15 min. Wash swollen powder with about 200 ml phosphate buffer on a sintered glass filter. Additional Protein A products include Affi-Gel Protein A Gel (BioRad) and Protein A Agarose (BRL). (See also, "Purification on Protein A Beads", Text Ref. 14, p. 309.)
(b) Pack Protein A-Sepharose into a small column.
(c) Dialyze ammonium sulfate-fractionated IgG against phosphate buffer and apply to column.
(d) Wash off unbound protein with phosphate buffer, pH 8.0, until absorbance at 280 is zero.
(e) Elute IgG with 1 M acetic acid.
(f) Collect eluted IgG fraction, neutralize immediately with 2 M Tris base, and dialyze against PBS.

---

[d] Protein G, from Streptococcus GX, which has broad affinity for IgG (F_c region) of diverse animal species, is now available commercially (e.g., Zymed Laboratories).

(4) Purification of IgG by dissociation from homologous antigen-antibody complex (See Text Ref. 32, p. 52).

(a) Dialyze IgG against distilled water using 2 changes of water.
(b) Adjust pH of IgG to 7.8 with 0.1 N NaOH.
(c) Prior knowledge of antigen stability over the specified pH range is required.
(d) Precipitate antigen from supernatant fluid 2 or more times until no further reaction is visible. Pool precipitates.
(e) Resuspend the precipitate in water, adjust pH to 2.9 with 0.1 N HCl. Pellet the antigen by high speed centrifugation.
(f) Neutralize the antibody in the supernatant fluid immediately.
(g) If antigen is adsorbed onto a solid phase [Cyanogen bromide (CNBr)-activated Sepharose 4B, Pharmacia] it can be separated from the released antibodies by centrifugation at 10,000 rpm - 15 min, thereby decreasing the time of exposure of antibodies to low pH (See Affinity Purified Antibodies, Text Ref. 14, p. 312).

(5) Elution of IgG from antigen-antibody complexes by sucrose density gradient centrifugation (24).

(a) To 5 ml of serum, add increment amounts of antigen (e.g., purified intact virus). Collect the precipitate after each addition.
(b) Combine precipitates and wash in PBS. Resuspend pellets in 2.0 ml of saline (0.85% NaCl) and adjust to pH 2.0 with 1 N HCl.
(c) After incubation, layer 1.0 ml on a linear sucrose density gradient column (7, 10, 10, 10 ml of 10, 20, 30, 40% sucrose in saline, pH 2.0) previously equilibrated overnight.
(d) Centrifuge 82,500 x g for 2.5 h. Fractionate gradient using fractionator and UV scanner. Antigens (not degraded by exposure to pH 2.0) are separated from IgG forming an absorbing zone characteristic of the protein and IgG remains at the top of the column.
(e) Collect the IgG zone and dialyze against PBS to remove sucrose and to restore neutral pH.
(f) The IgG can be concentrated by precipitation with 50% saturated ammonium sulfate and dialyzed in PBS as described previously.

(6) Preparation of $F(ab')_2$ fragments. The $F(ab')_2$ fragment of IgG is sometimes employed as coating antibody in indirect ELISA tests since it allows for the same antibody source to be used for trapping and detecting the antigen in DAS ELISA without the need to prepare an enzyme-antibody conjugate. In the test, the second antibody binds to the trapped antigen. The sandwich antibody is probed with an enzyme conjugate Protein A that reacts specifically with the Fc portion of the IgG molecule. Since no Fc fragment is present in the coating $F(ab')_2$ antibody fragment, the conjugate only reacts with the second antibody.

(a) Prepare 0.7 M sodium acetate buffer, pH 4, containing 0.05 M NaCl. Prepare 0.1 M cysteine HCl in water.
(b) To 1-5 mg/ml IgG in acetate buffer, add pepsin at 45-50 μg enzyme/ml IgG. Incubate overnight, 37 C.

(c) Add 1 volume cysteine solution to 9 volumes enzyme mixture.
(d) Separate $F(ab')_2$ fragments from enzyme and other digestion products by chromatography on Sephadex G100 column, by rapidly desalting on Sephadex G25, or by dialysis against 3 changes of PBS (1). Recent applications of this procedure have been reported with cucurbit viruses (8), southern bean mosaic virus (28), maracuja mosaic virus (13), and blueberry red ringspot virus (15). (A commercial kit for preparing $F(ab')_2$ is available from Pierce.)

(7) Alternate IgG purification procedures. In addition to the above, various other procedures have been used to obtain pure immunoglobulin preparations. For example, the use of high performance liquid chromatography (HPLC) gives extremely pure preparations in a short period of time but instrumentation is costly. Ultrafiltration through size-specific membranes can also be used. The high molecular weight IgM class of molecules can readily be purified by gel filtration using gel matrices such as Sephacryl S300 or S500 (Pharmacia).

(8) Testing antibody purity. The degree to which antibody molecules (immunoglobulins) have been purified from other serum components is most readily assayed by polyacrylamide gel electrophoresis in sodium dodecyl sulfate (SDS-PAGE) according to the procedure of Laemmli (21). Standard procedures for protein analysis work well, resolving the immunoglobulins in two major bands consisting of the heavy and light chains, respectively. Using a minigel apparatus the procedure is as follows:

(a) Prepare 10 ml separating gel solution by adding to 3.35 ml distilled water, 2.5 ml 1.5 M Tris-HCl (pH 8.8), 100 μl 10% SDS, 4 ml 30% acrylamide/bis (29.6 g acrylamide and 0.8 g N'N'-Bis-methylene-acrylamide in 100 ml distilled water), 50 μl 10% ammonium persulfate, and 5 μl TEMED.
(b) Pour separating gel solution immediately into assembled slab gel apparatus (using 0.75 mm spacers) to within about 2 cm of the top of the assembly, cover with a layer of tert-butanol, and allow to polymerize for about 1 hr.
(c) After polymerization is complete, pour off butanol, rinse top of gel with distilled water, and wick away all traces of water with tissue paper.
(d) Prepare 10 ml of stacking gel solution by adding to 6.1 ml distilled water, 2.5 ml 0.5 M Tris-HCl (pH 6.8), 100 μl 10% SDS, 1.3 ml 30% acrylamide/bis, 50 μl ammonium persulfate, and 10 μl TEMED.
(e) Pour stacking gel solution on top of the separating gel and insert comb.
(f) After the stacking gel polymerization is complete, remove comb, and wash wells with distilled water.
(g) Mix immunoglobulin preparation with an equal volume of sample buffer (4 ml distilled water, 1 ml 0.5 M Tris-HCl pH 6.8, 0.8 ml glycerol, 1.6 ml 10% SDS, 0.4 ml 2-β mercaptoethanol, and 0.2 ml 0.05% bromophenol blue).

(h) Load wells with 2, 5, 10, and 15 μl of sample, and place gel sandwich into running tank.
(i) Fill lower and upper buffer reservoirs with running buffer (5x buffer stock contains 15 g Tris, 72 g glycine, and 5 g SDS per liter of distilled water).
(j) Attach electrodes and run at 60 V until the bromophenol blue tracking dye passes into the separating gel, about 15 min. Then run at 25 mA/gel constant current for about 1 hr or until the tracking dye is near the bottom of the gel.
(k) Carefully remove gel from gel sandwich and stain for 1/2 hr in Coomassie stain (0.1% Coomassie Blue R-250 in 40% methanol/10% acetic acid).
(l) Destain gel to remove background by placing in 40% methanol/10% acetic acid for 1-3 hr.

(9) <u>Cross-absorption for removing non-pathogen-specific antibodies</u>

(a) When initial evaluations of antisera indicate the presence of non-pathogen-specific antibodies, cross-absorption with healthy-plant extracts may be necessary. Several approaches and protocols are available for this purpose, both for whole antiserum and purified IgG.

[1] The preferred method for cross-absorption involves immobilized healthy-plant extract (i.e., acetone-extracted powder) to preclude addition of extraneous soluble complexes to the antiserum.

[a] Clarify healthy-plant sap by the procedure used to prepare immunogen.
[b] Following first low-speed centrifugation, pour off supernatant fluid and resuspend pellet in buffer (1/3 volume of buffer used for extraction/clarification). In fume hood, add 1/2 volume of acetone and mix in blender.
[c] Centrifuge preparation 10 min at 8,000 g and discard supernatant fluid. Wash pellet in buffer three times by repeated resuspension, stirring, and centrifugation.
[d] Repeat acetone extraction by twice resuspending pellet in buffer, adding equal volume of acetone, mixing, and centrifuging.
[e] Pour suspension onto filter paper in a Buchner funnel, with light suction applied. Wash residue with acetone and bring to partial dryness by funnel aspiration.
[f] Place filter with residue in a Petri dish and air dry in fume hood. Scrape dried residue into a mortar and grind to fine powder. Store residue in sealed vial at -20 C. **Example**: 100 g <u>Physalis</u> spp. leaves yield 0.6 g powder; 100 g <u>N. tabacum</u> leaves yield 1 g powder.
[g] For cross-absorption, add 40 - 50 mg of powder to 1 ml antiserum and mix. Incubate 2 hr at 22 C, with occasional mixing, and refrigerate overnight. Centrifuge to remove precipitate. Test cross absorption efficacy by appropriate serological test.

[2] Alternatively, remove non-pathogen-specific antibodies by passing antiserum through DEAE-cellulose to which plant antigens have been attached.

[a] Add 20 - 50 ml demineralized water to 1.0 g DEAE cellulose. Wash several times in appropriate buffer, using 2,000 x g centrifugation to separate cellulose from supernatant fluid.

[b] Clarify healthy-plant sap by the procedure used to prepare immunogen.

[c] Resuspend cellulose in 40 - 50 ml of clarified plant sap and stir 1 hr at 22 C. Wash with buffer until supernatant fluid is clear.

[d] Bring mix to 2.0% BSA to saturate unbound sites on cellulose. Bring mix to 0.25% gluteraldehyde to cross-link bound reagents.

[e] Pour treated cellulose into a column. Equilibrate column with buffer and apply antiserum sample. **Example**: When 10 ml PVY antiserum is added to a column prepared as above and equilibrated with PBS, a 6 ml void volume is followed by a collection of 20 ml of serum, yielding 49.5 mg of IgG. Cross reactions with healthy sap are removed by this process.

[f] If the column is to be subsequently used to again cross-absorb antiserum, it should be treated with 1.0 M glycine, pH 2.8, to dissociate antibodies, then re-equilibrated with an appropriate buffer before use.

[3] Cross-absorption with soluble plant protein. If use of soluble plant protein is preferred, healthy-plant sap is clarified as described in No.1, above, except that centrifuged supernatant fluid rather than pellet is saved. After cross-absorption and centrifugation of antiserum, virus-specific IgG is rescued by passage through a column containing Protein A-Sepharose CL-4B (Pharmacia). (See Purification of IgG by Protein A affinity chromatography, this Section.)

[4] Direct cross-absorption in ELISA

[a] DAS ELISA. Coat wells with primary antibody and add antigen as in standard procedure. Dilute IgG-enzyme conjugate with clarified healthy plant sap (1:400 to 1:1000 in buffer), and incubate mixture 20 min. Add sap-conjugate mix in the place of standard conjugate and complete test by the standard procedure.

[b] Indirect ELISA. Coat wells with antigen, as in standard procedure. Dilute IgG stock to desired concentration (e.g., 1 mg IgG/ml) with healthy plant extract (e.g., 10 to 100 ml buffer per g tissue). Place mix directly into wells of microtiter plate, incubate 1 hr at 37 C, and thoroughly wash wells. Proceed with remaining steps by standard procedure.

(b) Bacteria. Since most bacteria can be cultured on artificial media, bacterial cells free of host plant antigens can be obtained readily. Sometimes, however, polyclonal antisera prepared against a specific plant pathogenic bacterium may cross-react with other phytopathogens or saprophytic bacteria. Such cross-reaction is usually caused by the presence of shared antigenic determinants. Antiserum can thus be rendered more specific by removing antibodies to shared antigens by cross-absorption.

[1] Suspend 10^6 to 10^8 cells per ml of PBS from a 24- to 48-hr-old culture of the cross-reacting bacterium. Mix suspension 1:1 with antiserum and incubate 2 to 4 hr at 48 to 50 C.[e]

[2] Remove bacteria from antiserum by centrifugation 15 min at 7,000 g. Test cross-absorption efficacy by agar double diffusion serology.

[3] Repeat cross-absorption, as necessary, to complete the removal of non-pathogen-specific antibodies.

Comments: Antiserum to *Erwinia carotovora* subsp. *atroseptica* that cross-reacted with some strains of *E. carotovora* subsp. *carotovora* was rendered specific for subsp. *atroseptica* by cross-absorption.[f] However, antiserum to *Corynebacterium sepedonicum* was not rendered species-specific by cross-absorption, because this species shares antigens with many saprophytic bacteria.[g]

C.2. Chickens. The laying hen presents many advantages as a source of antibody. The bird can be immunized with relatively small amounts of antigen, blood can be collected via cardiac puncture (see footnote for Rabbits, blood collection) or veinal bleeding, but most importantly, eggs can be collected from which immunoglobulin G can be isolated. Yolk preparations yield IgG in concentrations that parallel the amount in blood serum. Whether a lipid solvent or polyethylene glycol (PEG) precipitation is used to extract egg yolk-IgG, the product is the same. An equal volume of PBS (0.01 M phosphate) is stirred into 2-4 yolks. Then, with stirring, an equal volume (to the yolk-mix) of chloroform is added. The emulsion is broken by centrifugation and the water phase containing IgG is removed. The antibodies can be further purified by $(NH_4)_2SO_4$ precipitation and passage through a DEAE-Sephacel column (17). With PEG extractions, 1 volume of PBS is mixed with 2 volumes of yolk. Pulverized PEG-6000 is brought to 3.5% and mixed until dissolved. After centrifugation, the clear supernatant layer is mixed with a 12% concentration of PEG. The IgG is pelleted

---

[e] Crowley, C. F., and De Boer, S. H. 1982. Nonpathogenic bacteria associated with potato stems cross-react with *Corynebacterium sepedonicum* antisera in immunofluorescence. Am. Potato J. 59:1-8.

[f] Edwards, P. R., and Ewing, W. H. 1972. Identification of Enterobacteriaceae. 3rd Edition. Burgess Publishing Co., Minn. 362 pp.

[g] Vruggink, H., and Maas Geesteranus, H. P. 1975. Serological recognition of *Erwinia carotovora* var. *atroseptica*, the causal organism of potato blackleg. Potato Res. 18:546-555.

by centrifugation, redissolved and the precipitation repeated. Finally, the pellet is taken up in PBS buffer at 1/2 the original volume of yolk. This product can be further purified as described previously (25).

A chicken immunized with a total of 150 μg of carnation etched ring virus produced egg yolk IgG with a titer of 1:25,000 by the eighth day of immunization and 1:162,000 after 30 days using indirect ELISA tests (16). In a study with the SGV strain of barley yellow dwarf virus, a hen immunized with less than 100 μg of virus produced some 216 antibody-yielding eggs over a 15 month study. While the antibody level was low, the yolk-IgG was effective in neutralization tests with aphids and in reacting specifically with virus in indirect ELISA tests. A yield of 5 mg IgG/ml yolk was reported (17). In other studies, where antigen purification and concentration was not a limiting factor, 6-12 mg IgG/ml yolk (25) and 25 mg IgG/ml yolk were reported (2), good yields of IgG considering 200 ml of yolk can be obtained in 20 days following immunization.

C.3. Mice. Mice are becoming an increasingly useful antibody producer for the plant pathologist in more ways than for production of monoclonal antibodies. Though they yield very little blood, mice are susceptible to an ascites tumor which produces abdominal fluid as rich in antibody as the blood serum. The advantages of using murine ascitic fluid are: 1) very small doses of the antigen are required for immunization, 2) the maximum total volume of fluid collected under ideal conditions is 20 ml (40 ml total from a single mouse has been reported) (10), 3) the ascitic fluid can be treated for purification of IgG and stored as described for immune serum, and 4) ascites antibodies can be used very effectively for coating or as "second" antibody in enzyme assays requiring a two-antibody system from genetically diverse serum sources.

Swiss Webster mice were given 4 intraperitoneal (IP) injections of cucumber mosaic virus in adjuvant and 2 IV injections in saline, all at weekly intervals. During the fifth week strain E-Erlich Lettre carcinoma cells were injected IP. Ascitic fluid was drained from the intraperitoneal cavity every 3 days. Two mice which received a total of 300 and 600 μg of antigen produced 10 and 20 ml, respectively, of ascitic fluid. The microprecipitin titers were 1:32-64 with #1 and 1:64-128 with #2 but were not significantly different in indirect ELISA tests (29). Three mice injected with a total of 0.12 mg of zucchini yellow mosaic virus per mouse yielded, respectively, 5 (mouse died early in experiment), 27, and 20 ml of ascitic fluid (7). Other recent reports involve the use of murine ascities antibodies to blueberry red ringspot virus (15), chickpea filaform virus (20) and barley yellow dwarf virus (34). Earlier use of ascitic fluid as well as chicken yolk IgG is reviewed by van Regenmortel (32, p. 46-48).

Other reviews for antibody preparation can be found in Bos (3), Johnston and Thorpe (19), and in deBokx (9).

Literature Cited, Text

1. Barbara, D. J., and Clark, M. F. 1982. A simple indirect ELISA using $F(ab')_2$ fragments of immunoglobulin. J. Gen. Virol. 58:315-322.
2. Bar-Joseph, M., and Malkinson, M. 1980. Hen egg yolk as a source of antiviral antibodies in the enzyme-linked immunosorbent assay (ELISA); a comparison of two plant viruses. J. Virol. Methods 1:179-183.
3. Bos, L. 1983. Introduction of Plant Virology. Chapter 6. Longman, NY. 160 pp.

4. Bouzar, H., Moore, L. W., and Schaad, N. W. 1986. Serological relationship between 50S ribosomal subunits from strains of Agrobacterium and Rhizobium. Phytopathology 76:1265-1269.
5. Bruck, C., Portetelle, D., Glineur, C., and Bollen, A. 1982. One-step purification of mouse monoclonal antibodies from ascitic fluid by DEAE affi-gel blue chromatography. J. Immunol. Methods 53:313-319.
6. Clark, M. F., Lister, R. M., and Bar-Joseph, M. 1986. ELISA techniques. Pages 742-766 in: Methods in Enzymology. Vol. 118. A. Weissbach and J. Weissbach, eds. Academic Press Inc., London. 829 pp.
7. Davis, R. F. 1986. Partial characterization of zucchini yellow mosaic virus isolated from squash in Turkey. Plant Dis. 70:735-738.
8. Davis, R. F., and Mizuki, M. K. 1987. Detection of cucurbit viruses in New Jersey. Plant Dis. 71:40-44.
9. DeBokx, J. A., and van der Want, J. P. H. 1987. Viruses of potatoes and seed potato production. Chapter 4. Pudoc Wageningen. The Netherlands. 259 pp.
10. Demski, J. W., and Boyle, J. S. 1969. Serological relationships of some virus isolates from sour cherry using antisera from mice and rabbits. Phytopathology 59:996-998.
11. Dickey, R. S., Zumoff, C. H., and Uyemoto, J. K. 1984. Erwinia chrysanthemi: serological relationships among strains from several hosts. Phytopathology 74:1388-1394.
12. Fackrell, H. B., and Sinha, R. C. 1983. Serological analysis of Pseudomonas syringae pv. tomato. Phytopathology 73:178-181.
13. Fribourg, C. E., Koenig, R., Lesemann, D. E. 1987. A new tabamovirus from Passiflora edulis in Peru. Phytopathology 77:486-491.
14. Harlow, E., and Lane, D. 1988. Antibodies. A Laboratory Manual. Cold Spring Harbor Laboratory. Cold Spring Harbor, N.Y. 726 pp.
15. Hepp, R. F., and Converse, R. H. 1987. Blueberry red ringspot virus detection in crude sap of highbush blueberry plants. Plant Dis. 71:536-539.
16. Hsu, H. T., and Lawson, R. H. 1985. Comparison of mouse monoclonal antibodies and polyclonal antibodies of chicken egg yolk and rabbit for assay of carnation etched ring virus. Phytopathology 75:778-783.
17. Hu, J. S., Rochow, W. F., and Dietert, R. R. 1985. Production and use of antibodies from hen eggs for the SGV isolate of barley yellow dwarf virus. Phytopathology 75:914-919.
18. Jensen, S. G., Palomar, M. K., Ball, E. M., and Samson, R. 1985. Factors influencing virus titer in maize dwarf mosaic virus-infected sorghum. Phytopathology 75:1132-1136.
19. Johnstone, A., and Thorpe, R. 1982. Immunochemistry in Practice. Blackwell Scientific Publications, Oxford, Great Britain. 298 pp.
20. Kaiser, W. J., Wyatt, S. D., Hannan, R. M., and Cody, Y. 1988. Chickpea filiform, a new viral disease of Cicer arietinum. Plant Dis. 72:70-74.
21. Laemmli, U. K. 1970. Cleavage of structural proteins during assembly of the head of bacteriophage T4. Nature 227:680-685.
22. Lindmark, R., Thoren-Tolling, K., and Sjoquist, J. 1983. Binding of immunoglobulins to protein A and immunoglobulin levels in mammalian sera. J. Immunol. Methods 62:1-13.
23. Malin, E., Belden, E. L., and Roth, D. 1985. Evaluation of radio-immunoassay, indirect enzyme linked immunosorbent assay, and dot blot assay for the identification of Xanthomonas campestris pv. phaseoli. Can. J. Plant Pathol. 7:217-222.

24. McLaughlin, M. R., Bryant, G. R., Hill, J. H., Benner, H. I., and Durand, D. P. 1980. Isolation of specific antibody to plant viruses by acid sucrose density gradient centrifugation. Phytopathology 70:831-834.
25. Polson, A., von Wechmar, M. B., and van Regenmortel, M. H. V. 1980. Isolation of viral IgY antibodies from yolks of immunized hens. Immunol. Comm. 9:475-493.
26. Powelson, M. L., and Apple, J. D. 1984. Soil and seed tubers as sources of inoculum of _Erwinia carotovora_ pv. _carotovora_ for soft rot of potatoes. Phytopathology 74:429-432.
27. Reif, A. E. 1969. Batch preparation of rabbit G globulin with DEAE-cellulose. Immunochemistry 6:723-731.
28. Ronald, E. P., Tremaine, J. H., and Mackenzie, D. J. 1986. Assessment of southern bean mosaic virus monoclonal antibodies for affinity chromatography. Phytopathology 76:491-494.
29. Smith, D. B., and Davis, R. F. 1986. Identification of cucumber mosaic virus from _Vinca minor_ in New Jersey. Plant Dis. 70:1056-1059.
30. Thaveechai, N., and Schaad, N. W. 1986. Serological and electrophoretic analysis of a membrane protein extract of _Xanthomonas campestris_ pv. _campestris_ from Thailand. Phytopathology 76:139-147.
31. Vaitukaitis, J. L. 1981. Production of antisera with small doses of immunogen: multiple intradermal injections. Pages 46-52 in: Methods in Enzymology, Vol. 73. Immunochemical Techniques, Part B. J. J. Langone and H. Van Vunakis, eds. Academic Press, New York. 739 pp.
32. van Regenmortel, M. H. V. 1982. Serology and Immunochemistry of Plant Viruses. Academic Press, N.Y. 302 pp.
33. Watanaabe, M., Ishii, T., and Nariuchi, H. 1981. Fractionation of IgG_1, IgG_{2a}, IgG_{2b}, and IgG_3 immunoglobulins from mouse serum by protein A-sepharose column chromatography. Japan. J. Expt. Med. 51:65-70.
34. Wyatt, S. D. and Seybert, L. J. 1988. Status of the barley yellow dwarf problem of winter wheat in eastern Washington. Plant Dis. 72:110-113.
35. Zola, H., and Brooks, D. 1982. Techniques for the production and characterization of monoclonal hybridoma antibodies. Pages 1-57 in Monoclonal Hybridoma Antibodies: Techniques and Applications. J. G. R. Hurrell, ed. CRC Press Inc. Boca Raton. 231 pp.

Literature Cited, Table

1. Berna, A., Briand, J.-P., Stussi-Garaud, C., Godefroy-Colburn, T. 1986. Kinetics of accumulation of the three non-structural protein of alfalfa mosaic virus in tobacco plants. J. Gen. Virol. 67:1135-1147.
2. Brakke, M. K., Ball, E. M., and Langenberg, W. G. 1988. A non-capsid protein associated with unencapsulated virus RNA in barley infected with barley stripe mosaic virus. J. Gen. Virol. 69:481-491.
3. Brakke, M. K., Ball, E. M., Hsu, Y. H., and Langenberg, W. G. 1987. Wheat streak mosaic virus cylindrical inclusion body protein. J. Gen. Virol. 68:281-287.
4. Bouzar, H., Moore, L. W., and Schaad, N. W. 1986. Serological relationship between 50S ribosomal subunits from strains of _Agrobacterium_ and _Rhizobium_. Phytopathology 76:1265-1269.
5. Calzolari, A., Bazzi, C., and Mazzucchi, V. 1982. Cross-reactions between _Corynebacterium sepedonicum_ and _Arthrobacter polychromogenes_ in immunofluorescence staining. Potato Res. 25:239-246.

6. De Boer, S. H. 1982. Cross-reaction of Corynebacterium sepedonicum antisera with C. insidiosum, C. michiganense, and an unidentified coryneform bacterium. Phytopathology 72:1474-1478.
7. De Boer, S. H., Copeman, R. J., and Vruggink, H. 1979. Serogroups of Erwinia carotovora potato strains determined with diffusible somatic antigens. Phytopathology 69:316-319.
8. deMejia, M. V. G., Hiebert, E., and Purcifull, D. E. 1985. Isolation and partial characterization of the amorphous cytoplasmic inclusions associated with infections caused by two potyviruses. Virology 142:24-33.
9. Dewey, F. M., Barrett, D. K., Vose, I. R., and Lamb, C. J. 1984. Immunofluorescence microscopy for the detection and identification of propagules of Phaeolus schweinitzii in infested soil. Phytopathology 74:291-296.
10. Duffus, J. E., Larsen, R. C., Liu, H. Y. 1986. Lettuce infectious yellows virus - a new type of whitefly-transmitted virus. Phytopathology 76:97-100.
11. El Nashaar, H. M., Moore, L. W., and George, R. A. 1986. Enzyme linked immunosorbent assay quantification of initial infection of wheat by Gaeumannomyces graminis var. tritici as moderated by biocontrol agents. Phytopathology 76:1319-1322.
12. Falk, B. W., Morales, F. J., Tsai, J. H., and Niessen, A. I. 1987. Serological biochemical properties of the capsid and major non-capsid proteins of Maize stripe, rice hoja blanca, and echinochloa hoja blanca viruses. Phytopatholgy 77:196-201.
13. Falk, B. W., Tsai, J. H., and Lommel, S. A. 1987. Differences in levels of detection for the maize stripe virus capsid and major non-capsid proteins in plant and insect hosts. J. Gen. Virol. 68:1801-1811.
14. Gingerly R. E., Nault, L. R., and Bradfute, O. E. 1981. Maize stripe virus: characteristics of a member of a new virus class. Virology 112:99-108.
15. Gleeson, M. L., Ghabrial, S. A., and Ferriss, R. S. 1987. Serological detection of Phomopsis longicola in soybean seeds. Phytopathology 77:371-375.
16. Glendloff, E. H., Ramsdev, D. C., and Burton, C. L. 1983. Fluorescent antibody studies with Eutypa armeniacae. Phytopathology 73:760-764.
17. Hearon, S. S. 1984. Comparison of two strains of kalanchoe latent virus, carlavirus group. Phytopathology 74:670-676.
18. Hellman, G. M., Thornbury, D. W., Hiebert, E., Shaw, J. G., Pirone, T. P., and Rhoads, R. E. 1983. Cell-free translation of tobacco vein mottling virus RNA. II. Immunoprecipitation of products by antisera to cylindrical inclusion; nuclear inclusion and helper component proteins. Virology 124:434-444.
19. Hiruki, C., Rao, D. V., Chen, M. H., Okuno, T., and Figueiredo, G. 1984. Characterization of sweet clover necrotic mosaic virus. Phytopathology 74:482-486.
20. Jensen, S. G., Palomar, M. K., Ball, E. M., and Samson, R. 1985. Factors influencing virus titer in maize dwarf mosaic virus-infected sorghum. Phytopathology 75:1132-1136.
21. Johnson, M. C., Pirone, T. P., Siegel, M. R., and Varney, D. R. 1982. Detection of Epichloe typhina in tall fescue by means of enzyme-linked immunosorbent assay. Phytopathology 72:647-650.
22. Leach, J. E., Ramundo, B. A., Pearson, D. L., and Claflin, L. E. 1987. Dot immunobinding assay for detecting Xanthomonas campestris pv. holcicola in sorghum. Plant Dis. 71:30-33.

23. Malin, E., Belden, E. L., and Roth, D. 1985. Evaluation of radioimmunoassay, indirect enzyme linked immunosorbent assay, and dot blot assay for the identification of Xanthomonas campestris pv. phaseoli. Can. J. Plant Pathol. 7:217-222.
24. Morimoto, K. M., Ramsdell, D. C., Gillett, J. M., and Chaney, W. G. 1985. Acquisition and transmission of blueberry shoestring virus by its aphid vector Illinoia pepperi. Phytopathology 75:709-712.
25. Nordin, J. H., Mason, T. L., Smith, L. L., Willmann, P. A., Richards, W. L., and Takai, S. 1987. Use of an enzyme-linked immunosorbent assay with murine ascitic antibodies to screen microorganisms for production of cerato-ulmin, a toxin of Ceratocystis ulmi. Phytopathology 77:96-100.
26. Quiot-Dovine, L., Purciful, D. E., Hiebert, E., and de Mejia, M. V. G. 1986. Serological relationships and in vitro translation of an antigenically distinct strain of papaya ringspot virus. Phytopathology 76:346-351.
27. Rao, A. L. N., and Hiruki, C. 1985. Clover Primary leaf necrosis virus, a strain of red clover necrotic mosaic virus. Plant Dis. 69:959-961.
28. Scott, S. W., and Barnett, O. W. 1984. Some properties of an isolate of broad bean wilt virus from Dogwood (Cornus florida). Plant Dis. 68:983-985.
29. Stobbs, L. W., and Van Schagen, J. G. 1987. Occurrence and characterization of a turnip mosaic virus isolate infecting Alliaria petiolata in Ontario, Canada. Plant Dis. 71:965-968.
30. Svircev, A. M., Jeng, R. S., and Hubbs, M. 1988. Detection of cerato-ulmi by immunocytochemistry and scanning electron microscopy. Phytopathology 78:322-327.
31. Thaveechai, N., and Schaad, N. W. 1984. Comparison of different immunogen preparations for serological identification of Xanthomonas campestris pv. campestris. Phytopathology 74:1065-1070.
32. Thaveechai, E., and Schaad, N. W. 1986. Serological and electrophoretic analysis of a membrane protein extract of Xanthomonas campestris pv. campestris from Thailand. Phytopathology 76:139-147.
33. Thaveechi, E., and Schaad, N. W. 1986. Immunological characterization of a subspecies-specific antigenic determinant of a membrane protein extract of Xanthomonas campestris pv. campestris. Phytopathology 76:148-158.
34. Van Alfen, M. K., McMillan, B. D., and Dryden, P. 1987. The multi-component extracellular polysaccharide of Clavibacter michiganense subsp. insidiosum. Phytopathology 77:496-501.
35. Yakrus, M., and Schaad, N. W. 1979. Serological relationships among strains of Erwinia chrysanthemi. Phytopathology 69:517-522.
36. Yeh, S. D., and Gonsalves, D. 1984. Purification and immunological analyses of cylindrical inclusion protein induced by papaya ringspot virus and watermelon mosaic virus. Phytopathology 74:1273-1278.
37. Zuiderma, D., Heaton, L. A., and Jackson, A. O. 1987. Structure of the nucleocapsid protein gene of sonchus yellow net virus. Virology 159:373-380.

Section II.B.2.

STRATEGY AND TECHNIQUES FOR THE PRODUCTION OF MONOCLONAL ANTIBODIES; MONOCLONAL ANTIBODY APPLICATIONS FOR VIRUSES

R. L. Jordan
USDA ARS, Florist & Nursery Crops Laboratory
Beltsville, Maryland 20705

Acknowledgments. Many thanks to Hei-ti Hsu for the introduction to hybridoma technology and continuing discussions, to Peggy Hall for manuscript typing, to Mal Shurtleff for the Procedures Figure, and to Mary Ann Guaragna for exceptional technical assistance.

I. Introduction

The introduction by Kohler and Milstein in 1975 (52) of the novel approach of immortalizing specific antibody-forming cells by their fusion with myeloma cells has provided a method for the production of homogenous and biochemically defined immunological reagents, namely monoclonal antibodies (McAbs). Hybridization of antibody-forming lymphocyte cells with malignant myeloma cells results in hybridomas which combine the parental traits of specific antibody secretion and continuous growth. Cloning and further selection of hybrids allows the development of McAbs of identical specificity and desired activity directed against a unique epitope of the immunizing antigen.

McAb production has recently been adapted to the plant sciences, including plant pathology (29,44,80). McAbs have been utilized for the detection and identification of plant pathogens, including viruses, bacteria, spiroplasmas, and mycoplasma-like organisms; for the characterization and analysis of variants and strains; for determining the localization and accumulation of pathogens and pathogen-encoded products; and for studying the structure of pathogens and their gene products. The successful production and use of (McAbs) in plant pathology is reflected in the over 80 published research articles reporting the use of McAbs as the key reagent (see references in 29,44).

When compared with conventional polyclonal antibody production, the ability to obtain practically unlimited quantities of the same antibody in a reproducible manner, and the ability to immortalize the production of such monospecific reagents by cryopreservation of the hybridomas for unlimited periods, are just two advantages of hybridoma-produced McAbs. One of the more striking advantages of the hybridoma approach is the ability to produce and select McAbs to almost any antigenic determinant, even when impure antigen or antigen mixtures are used as immunogen.

The methodology involved in the establishment of permanent cell lines producing McAbs is relatively simple, yet requires a large number of sometimes crucial steps, each of which may be carried out in many different ways. An illustration of a general protocol used for the fusion, selection and screening, cloning, and growth, maintenance and cryopreservation of hybridomas is shown in Fig. 1. The derivation of McAbs obviously depends not only on the generation of large numbers of hybridoma cell lines, but also on the successful immunization of the donor animal(s), knowledge and experience in basic cell culture techniques (including media production, cell line maintenance and preservation, and sterile culture techniques), and the development of preplanned, sensitive, reliable, and relatively fast and simple serological assays for identifying and characterizing antibodies.

In this chapter, readers are provided with details of the various techniques and the rationale used to develop the protocols and approaches. Preparation and characterization of McAbs to a wide spectrum of plant and plant-pathogen antigens, including viruses (2,40,41,42,43,46), spiroplasmas (12,45,53), and natural plant products (70) will be covered. The information presented was drawn from personal experiences, as well as from the literature (only the most recent sources of information on specific topics are cited). Several articles or books describing in more extensive detail the hybridoma technology and procedures for the preparation and characterization of McAbs are also cited (10,23,24,25,27,31,48,51,72,90,91). Other procedures related to cell culture, McAbs or antibodies in general are provided (7,57,58).

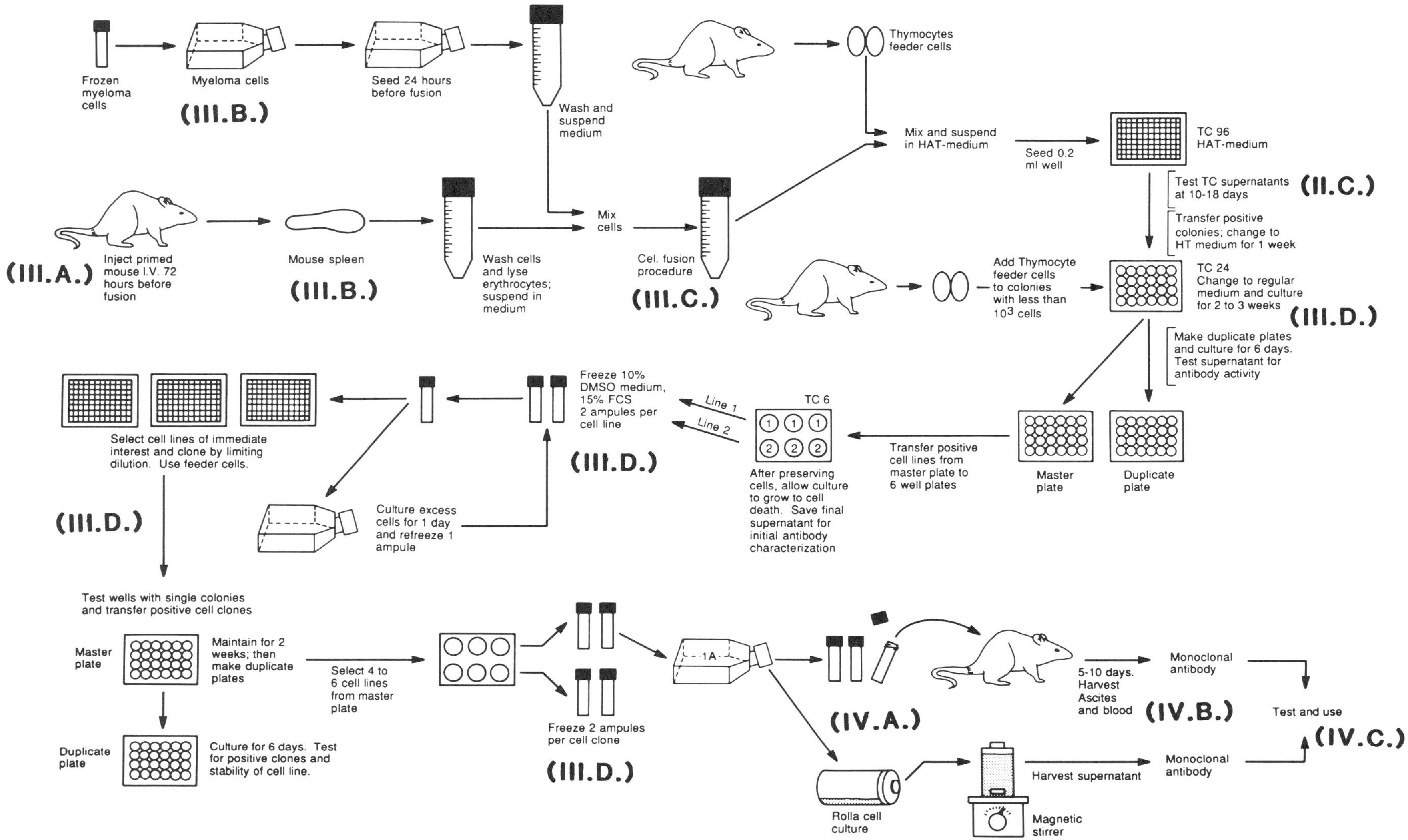

Fig. 1. Stylized flow diagram of the procedures used for the fusion, selection and screening, cloning, growth, and cryopreservation of monoclonal antibody-secreting hybridomas. Numbers and letters (e.g., III.A.) signify sections in text relevant to areas of diagram (diagram by courtesy of Dr. M. C. Shurtleff).

II. Basic Requirements for Fusion, Maintenance and Growth of Hybridomas

The researcher should have experience in sterile techniques, and knowledge and familiarity with the basic principles and techniques of cell culture. Hybridomas are rather fastidious cells which need to be evaluated microscopically to determine viability and rate of growth. Other important requisites to McAb production are the antigen and serological assay(s) to be used to identify and characterize the hybridoma antibodies. The assay systems used will depend upon the nature of the antigen and the projected use of the McAb.

II.A. Equipment and Plasticware

Hybridoma production requires various major and minor equipment and cell culture ware. While it is possible to successfully produce hybridomas in a general research laboratory, a distinct laboratory or laboratory area for hybridoma work is advantageous. Equipment listed below should be in close proximity and removed from normal traffic areas: 1) a laminar flow hood, 2) a humidified, temperature-controlled (37 C) carbon dioxide incubator, 3) an inverted microscope, and 4) a bench-top clinical centrifuge. The sterile hood and CO_2 incubator should be dedicated solely for hybridomas during hybridoma production. Outside the culture area, access to a 37-56 C water bath, -70 C freezer, autoclave and a liquid nitrogen storage container are required. Other general laboratory items such as a hemocytometer, single and 8- to 12-channel variable-volume (10-200 μl) automatic pipets and sterile tips, peristaltic pump refrigerators and freezers (-15 C) should be readily available. Such equipment dedicated entirely to hybridoma work would maximize efficiency.

Depending upon the nature of the assay used, additional equipment (and support materials) may also be needed: an ELISA reader, fluorescent microscope, fluorescent-activated cell sorter, scintillation counter, and spectrophotometer.

Animal holding facilities are needed to maintain the constant supply of mice or rats required for all stages of McAb production including immunization, production of ascitic fluid, and donation of thymuses and/or spleens for feeder cells. Cages, bedding, feed and adequate temperature control and ventilation are obvious necessities.

Required cell culture ware includes: 96-, 24-, and 6-well flat bottom sterile cell culture plates with covers; 25 cm^2, 75 cm^2, and 150 cm^2 sterile cell culture flasks; 1-, 5-, 10-, and 25-ml sterile plastic pipets; ampules for liquid nitrogen preservation; 5-, 15-, and 50-ml sterile screw-cap centrifuge test tubes. (Corning, Costar, Falcon, and Nunc are excellent sources for these items.) Glass, cotton-plugged reusable pipets may also be less expensive. Other necessary items include disposable plastic or glass Pasteur pipets, protective (vinyl) gloves, Petri dishes (square), sterile filters (syringe-type and 100-1000 ml reservoir-type or '1-10 liter'-type with positive pressure pump), 1 to 50 ml syringes, 26- to 16-gauge (3/4" to 1 1/2") needles, and surgical scissors.

II.B. Materials and Media

The establishment of hybridomas requires not only special attention, but also quality reagents, as it involves the growth of vulnerable cells that are sensitive to extremes in cell densities requiring continuous monitoring and nursing. Use of manufacturer-prescreened media and additives greatly aids the establishment, growth and stability of hybridomas, and avoids lost time and effort. (Commercial sources of specific reagents will be listed when appropriate.) All preparations are made with tissue-culture-grade, deionized, glass double-distilled water.

II.B.1. Media. The most commonly used media for hybridoma production are Dulbecco's Modified Eagle's Medium (DMEM) and Rosewell Park Memorial Institute (RPMI-1640) medium. The two media are very similar, the differences being that RPMI-1640 has no pyruvate and DMEM has no asparagine. Both media are usually buffered with carbonate/bicarbonate. RPMI-1640 (Gibco) will be the standard medium in this report; however, DMEM (Flow) performs equally well. Media may be purchased in liquid or powder form, the latter preferred because of longer shelf life. It is usually advisable to use prepared media that are less than 3 months old, especially when using purchased liquid media when exact dates of preparation are unknown. Glutamine is unstable and is added to 2 mM (from 200 mM stock; Gibco) to media older than 2 weeks. Some researchers also supplement the media with nonessential amino acids (to 1%; Gibco) and other growth media (i.e., NCTC 109 to 10%; Gibco). Prepared media should be stored (4 C) in the dark to prevent the production of highly toxic photoproducts (24,25).

2. Antibiotics. The most common cell culture contaminants are bacteria, yeast and fungi. The best control for these organisms is scrupulous use of sterile techniques. Most researchers add penicillin and streptomycin as a matter of routine (Gibco; 100X = 10,000 units/ml) to minimize bacterial contamination. When necessary, or as an alternative, gentamycin (at 100 μg/ml) may be used as a broad spectrum bacterial and mycoplasma antibiotic. Lincomycin and tylosin tartrate (100 μg/ml each) have been used, in conjunction with phagocytic macrophages, for removal of mycoplasma from important cell lines (31). Fungizone (Amphotericin B) at 5-10 μg/ml is useful against fungal contamination.

3. Sera. Fetal bovine serum (FBS) is used in nearly all hybridoma work. The quality of available sera for cell culture is now such that extensive prior testing of many different batches is usually unnecessary. Several companies, in fact, provide "hybridoma-tested" sera or sera developed primarily for hybridoma work (e.g., HyClone, Sigma). Even so, the serum chosen should be able to support the growth of myeloma cells at 1 cell/well. The serum can be kept frozen at -20 C for at least 1-2 yr. Most researchers heat the serum to inactivate complement, then store frozen. For use, the FBS is thawed at room temperature, then transferred to 56 C for 30-45 min with occasional mixing prior to aseptic dispensing into smaller one-time-use portions for -20 C storage.

4. Serum substitutes, supplements and serum-free media. Iron-supplemented bovine calf serum (38) and horse serum are sometimes substituted for FBS and are generally less expensive. However, both contain high levels of contaminating immunoglobulins (FBS has very low levels) which can interfere with many immunoassays. Several companies offer defined or processed serum replacements which reportedly have consistent chemical

and performance characteristics and can be used in place of FBS. Three supplements/substitutes can completely (NuSerum) or partially (Mito +; CPSR-3) replace FBS as the serum component (unpublished), thereby reducing overall serum costs without loss of hybridoma production efficiency. Sigma's CPSR-3 (Controlled Process Serum Replacement) contains components derived from adult bovine plasma. Collaborative Research offers an FBS alternative (NuSerum) which has a consistent formulation of newborn bovine serum (25%) supplemented with growth factors, cell growth supplements, hormones, vitamins, lipids, and trace elements. Mito + Serum Extender (Collaborative Research) is a defined concentrated formulation of hormones, growth factors, and minerals that can replace up to 80% of the standard serum supplements or can be used together with serum-supplemented medium for an "enhanced" medium.

Serum-free media that support the growth of several parent myeloma lines (and their fusion-derived hybridomas) have been recently described (47,54). These media are usually mixtures of RPMI and DMEM supplemented with hormones, transferrin, lipids, trace elements, and other factors. Unfortunately, none of the serum-free media are presently suitable for cultivating all hybridomas or myelomas.

II.B.5. Preparation of media and working solutions for fusion, maintenance and growth of hybridomas.

a. 100X HT (= 10 mM hypoxanthine, 1.6 mM thymidine): 136.1 mg hypoxanthine (Sigma) and 38.8 mg thymidine (Sigma) is dissolved in 100 ml water at 60-70 C, 0.2 μm filter sterilized, and stored frozen at -20 C in 5 ml aliquots.
b. 100X HAT (=10 mM hypoxanthine, 4 X 10^{-5} M aminopterin, 1.6 mM thymidine): Prepare 100X HT as above (but in 90 ml water) and add 1.8 mg aminopterin (Sigma) and 0.5 ml 5N NaOH to dissolve, then neutralize with 0.5 ml 5N HCl and adjust to 100 ml before filter sterilizing. Store at -20 C as for HT, protected from light.
c. Serum-free RPMI-1640: 50 ml 100 mM sodium pyruvate (Gibco), 10 g sodium bicarbonate (Sigma), 0.5 g gentamycin (Sigma) and a 5 liter package of powdered RPMI-1640 with glutamine (without bicarbonate) (Gibco) are thoroughly dissolved in 3600 ml of tissue-culture-grade water. The solution is dispensed by peristaltic-pump through a 0.2 μm filter (Millipore) into sterile 500 ml bottles (360 ml/bottle). The medium is stored at 4-6 C until use.
d. Complete RPMI-1640: (Serum-free RPMI-1640 containing 15% serum, 10% 'conditioned medium', 2 mM glutamine, 6 mM HEPES, and 2 x 10^{-5}M 2-mercaptoethanol). To 360 ml serum-free RPMI-1640 aseptically add 75 ml serum, 50 ml myeloma conditioned medium (see III.B.), 5 ml 100 mM glutamine (Gibco), 3 ml 1 M HEPES and 0.1 ml 0.1 M 2-mercaptoethanol (2MCE). (The glutamine, HEPES, and 2MCE are added to 5 ml serum-free RPMI-1640 and 0.2 μm filter sterilized as it is added to the complete medium.)
e. Complete medium + HAT (or HT): Add 5 ml 100X HAT (or 100X HT) to 500 ml complete medium.
f. Polyethylene glycol (PEG): At the time of fusion, a 45% solution is prepared by dissolving 2.0 g PEG (M_r 1000 to 4000 with 2.5 ml serum-free RPMI-1640 at 50-60 C (water bath). The slightly alkaline (pH 7.5-8.2; pink, not orange or purple) solution is filter sterilized

and held at 37 C until fusion.

g. "Freezing medium" (= Complete medium + HT containing ca. 25% serum and 8% DMSO): Aseptically add 5 ml serum and 4 ml dimethylsulfoxide (DMSO; do not try to filter sterilize as it will dissolve nitrocellulose membranes) to 40 ml complete medium + HT and store at 4-6 C.

h. Preparations of 50-500X HT and HAT, and PEG and DMSO components can be obtained as ready-to-use, pretested/screened stock solutions from several companies (e.g., American Type Culture Collection; Sigma; Gibco).

II.C. Guidelines for Screening Assays

The immunoassay selected for screening antibody activity in the hybridoma culture supernatant fraction is critical to hybridoma production. In principle, any assay capable of detecting low antibody concentrations can be used. There are many ways to screen hybridomas (10,27,57), and applicable methods are presented separately (**Section IV.E.3.** of this manual). The screening method must be reliable, sensitive enough to detect all McAb-secreting hybridomas of interest, simple and quick enough to test hundreds of samples at a time, and must be definitive.

In initial screens following fusion, 400-2000 cell lines may require assay. If 10% of these are antigen-specific, then 40-200 cell lines must be rescreened at least twice during expansion and freezing. Selected cell lines are then cloned, adding typically 30 clones per 'parent' cell line for screening. In essence, hundreds to thousands of assays may be required in the course of a single hybridoma project.

Alternatively, if the screening assay procedure is difficult, time consuming, or expensive, yet necessary to select a specific antibody activity, screening may be performed in two steps. The first step might involve a fast, non-specific assay, that would reduce the number of hybrid cultures to be screened in the second more specific assay. The non-specific first-phase screening might include testing for mouse or rat antibody in culture supernatant fraction with antiglobulin capable of detecting all desirable subclasses.

A second factor in selecting assays is the nature of the antigen; e.g., purity, abundance, and physical form (size, native conformation, fixed, denatured). Solid phase assays are suitable for most antigen preparations encountered in plant pathology. High-molecular weight, soluble, highly purified antigens such as viruses, as well as insoluble, complex antigens such as bacteria or bacterial cell walls, can be immobilized by binding to a solid phase matrix. Pre-treatment of the solid phase with gluteraldehyde or polylysine, for example, may be necessary with some antigens, whereas low-molecular-weight antigens such as peptides, enzymes or hormones, may need to be coupled to a suitable carrier for maximal binding (27). Immunofluorescence and immunocytochemical techniques are suitable (albeit labor intensive) when screening for McAbs to cellular or tissue antigens. McAbs to functional antigens (e.g., enzymes) must be able to precipitate, modify, inhibit, or neutralize the activity mediated by the specific antigen.

Lastly, the screening method should be appropriate for the intended use of the antibody. Accordingly, the condition of the antigen in the screening assay should approximate that of the final assay. McAbs tend to be very assay-specific, useful in one assay, but not in another. This is due primarily to the strict specificity of McAbs for their respective epitopes and the "accessibility" of those epitopes in the assay(s) (3). Proper selection of an assay can reduce the workload and ensure that the desired McAbs are identified.

II.D. Comments Concerning Myeloma Cell Lines and Immunogens

The choice of animal species as immune spleen donor depends mainly on the myeloma cell line to be used for fusion. Mouse and rat myeloma lines are readily available, with mice as the most common species used. The Balb/c strain of mice is preferred as all the murine myelomas in use were derived from this strain. Other inbred strains are used when low immune response is encountered in Balb/c. The LOU/C or LOU/M strains of rat are recommended when using rat-derived myelomas (23).

Myeloma cell lines are mutants deficient in either the enzyme hypoxanthine-guanine phosphoribosyl-transferase (HGPRT) (azaguanine-resistant) or thymidine kinase (TK) (bromodeoxyuridine-resistant). Such mutants die in the presence of aminopterin, which blocks the main DNA synthesis pathway. Selection of antibody-secreting hybridomas is possible because myeloma-spleen cell hybrids can survive in aminopterin when hypoxanthine and thymidine (components of HAT and HT-containing media) are present for use by the salvage pathway.

The choice of a myeloma is based on its overall performance, including growth rate, fusion efficiency, feeder cell dependence and cloning efficiency. Several cell lines currently in use yield high frequencies of hybridomas (27,90,91), including P3/NS1/1.Ag.4.1 (NS1), P3X63/Ag8.653 (P3) and Sp2/0.Ag14 (Sp2). All three cell lines have been used successfully in plant pathology. NS1 is an excellent choice because of its growth rate, lack of feeder cell dependence (R. L. J.) and excellent fusion and cloning efficiencies. However, it does produce endogenous intracellular K light chain. Neither P3 nor Sp2 produce endogenous antibody and Sp2 is less dependent upon feeder cells than P3. The recently introduced FOX-NY (81) cell line, an NS1-derived myeloma which also does not produce antibody, is deficient in HGPRT (located on the γ chromosome) and in the enzyme adenosine-PRT (located on chromosome 8). This cell line and use of HAT screening medium enables the survival of hybrids that have lost the chromosome encoding HGPRT, and which could consequently be lost in HAT selection media. This ensures the survival of Ig producers usually lost with other myeloma fusion partners when the X chromosome is lost.

Rat myeloma cell lines include YB2/0 and IR983/F (10). Rats are advantageous for production of larger volumes of ascitic fluid, but their myeloma fusion efficiencies are lower than the murine myelomas (10).

Most of the mouse and rat myelomas are available from Flow, Gibco, ATCC, the Institute for Medical Research, or NIGMS Cell Repository.

Some elements to consider in choosing and preparing an immunogen include the following: Immunogen purity is less crucial for McAb production than for polyclonal antibody production. If the target antigen is a poor immunogen or the screening assay cannot distinguish between target-specific and impurity-specific antibodies, an immunogen of greater purity is required. The immunogen should present an appropriate antigenic site conformation, i.e., McAbs generated from denatured antigen may not necessarily recognize native antigen, and vice versa. Low molecular weight antigens (less than 1K dalton) and peptides are generally poor immunogens and should be coupled to larger immunogenic carriers such as albumin or keyhole limpet hemocyanin. Aggregated and/or particulate antigens, such as bacteria and cell walls or membranes, are stronger immunogens than soluble antigens. Likewise, repeating antigenic determinants of a molecular complex (e.g., native virus particle) elicit stronger responses than monodispersed determinants (e.g., protein capsid subunit).

III. Production of Monoclonal Antibody-Secreting Hybridomas

Important elements in the production of McAbs include the immune state of the spleen donor animal, the ability to prepare and fuse myeloma and spleen cells and obtain viable hybrids, the screening method, the ability to clone and cryopreserve cells, and perseverance through hundreds of hours over a 3-6 month period.

III.A. Immunization Schedule

III.A.1. Principle

The immune state of the animal from which the spleen is taken is very important in determining the success of a hybridoma project. The degree of immunization is dependent upon the choices of animal, immunogen, and the immunization schedule (10). Immunization protocols and schedules vary considerably and should not be considered separately from the immunogen. A protocol that works well for a membrane antigen may not necessarily be satisfactory with a soluble protein. A generally satisfactory protocol follows.

III.A.2. Immunization protocol

a. Inject two 1-4 month old mice or rats intraperitoneally with 10 μg antigen emulsified in 300 μl complete Freund's adjuvant or 1 X 10^7 cells in buffered saline or Freund's incomplete adjuvant.
b. Inject 2 to 3 more times with the same dose in Freund's incomplete (or saline for cells) at 2- to 8-wk intervals.
c. Take serum samples from individual animals 10 to 14 da later, and determine the titer of the relevant antibodies in the same assay(s) that will be used later for screening hybridomas.
d. Give an additional 1-5 wk rest period (more than 3 wk from the last injection) to animals showing the highest antibody titers prior to administering a final intraperitoneal boost of the same or higher dose in aqueous solution.
e. Three to 5 days later, remove the spleen(s) and prepare for cell fusion.

III.A.3. Comments

a. Doses of 1-50 μg of molecular antigen or 1 X 10^6 - 5 X 10^7 cells per rodent per injection is usually sufficient to elicit good responses if repeated two or three times at intervals of 2-8 wk.
b. Immunogenicity of soluble antigens is enhanced by the use of an adjuvant (e.g., Freund's) using Freund's complete in the first injections, and Freund's incomplete in subsequent injections. When

cells are used as immunogen, adjuvant is not usually required.

c. The site of injection is probably not critical. Subcutaneous, intramuscular and intraperitoneal immunization routes are adequate to stimulate an immune response. Intradermal injections at multiple sites can cause painful ulcers and the other routes are as efficient for priming (10,27).

d. Volumes of 50-300 μl per mouse and 1-2 ml per rat are typical.

e. Many laboratories require the immunized animal to have a specified serum titer before its spleen will be used in a fusion (23,90). Generally, it is wise to select animals with high titers of circulating antigen-specific antibodies. Note, however, that the spleen cells that generate a high frequency of antigen specific hybridomas are the rapidly dividing plasmablasts which do not themselves secrete large amounts of antibody (10).

f. Regardless of serum titer 'requirements', a final boost 3-5 da prior to fusion generates a maximal number of plasmablasts (24,25).

III.B. **Preparation of Parental Cells**

III.B.1. Principle

A critical component in the production of hybridomas is the preparation of cells for fusion. Both spleen B-lymphocyte cells and myelomas need to be in the optimum cell cycle stage and condition for successful hybridization and subsequent hybrid proliferation. A successful immunization schedule, including a 3-5 da pre-fusion antigen injection, should produce sufficient numbers of stimulated lymphoblasts in the donor spleen for use in the lymphocyte-myeloma cell fusion procedure. After the spleen is removed from the animal it is squeezed, teased, smashed, minced, strained, and/or injected with a solution to release the lymphocytes, which are then pelleted, washed and counted before fusion with the myeloma cells. The procedure used to prepare the spleen cell suspension must be gentle to preserve cell viability and function. A mouse spleen will typically contain about 10^8 cells, and a rat spleen about 5-10 X 10^8 cells (27). Some researchers incubate the cell suspension in a lysis medium to remove red blood cells (90). The lysis method can give variable results and adequate numbers of hybrids are obtained without attempting to remove non-lymphatic cells (10,25,51).

III.B.2. Preparation of spleen cells

a. Kill immunized mouse (mice) by cervical dislocation or CO_2 asphyxiation.

b. Pin feet to corkboard, saturate abdomen and chest with alcohol. Cut skin to expose abdominal wall. Saturate with alcohol.

c. Cut through abdominal muscles and gently remove the spleen. Dip it in 70% alcohol 10 sec, then transfer it to a petri dish containing serum-free RPMI-1640. Remove any connective tissue and place the spleen in a centrifuge tube containing serum-free medium.

d. Cut open chest cavity, snip heart and collect blood in chest cavity around heart and lungs. (This is positive control mouse polyclonal antibody source.)

e. Place mouse in autoclave bag and freeze (until time for autoclaving and incineration).

f. Prepare a single cell suspension by successive sac injection and smashing as follows: Place the spleen in a sterile Petri dish and inject with 20 ml serum-free RPMI-1640 with a syringe and needle, forcing the cells out of the spleen sac. Crush the spleen between two sterile frosted glass slides to release remaining cells into the medium.
g. Harvest the cells by centrifugation (400 g, 10 min) and wash three times in serum-free RPMI-1640.
h. Resuspend in 10 ml RPMI-1640 and count the cells by diluting a sample 1:10 in 0.1% trypan blue (in saline). (Live cells will exclude the dye and dead cells will be blue.) Cells are only used in hybridization if the viability is greater than 90%.

III.B.3. Preparation of myeloma cells

a. Prepare from stock frozen cultures a viable, actively growing (log phase) myeloma cell culture (2-10 75 cm^2 flasks) at least 10 da prior to fusion. (See III.D., this chapter, for information on reviving frozen stock cells.)
b. The myeloma cells can be grown in complete RPMI-1640, containing 10% serum and without supplemented conditioned medium. Split and feed the cells every other day.
c. For 3 da prior to fusion, split and feed the cells every day.
d. Harvest the myeloma cells by centrifugation (400 g, 10 min) and wash three times in serum-free RPMI-1640. Count cells and determine viability as above.
e. Use as many flasks as necessary so that final pooled myeloma cell pellet is the same volume as the packed spleen cells (for NS1 or FOX-NY myelomas this is equal to a final spleen to myeloma cell ratio of 5:1 for fusion).
f. The spent myeloma culture fluid is saved and filtered (0.22 μm) for later use as conditioned medium.

III.B.4. Comments

a. Regardless of the choice of myeloma to be used for fusion, the cells should be maintained at high viability and in logarithmic growth prior to fusion (25,51). Good culture conditions of the myeloma cell line improves the fusion frequency tremendously.
b. To avoid revertants, do not keep the same culture going for long periods of time (months). Cells should be thawed from frozen stock, scaled up (and fresh vials replaced as soon as possible), and propagated in early log growth for at least a week prior to fusion.
c. If the cells have been passaged through medium containing, for example, 8-azaguanine (30 μg/ml) to ensure aminopterin sensitivity, they need to be passaged at least 3 times in medium without 8-azaguanine prior to fusion.
d. Hybridoma cells are often intolerant of dilution. Feeder cells or conditioned medium of some sort increase the ability of cultured cells to grow at very low densities. If culture conditions are optimal, feeder cells may make little difference other than reducing variability between fusions (24). Commonly used feeder cells (10^6/ml) include normal spleen cells (23), thymocytes (25,51),

irradiated fibroblasts (10), and peritoneal macrophages (48). The presumed functions of feeder cells include: the addition of soluble growth factors to the medium (all feeder cells), removing waste products and dead cells (phagocytic macrophages only), or giving emerging hybridomas or clones cell contact at otherwise low cell density (all feeder cells). Normal spleen cell feeders are prepared as per the parental spleen cells (as described above). Procedures for preparing other types of feeder cells are described elsewhere (10,27,90,91).

e. Conditioned medium supports actively growing cells. The cells are removed and the medium is filtered and used to supplement (at 10-50%) the medium used to grow hybridomas.

f. It is objectionable and often inconvenient to use cells from specially sacrificed animals either as feeders or for conditioned medium. Commercially available cell lines (i.e. mouse L929 and 3T3 fibroblast cell lines; ATCC), or conditioned medium (thymocyte, Hana Biologics, and endothelial cell growth supernate, Collaborative Research) are good alternatives (75). The spent medium from the log growth phase myeloma fusion partner is generally a good source for conditioned medium.

III.C. **Cell Fusion**

III.C.1. Principle

The next critical step in the production of hybridomas is the hybridization or fusion procedure. However, before initiating cell fusion, healthy and myeloma cells should be visually characterized to facilitate distinction between hybrid cells and those of parent cell types. Likewise, optimal cell concentration (cells/ml) must be previously standardized.

A successful fusion protocol should bring parent cells together and allow fusion to occur at a sufficiently rapid rate without causing more than minimal damage to the cells. The first hybridomas were made using Sendai virus as the fusing agent, whereas nearly all hybridomas are now generated with polyethylene glycol (PEG) as the fusion inducer (27). The conditions which favor successful fusions are not unique. Several modifications of the original PEG-induced fusion procedure (25,51) have been described, and all variations seem to work equally well. The more important variables include: the concentration and pH of the PEG mixture, the duration of exposure to PEG, and the physical handling and processing of the cells during and after fusion. The mechanism of fusion is complex, involving cell agglutination, cell swelling, and membrane fusion and the optimal conditions for the three processes are often at odds (10). Membrane fusion efficiency, but also toxicity, increases with increasing PEG concentration (30-50%), time of exposure (2 to 10 min), and temperature 20-40 C). All fusion protocols operate under these ranges, compromising conflicting needs to achieve sufficient fusion while keeping cell damage to an acceptable level. Fusion frequency is also dependent upon pH, with maximal numbers of hybrids occurring at pH 7.8-8.2. The dilution and manipulation procedures are also very critical. Fast addition of medium and rough handling are harmful as the aggregated cells which will eventually fuse can be dissociated by vigorous treatment. We have obtained moderate success using the following protocol, which is a compilation of several protocols (56,72,88).

III.C.2. Fusion protocol

a. Mix the resuspended spleen and myeloma cells in one tube (after keeping aside a sample of both for use as controls). Centrifuge at 400 g, 10 min and remove ALL the supernatant liquid by careful aspiration.
b. Loosen the pellet by gently tapping the tube.
c. Using a 5 ml pipet, gently stir the cell pellet while adding, dropwise, 1.5 ml PEG (37 C) over a 45-sec period.
d. Incubate with gentle swirling for 30-75 sec (essentially the time it takes to prepare for step e).
e. Slowly add 2 ml serum-free medium with stirring over a 60-sec period.
f. Incubate and swirl 60 sec.
g. Repeat e and f.
h. Add 35 ml serum-free medium over the next 3-5 min.
i. Total time = about 9 min.
j. Centrifuge cells 400 g, 10 min. Aspirate and discard supernatant liquid.
k. GENTLY, with much patience, resuspend cells in 20 ml complete medium + HAT by tapping base of the tube. Do not pipet cells with anything smaller than a large-bored 25 ml pipet (i.e., about 3 mm i.d. at the tip). Cells are very fragile clumps at this stage.
l. Transfer cells to a 75 cm^2 flask containing 100-180 ml complete medium + HAT (resuspend in volume calculated to give a cell concentration of less than 1 X 10^5 myeloma cells/ml, based on original cell counts).
m. Put flask in incubator (37 C, 5-10% CO_2), standing up with loosened cap. Let stand at least 1 hr. (Fibroblast cells and large clumps/debris will generally stick to the bottom of the flask.)
n. Carefully pour suspended fused cells to a new flask. Mix by gentle pipetting up and down with a 25 ml pipet.
o. Pipet cells (gently) into a Petri dish (square) and dispense 200 μl/well into 96-well cell culture plates with an 8- or 12-channel microtiter pipet (usually 10-15 plates per spleen). If feeder cells are used, they are usually plated the day before (100 μl/well) and fused cells are then added at 100 μl/well (at a cell concentration of 2 X 10^5 myeloma cells/ml).
p. Plate out saved spleen and myeloma control cells in complete medium + HAT separately at cell densities similar to the fused cells.
q. Incubate at 37 C in 5-10% CO_2 in a humid sterile incubator.

III.C.3. Comments

a. Other factors which might influence the fusion efficiency under suboptimal conditions include the molecular weight of the PEG (range from 1000 to 6000), the source of PEG [Merck 4000, Kodak 1450, Baker 3350, and Sigma 4000 are adequate for high fusion efficiency (56)], the ratio of spleen cells to myeloma cells (1:1, 2:1, 5:1, 10:1 are common ratios), the presence or absence of serum or red blood cells (90,91), and the addition of DMSO with the fusogen [doesn't hurt and might be beneficial (24)].

b. Another area of variation between researchers concerns the size of the initial cultures (i.e., post fusion plating densities). The two extremes include plating at high cell density (10^8/ml) in 1 to 5 24-well plates (1 ml/well) (25,51,72) and plating at low cell density (10^5/ml) in 10 to 20 96-well plates (0.2 ml/well) (10,27). Plating at relatively low numbers of cells per well has the advantage that "assay positive" wells will probably contain one hybrid per well and thus McAbs from the beginning.

c. A standardized, reliable, fast and sensitive screening assay should be developed before proceeding with cell fusion samples. 1000-2000 wells require screening for antibody production within two weeks of hybridization and cells will be growing at a fast rate and cannot wait for the development of a screening assay. A commitment of 3-8 wk of continuous work must also be made at this time.

III.D. **Selection, Propagation, Cryopreservation and Cloning of Hybrid Cells**

III.D.1. Principle

After fusion the hybrids are placed into a medium that allows hybrids to grow and prevents the growth of unfused parent cells. The cultures are monitored daily (with the microscope) to assess the progress of the hybridomas, to decide upon the necessity of feeding, or testing supernatant fluids for the presence of antibody, and to remove any contaminated cultures. This stage in the production of McAbs can be difficult and time consuming. Hundreds of colonies, at various times after fusion, will be ready for screening. Once antigen positive antibody secreting hybridomas have been identified, they are transferred to larger vessels, rescreened, expanded again, rescreened and frozen, cloned, screened, recloned, rescreened and bulk cultured for antibody production.

III.D.2. Propagation and maintenance procedures after fusion

a. Do not disturb the culture plates for the first 3 days after fusion. Don't even open incubator door.

b. Day 3: Examine plates for contamination. Individual contaminated wells can be aspirated and treated with NaOH or $CuSO_4$ and then left empty. Gross contamination of the 96-well plates is almost impossible to cure and it is best to discard a whole plate rather than risk the spread of the contaminant. Feed all wells by first aspirating half of the medium in each well, using a 12 channel microtiter pipet, and replacing with about 100 μl of fresh medium + HAT.

c. Days 7-14: Begin screening assays. Aseptically (e.g., with a sterile 12 channel microtiter pipet) remove culture fluid for screening when the bottom of the well is 30-50% covered by hybridoma cells and the medium has changed color to yellow. Replace with fresh medium. Screen and feed every 3-4 days. Plan to spend all day working with the cells and screen. Transfer positive cell lines to 24-well plates by gently aspirating the media and cells up and down in the well with a sterile 12 channel micrototer pipet (about 3-4 times to mix the cells with the media), and adding this suspension to 1 ml fresh medium in the new wells.

d. Day 14: Control myeloma cells should have been dead by day 10. Begin feeding with complete medium + HT. Continue feeding every 3-4 da.
e. Days 14-30: Allow sufficient time for proliferation of transferred cells and rescreen. Transfer positive cell lines to 25 cm^2 flasks (5-10 ml complete medium + HT per flask). Freeze as soon as possible. Continue to monitor 96-well plates, screen and transfer to larger-volume flasks accordingly.
f. Days 30-45: About 80-90% of all the antigen positive hybridomas will have been selected in the first 3 wk of screening. The remaining 10-20% will be selected in the next 2-3 wk.
g. When all antigen-specific antibody-secreting cell lines have been placed in frozen storage, retained final culture supernatant fluids are evaluated in detail in a variety of assays (see IV.C. this chapter) to identify antibodies of interest and to ascertain which cell lines to carry through the cloning process and antibody production stage.

III.D.3. Comments

a. Laboratory protocols vary widely in the post fusion care of hybridomas. Four to five days after fusion there should be massive cell death. If there are living cells in the myeloma control wells, then the aminopterin stock may not be any good or the myeloma cell line has reverted to aminopterin resistance. All myelomas and parent spleen cells should be dead by day 10.
b. Some protocols do not allow cells (hybrids) to be fed any sooner than 7-10 da postfusion, whereas others suggest feeding as early as 3 da postfusion. Every manipulation increases the risk of contamination; therefore, feeding should be as infrequent as possible.
c. Feeding is accomplished by aseptically aspirating about half of the culture supernatant and replacing it with fresh medium. Early feeding of cultures actually serves several purposes. One reason for feeding is to remove waste products from the predominantly dead cultures more than to replenish nutrients. A second reason for early feeding is to gradually dilute out any antibody made by unfused antibody-secreting spleen cells. (It is not unusual to have 100% of the fusion wells to be antigen-positive, without any hybridoma growth, because of the presence of antibodies secreted by the immune lymphocytes surviving in the fusion medium.)
d. Later feedings are determined according to hybrid cell growth. Ten to 14 da after fusion, HAT medium is gradually replaced with HT medium. Because of the sensitivity of normal cells to aminopterin (without HT), cells should be slowly weaned from HT to complete medium without HT. (Once weaned of HAT, all subsequent operations are routinely carried out in medium + HT.)
e. To allow for adequate accumulation of antibody, cells should not be fed for at least 48 hr before the removal of culture fluid for screening. Hybrids are usually screened and fed in the same step; i.e., culture removed for screening assay is replaced with fresh medium. Antigen-positive wells are transferred to the next larger size vessel (96-well plate to 24-well plate to 25 cm^2 flask). Cells remaining in the original well are diluted with fresh medium and serve as the backup culture for the transferred cells.

f. A general routine involves removal of culture fluids on Mondays and Thursdays for antibody screening and transfer of antibody positive cell lines on Tuesdays and Fridays. Extremely fast-growing hybrids are simply moved up first and then screened in the next round of assays. Hybrids are screened at each expansion stage and those secreting the desired antibodies should be frozen as soon as possible.

III.D.4. Procedures for the freezing and thawing of cells

a. Harvest the cells by repeated gentle aspiration of the medium and cells across the bottom of the 25 cm^2 flask (or up and down in the well if harvesting from a 6-well plate) and centrifugation at 400 g, 10 min.
b. Remove and retain the supernatant fluid for analysis (5-10 ml from flask; 2-4 ml from plate wells).
c. Resuspend cells in about 1 ml freezing medium and aseptically dispense aliquots to four sterile 2-ml cryovials (2 vials for 6-well plate samples). Tighten the caps securely.
d. Immediately place the cryovials into a cooling chamber where the rate of cooling is approximately -1 C/min. [Adequate cryopreservation has been obtained when the vials have been placed in a 2" X 5" X 5", 81-space, cardboard freezer storage box (Queue, Revco) inside a foam insulated styrofoam box and left in a -70 C or -135 C freezer for 18-24 hr.]
e. Once the vials have reached a temperature below -60 C, transfer the vials to liquid nitrogen or -135 C mechanical freezer storage.
f. To retrieve cells from cryostorage, thaw frozen cells immediately and quickly in a 37 C water bath (do not submerge).
g. As soon as the vial contents have thawed, transfer the cells to 10 ml complete medium + HT and centrifuge 400 g, 10 min.
h. Resuspend the cells in 2 ml complete medium + HT. Transfer cells to a top well of a 24-well plate (e.g., well 2A). Take 1 ml from this well and add it to 1 ml complete medium + HT in the next well down (well 2B) Repeat this dilution series down the plate (i.e., well 2B to well 2C to well 2D).
i. Examine the cultures after 24 hr growth at 37 C, 5-10% CO_2. Propagate, refreeze and/or clone from the well containing the most viable and best growing cells.

III.D.5. Comments

a. Cryopreservation of myeloma stock cells and stocks of each of the antigen reactive hybridoma 'parent' cell lines and clones is an essential safeguard and a very important part of hybridoma work. Parent myelomas and established hybridomas can be kept frozen with little difficulty for many years in liquid nitrogen with good recovery. More recently, mechanical freezers that can maintain temperatures less than -130 C have become available (Queue, Harris).
b. Aliquots of cells are resuspended (10^7/ml) in medium containing 20-90% serum and 5-10% DMSO. Aliquots are frozen slowly (about 1 C/min) to less than -70 C before transfer to permanent storage. Retrieval and thawing of cells are, in contrast, done quickly at 37 C before removal from the DMSO and subsequent culture.

c. During the propagation, expansion, and cryopreservation stages of hybridoma production, no back-up cultures should be discarded without first 'test-thawing' a cryovial to be sure that the frozen 'stock' cells can be recovered in a viable (and sterile) state.
d. A procedure has been described for freezing whole tissue culture plates containing hybrids (87). This can be advantageous for those 'highly successful' fusions.
e. As the cell lines are prepared for frozen storage, the culture supernatant fluids are retained and used in more extensive serological assays.

III.D.6. Protocol for cloning hybridomas by limiting dilution

a. If feeder cells are required by the cell line to be cloned, they should be prepared 1-2 da ahead of cloning and plated out, 100 μl/well.
b. A sub-sample of the chosen cell line (growing in log phase) is serially diluted across a row in a 96-well plate. The wells are inspected microscopically. The well containing 30-50 cells is resuspended in 20 ml (10 ml if feeders are used) complete medium + HT and plated out in 200 μl (100 μl if feeders are used) sub-samples into a 96-well plate.
c. The plate is incubated at 37 C, in an atmosphere of 5-10% CO_2, for 1-3 wk until visible clones appear. Clones are usually tested 10-14 da after plating but the time required may vary among clones. Screen for antigen-specific clones only if the plate shows growth in no more than 60 wells (=63%). If no antigen-specific positives occur, return to the frozen stock and again establish clone growth.
d. Four to 6 antigen-positive clones per parent cell line are usually selected, transferred, and expanded in culture, as for the original fusion.

III.D.7. Comments

a. Once antigen-positive antibody-secreting hybridomas have been identified they need to be cloned as soon as practicable. This is to ensure that a given culture contains only one cell type and that it is producing only one immunoglobulin specificity; i.e., true monoclonality.
b. Cloning is achieved by seeding single cells into culture wells (or on agar) and allowing them to grow into colonies. Hybridomas placed in culture at very high dilution have a tendency to die out, however. Cloning efficiency can be improved by using feeder cells, conditioned medium, or cell growth supplements (as discussed earlier for fusion).
c. Limiting dilution is the most widely used method of cloning and relies on diluting cells to less than one cell per well in a culture plate. Dispensing single cells into multiple wells is a random process and is best achieved by seeding large numbers of cells on several plates at 0.3 cells/well or lower. In practice, seeding 30-50 cells onto 1 96-well plate is sufficient.
d. Statistically, according to the Poisson distribution of probabilities, a single clone per well can be concluded if no more than 63% of the wells contain growing cells and if viable colonies are randomly distributed in the plate (11).

IV. **Characterization of Monoclonal Antibodies**

The characterization of the McAbs can be one of the most exciting aspects of McAb production. It can also be a long and complex procedure depending on the intended use of the antibodies. The most interesting characteristic of a McAb is, of course, its specificity. Other characteristics of the antibody that may also be useful in evaluating the McAb's utility include its immunoglobulin class and subclass, specific titer, storage stability, purification properties, and affinity.

IV.A. **Antibody Production and Purification**

IV.A.1. Principle

Large amounts of McAb can be produced either by culturing cells in vitro or growing them as solid or ascites tumors in vivo. The McAb is secreted and is accumulated in the spent medium of the cultured cells (at 10-100 μg/ml) and in the body fluids and serum (at 2-20 mg/ml) in tumor-bearing animals. For many purposes, purification of an antibody is not necessary and culture or ascitic fluid may be used directly. However, many methods used to characterize a McAb (or for using the McAb) involve labeling it with enzymes or isotopes and a pure antibody is more suitable for this purpose. Depending on the purity required, the antibody may be precipitated with ammonium (or sodium) sulfate. A higher degree of purity may be obtained by ion-exchange chromatography or affinity separation using Protein A, anti-immunoglobulin or by antigen affinity columns.

IV.A.2. Procedure for the production of monoclonal antibody in cell culture

a. Expand desired cell line(s) from 25 cm^2 to 150 cm^2 flasks (=about 50 ml medium). Hybridomas may be adapted to low serum growth in the transfer stages; i.e., 20% serum in 25 cm^2 flask to 10% serum in 75 cm^2 and 2-5% serum in 150 cm^2 flask.
b. Allow cells to reach log phase growth in maximum amount of medium per vessel and then to exhaustion (death), usually an additional 48-72 hr.
c. Harvest antibody containing culture supernatant fluid and discard cells.
d. For larger quantities of antibody, cells can be successively transferred up to 1- to 5-liter roller or spinner bottles.

IV.A.3. Procedure for the production of monoclonal antibodies in animals

a. Prime recipient mice by intraperitoneal injection with 0.1-0.3 ml pristane (Sigma) 10 days before hybridomas are used.
b. Harvest cells from a vigorously growing culture and resuspend in serum-free RPMI-1640 at 1-10 X 10^6 cells/ml.
c. For ascitic fluid production, inject 0.3-0.5 ml cells intraperitoneally (20-22 G 1-1/2 inch needle) into mice pristane-primed 10 da earlier.
d. After about 10-14 da (sometimes longer) mice will show abdominal swelling. Ascitic fluid is removed by "tapping" the mouse. Insert an 18-20 G 1-1/2 inch needle intraperitoneally (in the lower right or

left side of the abdomen) and collect the milky solution in a 15 ml tube.

e. Mice are observed daily and tapped as necessary, usually every 1-3 da until death. A total of 1 to 30 ml can be collected from each mouse. Cells and the fibrin clot are sedimented by centrifugation and the clear ascitic fluid is then stored frozen (as small aliquots) at -20 C or at 4 C with added preservative (e.g., 0.2% thimerosal).

IV.A.4. Comments

a. Culture in vitro provides a more pure preparation of antibody. The protein impurities are the medium components (FBS, etc) and the only mouse (or rat) immunoglobulin present is the McAb. In contrast, ascites and serum from hybridoma-injected mice contain 1-10% non-hybridoma mouse antibodies.
b. For McAb production in vivo, mice must be histocompatible with the parent cells and should also be primed by injection with pristane (tetramethylpentadecane) 10 da before injection with hybridoma cells (33).

IV.A.5. Procedure for the concentration of antibody from culture fluid or ascitic fluid by ammonium sulfate precipitation

a. Use culture fluid as is, but dilute ascitic fluid 1:2 with saline.
b. Add saturated ammonium sulfate (pH 7.4 with NH_4OH at time of use) to a final concentration of 45% (some McAbs require 50%) by dropwise addition and stirring.
c. Stir at room temperature 30 min; centrifuge 10,000 g, 10 min. Resuspend pellets in phosphate- or Tris-buffered saline, pH 7.5, at 1/10th to 1/100th original volume.
d. Dialyze at 4 C against at least 3 changes of appropriate buffer. Clarify by centrifugation.

IV.A.6. Procedures for column chromatography

a. Apply ammonium sulfate concentrated sample to a Sephadex G-200 column equilibrated in buffered saline and elute with the same buffer.
b. Monitor eluants at 280 nm. The excluded peak contains IgM (and lipids), and the included peaks contain IgGs and albumin.
c. IgMs are further purified by dialysis against 5 mM Tris-HCl, pH 7.5 or 2 mM phosphate, pH 6.0. The precipitated IgM is collected by centrifugation and resuspended in buffered saline.
d. IgGs are further purified by general Protein A column procedures (25,27), DEAE-cellulose chromatography (10,73) and/or DEAE-Affi Gel Blue column chromatography (9).

IV.A.7. Comments

a. Affinity purification of antibodies is usually accomplished with Protein A. However Protein A does not bind to all McAbs, e.g., does not bind to IgM or IgA; binds to IgG; binds with greatest affinity to subclass IgG2a. Protein A binds to all mouse IgG in the presence of buffers > pH 8.2. Some McAbs are sensitive to low-pH elution buffers required for dissociation from Protein A.

b. DEAE cellulose has also been used for the purification of IgG antibodies. The main advantage of DEAE over Protein A is that extremes of pH are not necessary for purification and ionic strength gradients are used instead. DEAE-Affi Gel Blue columns have been used to successfully purify a variety of different hybridoma antibodies of all subclasses (9).
c. Conventional antibody purification procedures are generally based upon the properties of a majority of antibody molecules in a polyclonal mixture. A given McAb usually has distinctive physical properties that are not those of the majority. It is therefore better to try to use procedures that separate on the basis of properties that do not vary greatly between different antibodies. (For example, IgMs are easily purified by dialysis against low ionic strength buffers - IgMs will exclusively precipitate.)
d. All stages of antibody purification should be monitored with an antigen-specific serological assay.

IV.B. **Comments Concerning Antibody Storage Stability, and Class and Subclass Determinations**

Conventional polyclonal sera are generally stable. This is because polyclonal sera contain many different 'monoclonal' antibodies, each with different stability. If only some of the antibodies are sensitive to a certain treatment, the activity of the preparation, as a whole, may not be seriously impaired by that treatment. This 'sensitivity' becomes very critical to a monoclonal reagent and is of practical significance. Most McAbs are stable and can be stored the same as conventional antisera. It is recommended, however, that each McAb's sensitivity to freeze-thawing be tested. Regardless, it is best to make small aliquots of the McAb for storage at -20 C (or better yet -80 C). Some antibodies are sensitive to freeze-drying, others to ammonium sulfate fractionation. The best approach is to keep in mind that the methods of production, purification and storage of a McAb can greatly influence its overall stability, reactivity, and specificity (84).

Determination of the class and subclass of a McAb is useful in predicting specific properties of the antibody, e.g., affinity for Protein A and choice of optimal purification procedure. Isotyping the antibody can also be an indication of monoclonality (or lack of it). If an Ab, for instance, is found to have both IgG1 and IgA antibodies, the cell line is probably a mixture of at least 2 cell lines, although having a single isotype does not prove the hybridoma is truly monoclonal (could be 2 McAbs with the same isotype).

Antibody class is most readily determined by the use of class specific antibodies by Ouchterlony double-diffusion assay or ELISA. Sets of antisera are commercially available from several sources (e.g., Litton, BioRad, HyClone) for the identification of IgG1, IgG2a, IgGb, IgG3, IgA and IgM. Class and subclass should be determined on culture fluid rather than ascitic fluid because the latter contains other classes and subclasses originating from the mouse, rather than the hybridoma.

IV.C. Determining Antibody Specificity

Once a number of positive hybridoma cell lines have been identified by a simple rapid screen, a more vigorous characterization of the McAbs in terms of specificity (and maybe affinity) is often desirable. The most interesting aspect of a McAb is indeed its specificity. One of the most relevant aspects of epitope analysis relates to the final use of the McAb. Clearly the specificity of the antibody needs to be assessed in order to evaluate the usefulness of the antibody. In addition to the initial screening assay, a variety of solid-phase and solution binding assays are utilized. These include ELISA, dot-blot, Western blot and immune precipitation assays (details of which can be found elsewhere in this manual). Fine specificity is further analyzed by labeled antigen and/or labeled antibody competition assays. The 'status' of the antigen as presented in each assay is also important in revealing the fine specificity of each tested antibody. For example, a conformation-dependent antibody reactive with an epitope present only on the native protein may not react with that same protein under denaturing conditions. In addition, the pH and ionic conditions of an assay, as well as the conditions for coating an antigen to solid phase, can structurally alter the antigen and affect the ability of a McAb to react with it. The antibody itself can be affected by the assay conditions or become inactive when coupled to an enzyme (29,40). Also, unless the antigen molecule has repeating antigenic determinants a McAb will not precipitate the antigen in double-diffusion or microprecipitin assays. In other words, care should be exercised in determining antibody specificity as changes in the environment of the antigenic determinant, or in the way the antigen (or antibody) is presented, could alter results.

V. Additional Alternative and New Approaches in Hybridoma Production

Hybridoma technology has made it possible to develop McAbs in almost every situation where polyclonal antibodies have been unsatisfactory. Large numbers of hybridomas (high efficiency of fusion) have been developed from fusions using cells and viruses as immunogen, and a large proportion of those hybridomas secreted the desired antibodies specific for the immunogen (high specific efficiency). However, many fusions have had specific efficiencies of close to zero. Realization of the potential of antigen-specific McAbs is often limited by failure to stimulate adequate numbers of antigen-specific B-lymphocytes during the in vivo immunization procedure. This failure can be due to any number of factors. Some of these include: low levels of antigen or 'poorly' immunogenic antigen (resulting in low splenic content of the appropriate cells), an antigen hierarchy response (selective responsiveness to one or a few components of the immunogen preparation, not necessarily the right one), or tolerance (antigen-specific unresponsiveness) (77). Several techniques have recently been employed to improve the yield of antigen-specific hybridomas. Some of these procedures will be introduced here.

Intrasplenic [direct injection of the spleen in the animal (79)] and in vitro [sensitization of cultured spleen cells with antigen (77)] immunization protocols have been found to increase the specific efficiency compared to in vivo immunization. These procedures are extremely useful when very low amounts of antigen are available. The 'immunization' schedule is reduced to 4-10 da. However, most of the McAbs obtained are concomitantly of the IgM isotype. An increase in the IgG response can be achieved by using the in vitro immunization as the secondary (or tertiary) immunization (41,78). Adoptive transfer of spleen cells from immunized animals to X-irradiated syngeneic recipients followed by in vivo antigen boosting has resulted in a 10- 50-fold increase in the percentage of antigen-specific hybridomas (78).

Other immunization-related techniques include using more highly immunogenic adjuvant peptides (e.g., N-acetylmuramyl-L-alanyl-D-iso-glutamine) (77) and injecting suspensions of antigen-bearing particles derived from antigen-spotted or -blotted nitrocellulose (1,55).

Procedures relating to the selection and/or enhancement of antigen-specific B-lymphocytes or to the suppression or induced tolerance of nontarget antigen-specific lymphocytes have also been described. Separation of splenocytes prior to fusion over a Percoll gradient, to prevent overgrowth of hybrids by macrophages and fibroblasts and increase fusion efficiency, has led to a higher yield of antigen-specific hybridomas (85). Solid-phase immunoadsorption removal of nontarget antigen-specific lymphocytes using anti-nontarget antigen antibodies has also been shown to be a viable protocol for increasing the yield of target-specific hybridomas (26,67). McAb production by receptor-mediated electrically induced cell fusion (64) is another form of the B-lymphocyte enrichment procedure. Suppression of immune response by cyclophosphamide (10) or induction of tolerance (32,36) has also been used. High doses of an identified nontarget antigen presented neonatally can suppress the ability of the immune system to recognize that nontarget antigen later in life when both target and nontarget antigen are present in the immunogen mixture (32,36).

The successful production of stable rabbit-mouse hybridomas that secrete complete rabbit Ig McAb of defined specificity (76) should enable the production of McAbs to antigens that were not immunogenic in mice. The procedure includes fusing immunized rabbit spleen cells with mouse myelomas, culturing the rabbit-mouse hybrids in the presence of normal rabbit serum and gradual adaptation of the rabbit Ig-secreting hybridomas to growing in medium containing FBS rather than rabbit serum (76).

II.B.2.a. ADDENDUM I. APPLICATIONS OF MONOCLONAL ANTIBODIES, VIRUSES

R. L. Jordan
U. S. Department of Agriculture, ARS
Beltsville, Maryland

As mentioned in the introduction, the hybridoma technology for the production of McAbs has been applied to plant virology. McAbs have been shown to be very useful for the identification and quantitative assay of plant viruses, for determining the degree of similarity between members of the various groups of viruses, as well as for studying the structural and functional aspects of structural and nonstructural gene products. McAbs have been produced to over 40 plant viruses representing ten taxonomic groups since 1981 (29,44,80). A few specific examples of the use of McAbs in studying plant viruses will be presented here.

The discriminatory properties of hybridoma antibodies has made them ideal reagents for the investigation of antigenic relationships between viral proteins at group, virus, strain and serotype levels. McAbs directed to the common strain of tobacco mosaic virus were able to differentiate between 10 different tobamoviruses (8). McAbs differentiating strains of barley yellow dwarf luteovirus have been produced in several different studies (19,20,34). The reactivity patterns of 10 McAbs produced to potato virus X (PVX) were useful in differentiating "resistance-breaking" phenotypes (82). Ilarvirus-specific McAbs were able to define 5 serotypes of apple mosaic virus and 3 serotypes of Prunus necrotic ringspot virus from a collection of 21 strains (30). Three of seven McAbs produced to an isolate of tobacco etch virus (TEV) were specific for TEV isolates (22), whereas the remaining seven had differential reactions with TEV and at least 6 other potyviruses. Twenty of 43 McAbs produced to a mixture of potyviruses recognized epitopes found only on strains of bean yellow mosaic virus (42), whereas the remaining 23 McAbs react with at least one of 30 different potyviruses. One of these pan-reactive McAbs has been able to detect all potyviruses so far tested (over 95 isolates representing 36 definitive potyviruses) [(43), Jordan and Hammond, U.S. Patent Application No. 07/082,591].

The molecular homogeneity of a McAb preparation ensures that only one epitope is investigated at any one time. This allows antigenic determinants to be studied in a defined and reproducible manner, i.e., epitopes can be identified, delineated and mapped, and their structural properties characterized using McAbs. McAbs have been used, for example, to study the structure of southern bean mosaic virus (65,83) and tobacco mosaic virus (4) and to localize specific viral epitopes on potyviruses (22,43), ilarviruses (2,40) and tobamoviruses (21).

The use of McAbs for routine viral diagnosis has been illustrated with potato virus Y-specific McAbs (28) and soybean mosaic virus (SMV) McAbs (18), in addition to many of the aforementioned references (see also 29,35,44,80). McAbs to SMV have also been used in immunoaffinity chromatography to purify the virus (17).

More recently, McAbs have been generated to nonstructural viral proteins. McAbs were produced to a cloned nonstructural protein from cucumber mosaic virus and used to localize the native protein in virus-infected plants (66). To further study helper component-mediated aphid transmission of potyviruses, McAbs have been generated to tobacco vein mottle and potato virus Y helper components (Jordan, Thornbury, and Pirone, unpublished). McAbs have also been produced to tobacco etch virus nuclear inclusion (Slade, Johnston, Carrington and Dougherty, unpublished).

McAbs have been produced that are specific to synthetic and viral double-stranded RNAs (74). Antigen-mimicking anti-idiotype McAbs have been produced to soybean mosaic virus-specific McAbs, which have been shown to be useful as a 'virus-standard' (in lieu of virions) in ELISAs (71).

II.B.2.b. ADDENDUM II. APPLICATIONS OF MONOCLONAL ANTIBODIES, BACTERIA

S. H. De Boer
Agriculture Canada Research Station
Vancouver, B.C., Canada

Monoclonal antibodies (McAbs) can be used in serological procedures for taxonomic classification, identification, detection of plant pathogenic bacteria, disease diagnosis, and monitoring bacterial metabolites. Because of their specificity, McAbs have been particularly useful for rapid identification of bacteria. McAbs to *Xanthomonas campestris* pv. *campestris* were used to identify bacteria isolated from black rot infected crucifers (89). Alvarez et al. (6) produced a series of 15 different McAbs whereby the genus *Xanthomonas* could be identified, as well as strains of *X. campestris* pv. *campestris* differentiated into six groups. Similarly, a panel of McAbs was used to identify a unique strain of *X. campestris* pv. *citri* causing Mexican citrus bacteriosis (5). Serogroup I of *Erwinia carotovora* subsp. *atroseptica* could also be differentiated from other *E. carotovora* strains on the basis of reactivity with a McAb (13).

Performance of serological procedures such as ELISA and immunofluorescence for disease diagnosis and detection of plant pathogenic bacteria can often be enhanced by McAbs because of their greater specificity. McAbs have been used in immunofluorescence tests to detect *Corynebacterium sepedonicum* (14,15), *Erwinia amylovora* (62), and *E. carotovora* subsp. *atroseptica* (13) directly in diseased plant tissue. Similarly, ELISA also could be used to diagnose the bacterial ring rot and blackleg diseases of potato with McAbs (13,16).

McAbs have been particularly useful for detecting fastidious prokaryotes including spiroplasmas and mycoplasma-like organisms (MLOs) because they cannot be as readily cultured as other phytopathogenic bacteria or in the case of MLOs be cultured at all. McAbs have been produced to the plum leaf scorch bacterium that causes disease in at least seven different woody hosts (37). Nine species-specific McAbs have been produced to *Spiroplasma citri* and seven to *S. kunkelii*, (59,60). In a separate study in which a mixture of *S. citri* and *S. kunkelii* was used as immunogen, and as antigen for screening hybridomas, 17 *S. citri*-specific, 17 *S. kunkelii*-specific, and 12 group I spiroplasma cross-reactive McAbs were produced (45). Three of the cross-reactive McAbs were shown to detect all 26 group I isolates tested (representing all 8 subgroups). Many of the species-specific and cross-reactive McAbs have been utilized in immunofluorescent studies of helical morphology (53), and detection of spiroplasmas in plants and insects (12, and Panday, Davis, and Jordan, unpublished). Aster yellows (AY) MLO-specific McAbs have been produced and used for the detection of the AY-MLO as well as for further purification of AY-MLO by immunoaffinity chromatography (39,61). More recently, McAbs have been produced to the tomato big bud MLO with an immunosuppression/tolerance technique using infected tissue extracts as immunogen (Hsu, Lee and Davis, submitted).

Often McAbs can be produced against whole, intact prokaryotic cells, and the McAbs with the desired specificity and reactivity selected for use in a serological assay procedure. The nature of the antigen that reacts with the McAb will, however, have a bearing on the usefulness of the McAb. McAbs directed to a heat labile antigen, for example, may not react with bacterial cells that have been heat fixed to microscope slides. A McAb directed to an extracellular polysaccharide antigen of *C. sepedonicum* also did not react with intact cells fixed to microscope slides, although it reacted very well in ELISA with *C.*

sepedonicum supernatants and ring rot infected plant extracts (16). In contrast, a McAb that reacted with a *C. sepedonicum* cell wall component was ideal for immunofluorescence but gave only a poor response in ELISA in which whole cells were not specifically fixed to the solid phase. A McAb against the lipopolysaccharide of *E. carotovora*, on the other hand, was useful for both immunofluorescence and ELISA because the antigen was not only an integral component of the outer membrane but also occurred in cultural supernatant and infected plant extracts.

McAbs are potentially very powerful tools to detect and monitor the presence of bacterial metabolites. They have already been used extensively for metabolites of non-plant pathogenic bacteria, and their role in elucidating mechanisms of pathogenicity and host specificity of phytopathogens is anticipated to be significant. McAbs to *Erwinia* pectate lyase has already been used to detect production of pectate lyase in culture (63) and in potato tubers (68,69), to compare pectate lyases in different strains by Western blot analysis (86), and to detect *Erwinia* by dot blot assays (49,50).

VI. Literature Cited

1. Abou-Zeid, C., Filley, E., Steele, J., and Rook, G. A. W. 1987. A simple new method for using antigens separated by polyacrylamide gel electrophoresis to stimulate lymphocytes in vitro after converting bands cut from Western blots into antigen-bearing particles. J. Immunol. Methods 98:5-10.
2. Aebig, J. A., Jordan, R. L., Lawson, R. H., and Hsu, H. T. 1987. Immunological and biological properties of a mouse monoclonal antibody reactive to Prunus necrotic ringspot ilarvirus. Intervirology 28:57-64.
3. Al Moudallal, Z., Altschuh, D., Briand, J. P., and van Regenmortel, M. H. V. 1984. Comparative sensitivity of different ELISA procedures for detecting monoclonal antibodies. J. Immunol. Methods 68:35-43.
4. Altschuh, D., Al Moudallal, Z., Briand, J. P., and Van Regenmortel, M. H. V. 1985. Immunochemical studies of tobacco mosaic virus - VI. Attempts to localize viral epitopes with monoclonal antibodies. Mol. Immunol. 22:329-337.
5. Alvarez, A. M., Benedict, A. A., and Civerolo, E. L. 1986. Unique strains of _Xanthomonas campestris_ pv. _citri_ associated with Mexican citrus bacteriosis. (Abstract) Phytopathology 76:1078.
6. Alvarez, A. M., Benedict, A. A., and Mizumoto, C. Y. 1985. Identification of _Xanthomonads_ and grouping of _Xanthomonas campestris_ pv. _campestris_ strains with monoclonal antibodies. Phytopathology 75:722-728.
7. Atassi, M. Z. 1986. Preparation of monoclonal antibodies to preselected protein regions. Methods Enzymol. 121:69-95.
8. Briand, J. P., Al Moudallal, Z., and Van Regenmortel, M. H. V. 1982. Serological differentiation of tobamoviruses by means of monoclonal antibodies. J. Virol. Methods 5:293-300.
9. Bruck, C., Drebin, J. A., Glineur, C., and Portetelle, D. 1986. Purification of mouse monoclonal antibodies from ascitic fluid by DEAE Affi-Gel Blue chromatography. Methods Enzymol. 121:587-596.
10. Campbell, A. M. 1984. Monoclonal Antibody Technology: The Production and Characterization of Rodent and Human Hybridomas. Elsevier, New York. 265 pp.
11. Coller, H. A., and Coller, B. S. 1986. Poisson statistical analysis of repetitive subcloning by the limiting dilution technique as a way of assessing hybridoma monoclonality. Methods Enzymol. 121:412-417.
12. Davis, R. E., Lee, I. M., Jordan, R. L., and Konai, M. 1987. Detection of corn stunt spiroplasma in plants by ELISA employing monoclonal antibodies and by isolation of the pathogen and its phages in serum-free medium. Pages 306-312 in: Plant Pathogenic Bacteria. E. L. Civerolo, et al, eds. Martinus Nijhoff, Boston. 1050 pp.
13. De Boer, S. H., and McNaughton, M. E. 1986. Production of monoclonal antibodies to the lipopolysaccharide _Erwinia carotovora_ subsp. _antroseptica_ serogroup I. Phytopathology 77:828-832.
14. De Boer, S. H., and McNaughton, M. E. 1986. Evaluation of immunofluorescence with monoclonal antibodies for detecting latent bacterial ring rot infections. Am. Potato J. 63:533-643.
15. De Boer, S. H., and Wieczorek, A. 1984. Production of monoclonal antibodies to _Corynebacterium sepedonicum_. Phytopathology 74:1431-1434.
16. De Boer, S. H., Wieczorek, A., and Kummer, A. 1988. An ELISA test for bacterial ring rot of potato with a new monoclonal antibody. Plant Dis. 72:874-878.

17. Diaco, R., Hill, J. H., and Durand, D. P. 1986. Purification of soybean mosaic virus by affinity chromatography using monoclonal antibodies. J. Gen. Virol. 67:345-351.
18. Diaco, R., Hill, J. H., Hill, E. K., Tachibana, H., and Durand, D. P. 1985. Monoclonal antibody-based biotin-avidin ELISA for the detection of soybean mosiac virus in soybean seeds. J. Gen Virol 66:2089-2094.
19. Diaco, R., Lister, R. M., Hill, J. H., and Durand, D. P. 1986. Demonstration of serological relationships among isolates of barley yellow dwarf virus by using polyclonal and monoclonal antibodies. J. Gen. Virol. 67:353-362.
20. Diaco, R., Lister, R. M., Hill, J. H., and Durand, D. P. 1986. Detection of homologous and heterologous barley yellow dwarf isolates with monoclonal antibodies in serologically specific electron microscopy. Phytopathology 76:225-230.
21. Dore, I., Weiss, E., Altschuh, D., and Van Regenmortel, M. H. V. 1988. Visualization by electron microscopy of the location of tobacco mosaic virus epitopes reacting with monoclonal antibodies in enzyme immunoassay. Virology 162:279-289.
22. Dougherty, W.G., Willis, L., and Johnston, R. E. 1985. Topographic analysis of tobacco etch virus capsid protein epitopes. Virology 144:66-72.
23. Eshhar, Z. 1985. Monoclonal antibody strategy and techniques. Pages 3-41 in: Hybridoma Technology in the Biosciences and Medicine. T. A. Springer, ed. Plenum Press, New York. 602 pp.
24. Fazekas de St. Groth, S., and Scheidegger, D. 1980. Production of monoclonal antibodies: Strategy and tactics. J. Immunol. Methods 35:1-21.
25. Galfre, G., and Milstein, C. 1981. Preparation of monoclonal antibodies: Strategies and procedures. Methods Enzymol. 73:3-46.
26. George, R. A., and Converse, R. H. 1988. Methods for the enrichment of desired B-cell populations before anti-cauliflower mosaic virus hybridoma formation. Phytopathology 78:1631-1636.
27. Goding, J. W. 1983. Monoclonal Antibodies: Principles and Practice. Production and Application of Monoclonal Antibodies in Cell Biology, Biochemistry and Immunology. Academic Press. New York, NY. 276 pp.
28. Gugerli, P., and Fries, P. 1983. Characterization of monoclonal antibodies to potato virus Y and their use for virus detection. J. Gen. Virol. 64:2471-2477.
29. Halk, E. L., and De Boer, S. H. 1985. Monoclonal antibodies in plant disease research. Annu. Rev. Phytopathol. 23:321-350.
30. Halk, E. L., Hsu, H. T., Aebig, J., and Franke, J. 1984. Production of monoclonal antibodies against three ilarviruses and alfalfa mosaic virus and their use in serotyping. Phytopathology 74:367-372.
31. Hammerling, G. J., Hammerling, U., and Kearney, J. F., eds. 1981. Monoclonal Antibodies and T Cell Hybridomas, Perspectives and Technical Advances. Elsevier/North-Holland, New York. 569 pp.
32. Hockfield, S. 1987. A Mab to a unique cerebellar neuron generated by immunosuppression and rapid immunization. Science 237:67-70.
33. Hoogenroad, N. J., and Wraight, C. J. 1986. The effect of pristane on ascites tumor formation and monoclonal antibody production. Methods Enzymol. 121:375-381.
34. Hsu, H. T., Aebig, J., and Rochow, W. F. 1984. Differences among monoclonal antibodies to barley yellow dwarf viruses. Phytopathology 74:600-605.
35. Hsu, H. T., and Lawson, R. H. 1985. Comparison of mouse monoclonal antibodies and polyclonal antibodies of chicken egg yolk and rabbit for assay of carnation etched ring virus. Phytopathology 75:778-783.

36. Hsu, H. T., Wang, Y. C., and Lawson, R. H. 1988. Induction of immunological tolerance to enhance the production of virus-specific hybridomas. (Abstract) Phytopathology 78:1537.
37. Hung, H. Y., Wells, J. M., and Chen, T. A. 1987. Monoclonal antibodies for differentiating some strains of the fastidious, gram-negative bacteria from plants. Pages 352-356 in: Plant Pathogenic Bacteria. E. L. Civerolo, et al, eds. Martinus Nijhoff, Boston. 1050 pp.
38. HyClone Laboratories, Inc. 1988. Iron-supplemented bovine calf serum: an excellent substitute for fetal bovine serum. Art to Science 7:1-4, 6.
39. Jiang, Y. P., Lei, J. D., Chen, T. A. 1987. Purification of aster yellows agent from diseased lettuce using affinity chromatography. (Abstract) Phytopathology 77:1726.
40. Jordan, R. L., Aebig, J., and Hsu, H. T. 1985. Reactivities of apple mosaic virus (ApMV)/Prunus necrotic ringspot virus (NRSV) monoclonal antibodies with ApMV, NRSV and other ilarviruses. (Abstract) Phytopathology 75:1353.
41. Jordan, R. L., Elliott, N., and Hsu, H. T. 1984. Production of monoclonal antibody secreting hybridoma cell lines to prune dwarf virus by _in vivo_ or _in vitro_ immunization. (Abstract) Phytopathology 74:862.
42. Jordan, R. L., and Hammond, J. 1986. Analysis of antigenic specificity of monoclonal antibodies to several potyviruses. (Abstract) Phytopathology 76:1091.
43. Jordan, R. L., and Hammond, J. 1988. Epitope specificity of strain-, virus, -subgroup-specific and potyvirus group cross-reactive monoclonal antibodies. (Abstract) Phytopathology 78:1600.
44. Jordan, R. L., and Hsu, H. T. 1990. Immunological detection and epitope analysis of plant pathogens and their gene products: Monoclonal antibodies in plant pathology. Acta Hort. (In Press).
45. Jordan, R. L., Konai, M. I., Lee, I. M., and Davis, R. E. 1989. Species-specific and cross-reactive monoclonal antibodies to the plant pathogenic spiroplasmas _Spiroplasma citri_ and _S. kunkelii_. Phytopathology 79:880-887.
46. Jordan, R. L., Nuss, D., and Hsu, H. T. 1985. Production of monoclonal antibodies to wound tumor virus using extracts from cultured virus-infected leafhopper cells. (Abstract) Phytopathology 75:1356.
47. Kawamoto, T., Sato, J. D., McClure, D. B., and Sato, G. H. 1986. Serum-free medium for the growth of NS-1 mouse myeloma cells and the isolation of NS-1 hybridomas. Methods Enzymol. 121:266-277.
48. Kennett, R. H., McKearn, T. J., and Bechtol, K. B., eds. 1980. Monoclonal Antibodies: Hybridomas, a New Dimension in Biological Analysis. Plenum Press, New York. 423 pp.
49. Klopmeyer, M. J., and Kelman, A. 1988. Use of monoclonal antibodies specific for pectate lyase as serological probes in the identification of soft rot _Erwinia_ spp. Phytopathology 78:1430-1434.
50. Klopmeyer, M. J., Livingston, R. S., and Kelman, A. 1986. Differentiation of soft rot bacteria using monoclonal antibodies. Phytopathology 76:1077.
51. Kohler, G. 1981. The technique of hybridoma production. Pages 285-298 in: Immunological Methods, Vol. 2. I. Lefkovits and B. Pernis, eds. Academic Press, New York. 316 pp.
52. Kohler, G., and Milstein, C. 1975. Continuous cultures of fused cells secreting antibody of predefined specificity. Nature 256:495-497.
53. Konai, M. I., Davis, R. E., and Jordan, R. L. 1986. Immunofluorescence microscopy using monoclonal antibodies to study _Spiroplasma_ with preserved helical cell shape. (Abstract) Phytopathology 76:1083.
54. Kovar, J., and Franck, F. 1986. Serum-free medium for hybridoma and parental myeloma cell cultivation. Methods Enzymol. 121:277-292.

55. Knudsen, K. 1985. Proteins transferred to nitrocellulose for use as immunogens. Anal. Biochem. 147:285-288.
56. Lane, R. D., Crissman, R. S., and Ginn, D. 1986. High efficiency fusion procedure for producing monoclonal antibodies against weak immunogens. Methods Enzymol. 121:183-192.
57. Langone, J., and Van Vunakis, H., eds. 1980-86. Immunological techniques (Parts A,B,C,D,E,F,G,H,I) in: Methods in Enzymology, Vols. 70,73,74,84,92, 93,108,116,121. Academic Press, New York.
58. Lefkovits, I., and Pernis, B., eds. 1981. Immunological Methods, Vols. 1 and 2. Academic Press, New York.
59. Lin, C. P., and Chen, T. A. 1985. Production of monoclonal antibodies against Spiroplasma citri. Phytopathology 75:848-851.
60. Lin, C. P., and Chen, T. A. 1985. Monoclonal antibodies against corn stunt spiroplasma. Can. J. Microbiol. 31:900-904.
61. Lin, C. P., and Chen, T. A. 1986. Comparison of monoclonal antibodies and polyclonal antibodies in detection of the aster yellows mycoplasma-like organism. Phytopathology 76:45-50.
62. Lin, C. P., Chen, T. A., Wells, J. M., and van der Zwet, T. 1987. Identification and detection of Erwinia amylovora with monoclonal antibodies. Phytopathology 77:376-380.
63. Livingston, R. S., Galuska, D. A., Klopmeyer, M. J., and Kelman, R. 1986. Production and characterization of monoclonal antibodies for pectic lyases. (Abstract) Phytopathology 76:1138.
64. Lo, M. M. S., Tsong, T. Y., Conrad, M. K., Strittmatter, S. M., Hester, L. D., and Snyder, S. H. 1984. Monoclonal antibody production by receptor-mediated electrically induced cell fusion. Nature 310:792-794.
65. MacKenzie, D. J., and Tremaine, J. H. 1986. The use of a monoclonal antibody specific for the N-terminal region of southern bean mosaic virus as a probe of virus structure. J. Gen. Virol. 67:727-735.
66. MacKenzie, D. J., and Tremaine, J. H. 1988. Ultrastructural location of non-structural protein 3A of cucumber mosaic virus in infected tissue using monoclonal antibodies to a cloned chimeric fusion protein. J. Gen. Virol. 69:2387-2395.
67. Mage, M. G., McHugh, L. L., and Rothstein, T. L. 1977. Mouse lymphocytes with and without surface immunoglobulin. Preparative scale separation in polystyrene tissue culture plates coated with specifically purified anti-immunoglobulin. J. Immunol. Methods 15:47-56.
68. Maher, E. A., and Kelman, A. 1988. Expression of pectate lyase by Erwinia carotovora subsp. carotovora in potato tuber tissue. (Abstract) Phytopathology 17:1551.
69. Maher, E. A., Livingston, R. S., and Kelman, A. 1986. Recognition of pectate lyase in western blots by monoclonal antibodies. (Abstract) Phytopathology 76:1101.
70. Mehta, A. M., Jordan, R. L., Anderson, J. D., and Mattoo, A. K. 1988. Identification of a unique isoform of 1-aminocyclopropane-1-carboxylic acid synthase by monoclonal antibody. Proc. Natl. Acad. Sci. 85:8810-8814.
71. Mernaugh, R. L., Durand, D. P., and Hill, J. H. 1987. Use of an anti-idiotype monoclonal antibody in lieu of a virus reference standard in ELISAs. (Abstract) Phytopathology 77:1765.
72. Oi, V. T., and Herzenberg, L. A. 1980. Immunoglobulin-producing hybridoma cell lines. Pages 351-372 in: Selected Methods in Cellular Immunology. B. B. Mishell and S. M. Shiigii, eds. Freeman, San Francisco. 486 pp.
73. Parham, P. 1983. Monoclonal antibodies against HLA products and their use in immunoaffinity purification. Methods Enzymol. 92:110-138.

74. Powell, C. A. 1987. Preparation of monoclonal antibody specific for double-stranded RNAs. Phytopathology 77:1743.
75. Ransom, J. H. 1986. Endothelial cell growth supplements for promoting the growth of monoclonal antibody producing hybridoma cells. Methods Enzymol. 121:293-295.
76. Raybould, T. J. G., and Takahashi, M. 1988. Production of stable rabbit-mouse hybridomas that secrete rabbit mab of defined specificity. Science 240:1788-1790.
77. Reading, C. L. 1986. In vitro immunization for the production of antigen-specific lymphocyte hybridomas. Methods Enzymol. 121:18-27.
78. Siraganian, R. P., Fox, P. C., and Berenstein, E. H. 1983. Methods of enhancing the frequency of antigen-specific hybridomas. Methods Enzymol. 92:17-26.
79. Spitz, M. 1986. "Single-shot" intrasplenic immunization for the production of monoclonal antibodies. Methods Enzymol. 121:33-42.
80. Stern, N., and Gamble, H. R. 1984. Hybridoma Technology in Agricultural and Veterinary Research. Rowman and Allenheld, New Jersey. 318 pp.
81. Taggart, R. T., and Samloff, I. M. 1983. Stable antibody producing murine hybridomas. Science 219:1228-1230.
82. Torrance, L., Larkins, A. P., and Butcher, G. W. 1986. Characterization of monoclonal antibodies against potato virus X and comparison of serotypes with resistance groups. J. Gen. Virol. 67:57-67.
83. Tremaine, J. H., MacKenzie, D. J., and Ronald, W. P. 1986. Determination of affinity constants and the number of binding sites of monoclonal antibodies specific for southern bean mosaic virus. Virology 155:452-459.
84. Underwood, P. A., and Bean, P. A. 1985. The influence of methods of production, purification and storage of monoclonal antibodies upon their observed specificities. J. Immunol. Methods 80:189-197.
85. van Mourik, P., and Zeijlemaker, W. P. 1986. Improved hybridoma technology: spleen cell separation and soluble growth factors. Methods Enzymol. 121:174-182.
86. Ward, L. J., and De Boer, S. H. 1989. Characterization of a monoclonal antibody against active pectate lyase from Erwinia carotovora. Can. J. Microbiol. 35:651-655.
87. Wells, D. E., and Bibb, W. F. 1986. A method for freezing hybridoma clones in 96-well microculture plates. Methods Enzymol. 121:417-422.
88. Westerwoudt, R. J., Naipal, A. M., and Harrisson, C. M. H. 1984. Improved fusion technique. II. Stability and purity of hybrid clones. J. Immunol. Methods 68:89-101.
89. Yuen, G. Y., Alvarez, A. M., Benedict, A. A., and Trotter, K. J. 1987. Use of monoclonal antibodies to monitor the dissemination of Xanthomonas campestris pv. campestris. Phytopathology 77:366-370.
90. Zola, H. 1987. Monoclonal Antibodies: A Manual of Techniques. CRC Press, Florida. 214 pp.
91. Zola, H., and Brooks, D. 1982. Techniques for the production and characterization of monoclonal hybridoma antibodies. Pages 1-57 in: Monoclonal Hybridoma Antibodies: Techniques and Applications. J. G. Hurrell, ed. CRC Press, Florida. 240 pp.

Section II.C.

PREPARATION OF ANTIBODY-ENZYME CONJUGATES

D. J. MacKenzie
Agriculture Canada, Vancouver Research Station
Vancouver, B.C. V6T 1X2

Introduction

Particularly since the report of Engvall and Perlmann on the use of antibody-enzyme conjugate for immunoassay (5), a number of additional techniques for conjugating enzymes to antibodies have been developed. In fact, antibody-enzyme linkage has become foundational to some of the most sensitive immunoassays in use today, including dot immunobinding assay, (DIA), enzyme-linked fluorescent assay (ELFA), enzyme-linked immunosorbent assay (ELISA), radioimmunoassay (RIA) and Western blotting.

Enzyme characteristics should be considered, however, before discussing antibody-enzyme linking procedures. The enzyme of choice should have a high specific activity, or turnover number, with a readily detectable substrate reaction product. The enzyme must have "exposed" reactive groups (e.g., amino, hydroxyl, sulfhydryl) to facilitate cross-linkage to antibody molecules. The reactive potential of both enzyme and antibody must be preserved in their cross-linked form and remain stable for reasonable periods. Enzymes fulfilling these criteria and in common use as conjugates with antibodies include alkaline phosphatase (AP), horseradish peroxidase (HRP) and β-galactosidase (BGD).

While numerous methods have been used to prepare protein-protein conjugates, relatively few have been consistently and successfully applied to the production of antibody-enzyme conjugates. Reagents for protein-protein coupling are generally non-specific in that they react with functional groups (primary amino, phenolic-hydroxyl, sulfhydryl or carboxyl) which are present on both protein participants. The balance between intra- and intermolecular cross-linking will depend not only on the total number, but also the distribution and availability for reaction of these groups on the surface of any two protein molecules. For this reason, optimal conditions for coupling any two proteins must be determined empirically on a trial and error basis, and methods developed for one particular combination of enzyme and antibody may not be directly applicable in the general case.

A. Antibody-Enzyme Conjugation Procedures

1. One-step, glutaraldehyde

This technique was originally introduced by Avrameas (1) and is the least technically demanding of all the conjugation methods. Glutaraldehyde is a homobifunctional aldehyde cross-linking reagent which reacts readily at neutral pH with the ϵ-amino groups of lysine residues present on the surface of most proteins. The resulting Schiff's base linkage is uncharacteristically stable, even in the absence of subsequent reduction, and is not likely the final reaction product. The high degree of stability observed in glutaraldehyde cross-linking may involve polymerization products of glutaraldehyde (12). Technical grade glutaraldehyde, which is presumably extensively polymerized, will often yield better results than highly purified glutaraldehyde (4). The enzymes most commonly employed are HRP and AP. Alkaline phosphatase and immuno-globulin G (IgG) in a ratio of $\geq$ 3:1 (w/w), will form efficient conjugates with very low concentrations of free IgG. Disadvantages of this technique include: (a) the antibody-enzyme conjugates produced are relatively heterogeneous in nature with many highly polymerized components; (b) excessive glutaraldehyde cross-linking may inactivate some antibody preparations. The latter may be of special concern with monoclonal reagents, some of which may be quite susceptible to glutaraldehyde inactivation. The protocol shown below for the preparation of AP conjugated antibody has been adapted from Engvall and Perlmann (5).

Procedure

a. Mix 3.0 mg of AP and 1.0 mg of purified antibody (see **Section II.B.1.**) in a total of 0.5 ml of PBS. It is important that both the enzyme and antibody preparations be free of nucleophilic amines (ie: Tris, $(NH_4)_2SO_4$, etc.) which can be conveniently removed by overnight dialysis against several changes of PBS at 4 C.
b. Add glutaraldehyde to a final concentration of 0.2% (ie: 10 μl of a 10% aqueous solution) and allow the reaction to proceed for 2 hr at room temperature with gentle stirring.
c. Dilute the reaction mixture to 1.0 ml with 10 mM Tris-HCl, pH 8.0, 0.15 M NaCl, 1 mM $MgCl_2$ and dialyze against this same buffer (2-3 changes; 1 liter) for 16-24 hr at 4 C.
d. Following dialysis, the conjugate should be stabilized by addition of bovine serum albumen (BSA) to 1% (w/v) and either dispensed as small aliquots or diluted with glycerol to 50% (v/v), and stored at -20 C, where it should remain stable for many months.

2. Two-step, glutaraldehyde

The two-step method differs from the one-step glutaraldehyde method shown above, in that one of the protein components (typically the enzyme) is activated with glutaraldehyde before addition of the antibody. This technique has been used primarily with HRP and takes advantage of the fact that this enzyme normally reacts with only one of the aldehyde groups of the cross-linker, with the second aldehyde being unable to react with the same or other HRP molecules. If HRP and IgG are incubated together with glutaraldehyde, IgG will effectively compete with HRP for available aldehyde groups, leading to extensive intermolecular cross-linking of the antibody with little coupling to the enzyme. Generally low molecular weight conjugates, consisting of one HRP and one IgG molecule are formed. A protocol for the preparation of HRP conjugated antibody (2) is shown below.

Procedure

a. Dissolve 5 mg of HRP in 0.2 ml of 0.1 M Na_3PO_4, pH 7.2, containing 1% glutaraldehyde.
b. React for 4-8 hr at room temperature with gentle agitation.

c. Dilute reaction mixture to 1.0 ml and dialyze extensively against 50 mM sodium carbonate, pH 9.2, (2-3 changes, 1 liter, 4 hr each) to remove excess glutaraldehyde. [Alternatively, glutaraldehyde may be removed by gel filtration chromatography over Sephadex G-25 (30 x .9 cm) equilibrated and eluted with 50 mM sodium carbonate, pH 9.2, 0.15 M NaCl. Fractions containing HRP should then be pooled and concentrated to 1.0 ml using an ultrafiltration membrane.]

d. Add 2.0 mg of purified antibody dissolved in 0.2 ml of 50 mM sodium carbonate, pH 9.2, and allow to react for 12-16 hr at 4 C.

e. Unreacted aldehyde groups are then quenched by addition of 0.1 ml of 0.2 M lysine and further 2 hr incubation at 4 C followed by dialysis against PBS at 4 C.

f. The conjugate may be separated from free antibody by gel filtration chromatography over Sephadex G-200 (50 x 0.9 cm) eluted with PBS. Fractions containing purified conjugate are then pooled, supplemented with BSA to 1% (w/v) and glycerol to 50% (v/v) and stored at -20 C.

3. Periodate-oxidation

This method has been applied primarily with HRP but can also be used for preparing conjugates with AP. The method takes advantage of the fact that vicinal hydroxyl groups present on the carbohydrate moiety of the enzyme can be oxidized, by sodium periodate, to reactive aldehyde groups which are then susceptible to subsequent nucleophilic attack by free amino groups on the surface of the antibody molecule. The resulting Schiff's base linkage is then reduced by reaction with sodium borohydride. The reverse approach has also been utilized, whereby the carbohydrate moiety present on the Fc region of IgG is first oxidized with periodate prior to coupling with amino groups of the enzyme (9,10). Self coupling of activated enzyme molecules can be prevented by blockage of free amino groups of the enzyme by reaction with fluorodinitrobenzene or by maintaining the oxidized enzyme solution at pH 4.5 prior to addition of antibody. The following protocol is adapted from Henning and Nielson (6), who found that substitution of $NaBH_4$ with ascorbic acid for reduction of the Schiff's base linkage resulted in greater retention of HRP activity.

Procedure

a. Dissolve 5 mg of HRP in 1.25 ml of 0.1 M sodium acetate, pH 6.5.

b. Add 0.25 ml of 0.1 M $NaIO_4$ and allow to react for 1 hr at room temperature in the dark.

c. Dialyze against 2-3 changes of 1 mM sodium acetate, pH 4.5, for 12-16 hr at 4 C.

d. Add 2 mg of purified antibody, dissolved in 0.2 ml of 0.2 M sodium carbonate, pH 9.5, and allow to react for 2-3 hr at room temperature with gentle agitation.

e. Add 0.25 ml of $NaBH_4$ (4 mg/ml) and leave at 4 C for 3 hr to overnight. [Alternatively, add 0.25 ml of ascorbic acid (5 mg/ml) and leave 12-16 hr at 4 C.]

f. Dialyze against 2-3 changes of PBS for 16-24 hr at 4 C.

g. Conjugate may be purified from free IgG by chromatography over Sephadex G-200 equilibrated with PBS and pooling of HRP-labeled antibody fractions (the first peak).

4. m-Maleimidobenzoyl-N-Hydroxysuccinimide Ester (MBS)

MBS is a heterobifunctional protein cross-linking reagent which reacts selectively with both amino ($-NH_2$) and sulfhydryl ($-SH_2$) residues. The hydroxysuccinimide active ester reacts readily with free amino groups while the maleimide moiety reacts specifically with sulfhydryl groups. This reagent has been used successfully to prepare antibody conjugates of β-galactosidase in a two step procedure (11). The first step involves reaction of the hydroxysuccinimide ester with amino groups of the antibody. Since IgG molecules do not possess free sulfhydryl residues there is no polymerization of the antibody and the incorporated maleimide groups are available for reaction with sulfhydryl groups on the enzyme molecule in the second step of the reaction. The efficiency of coupling will depend on the number of free sulfhydryl groups in commercial preparations of β-galactosidase, which generally ranges from 3-5 per molecule.

Procedure

a. Dissolve 1.5 mg of purified IgG in 1.5 ml of 0.1 M Na_3PO_4, pH 7.0, 0.05 M NaCl.
b. Add 15 μl of dioxan containing 0.32 mg of MBS and allow to react for 1 hr at 30 C with gentle agitation.
c. Excess MBS is removed by gel-filtration chromatography over Sephadex G-25 (0.9 x 30 cm) equilibrated with PBS containing 10 mM $MgCl_2$.
d. The void volume peak is collected, pooled, and mixed with 1.5 mg of β-galactosidase and maintained at 30 C for 1-2 hr. The reaction is terminated by addition of 1 M β-mercaptoethanol to a final concentration of 10 mM.
e. The conjugate may be purified by gel-filtration chromatography on Sephadex G-200 (0.9 x 90 cm) eluted with 10 mM Tris-HCl, pH 7.0, 50 mM NaCl, 10 mM $MgCl_2$ and 10 mM β-mercaptoethanol. Alternatively, good purification is achieved by ion-exchange chromatography on DEAE-Sephacel (0.9 x 15 cm). The column is equilibrated with 10 mM Tris-HCl, pH 7.0, containing 50 mM NaCl, 10 mM $MgCl_2$ and 10 mM β-mercaptoethanol, and washed with 50 ml of this same buffer following sample application. The antibody-enzyme conjugate is then eluted with equilibration buffer containing 0.2 M NaCl.

5. N-Succinimidyl 3-(2-pyridyldithio) proprionate (SPDP)

SPDP is a heterobifunctional cross-linking reagent (3) which has recently come into use for the preparation of antibody-enzyme conjugates. The principle of SPDP conjugation is to substitute proteins with 2-pyridyl disulfide moieties by reaction of amino groups on the surface of the protein with the N-hydroxysuccinimide active ester linkage. Then, one of the proteins (the enzyme) is treated with a reducing agent such as dithiothreitol (DTT) to expose free sulfhydryl groups on the added moieties. Both derivatized proteins are then combined and the newly exposed aliphatic thiol on the surface of the enzyme molecule will react with the 2-pyridyl disulfide moiety on the surface of the antibody. Since both protein participants in the conjugation are initially activated with SPDP, there are a number of parameters which can be controlled. These include the degree of substitution of both the enzyme and antibody with SPDP as well as the relative ratios of each protein in the final condensation reaction. This procedure has the theoretical advantage that the extent of homomolecular protein cross-linking can be minimized and the relative proportions of antibody and enzyme in the final conjugate may be more closely controlled. The SPDP method has been used to prepare antibody conjugates of both HRP (7) and AP (8). In the procedure described below, for the preparation of an AP conjugate, the following ratios have been used: SPDP/AP = 60; SPDP/IgG = 3; AP/IgG = 1.

Procedure

a. Dissolve 5 mg each of AP and IgG separately in 1 ml of PBS.
b. Add 5 μl and 110 μl, respectively, of SPDP (20 mM in absolute ethanol) to the solutions of IgG and AP. Incubate 30 min at room temperature.
c. Excess SPDP and released N-hydroxysuccinimide are removed by dialysis against PBS for IgG and against 0.1 M sodium acetate, pH 4.5, for AP.
d. Following dialysis, add 0.25 ml of freshly prepared 0.1 M DTT to the AP solution and let react for 30 min at room temperature.
e. Excess DTT and released pyridine-2-thione are removed from the enzyme sample by gel filtration chromatography over Sephadex G-25 (0.9 x 30 cm) equilibrated and eluted with PBS. The void volume peak is collected and pooled.
f. Modified IgG and AP are then mixed and incubated for 16-24 hr at room temperature with gentle agitation.
g. The final conjugate is then dialyzed against 10 mM Tris-HCl, pH 8.0, 0.15 M NaCl, 1 mM $MgCl_2$.

B. Purification and Storage of Conjugates

Most protein conjugation procedures give rise to products of heterogeneous size and composition. The final reaction mixture may thus be composed of free enzyme, free antibody, enzyme-enzyme or antibody-antibody conjugates, as well as antibody-enzyme complexes with different degrees of substitution. Of these, the presence of free antibody is the most critical since non-labeled antibody lowers the sensitivity of the assay. For those assays where the enzyme activity is measured on a solid phase support, removal of free unconjugated enzyme is not necessary. However, when large amounts of free enzyme are present, it becomes more critical to thoroughly rinse the solid phase to remove unbound enzyme; otherwise the specificity and precision of the assay are reduced. In many cases the amount of free antibody remaining following conjugation is negligible and subsequent purification is not necessary. The most common method for the purification of HRP and AP conjugates is gel filtration chromatography over Sephadex G-200 (0.9 x 60 cm) equilibrated and eluted with PBS. Other procedures include ion-exchange chromatography on DEAE-Agarose (11) and affinity chromatography on Con-A-Sepharose for purification of HRP conjugates (7). For long term storage, conjugates should be stabilized by addition of BSA to 1% (w/v) and should either be dispensed into single-use volumes, or mixed with glycerol to 50% (v/v), and stored at -20 C.

C. Comments

A complete review of all available protein-protein conjugation techniques is beyond the scope of this article; however, the protocols presented here represent those most commonly used. While it is not possible to state which method will reproducibly yield the best results under every set of circumstances, a few guidelines may be recommended. The one-step glutaraldehyde method generally gives good results with AP conjugates and, being the simplest procedure, should be the one tried initially with this enzyme. The best conjugates with HRP are generally obtained using the periodate oxidation method, although the two-step glutaraldehyde method also works well with this enzyme. The MBS method has generally been used primarily with β-galactosidase. The technology for producing protein-protein conjugates is continually evolving and may soon provide even more suitable procedures.

Literature Cited

1. Avrameas, S. 1969. Coupling of enzymes to proteins with glutaraldehyde. Use of the conjugates for the detection of antigens and antibodies. Immunochemistry 6:43-52.
2. Avrameas, S., and Ternynck, T. 1971. Peroxidase labelled antibody and Fab conjugates with enhanced intracellular penetration. Immunochemistry 8:1175-1179.
3. Carlsson, J., Drevin, H., and Axen, R. 1978. Protein thiolation and reversible protein-protein conjugation. Biochem. J. 173:723-737.
4. Engvall, E. 1980. Enzyme immunoassay ELISA and EMIT. Methods Enzymol. 70:419-439.
5. Engvall, E., and Perlmann, P. 1971. Enzyme-linked immunosorbent assay (ELISA). Quantitative assay of immunoglobulin G. Immunochemistry 8:871-874.
6. Henning, D., and Nielson, K. 1987. Peroxidase-labelled monoclonal antibodies for use in enzyme immunoassay. J. Immunoassay 8:297-307.
7. Jeanson, A., Cloes, J. M., Bouchet, M., and Rentier, B. 1988. Comparison of conjugation procedures for the preparation of monoclonal antibody-enzyme conjugates. J. Immunol. Methods 111:261-270.
8. Jeanson, A., Cloes, J. M., Bouchet, M., and Rentier, B. 1988. Preparation of reproducible alkaline phosphatase-antibody conjugates for enzyme immunoassay using a heterobifunctional linking agent. Anal. Biochem. 172: 392-396.
9. Murayama, A., Shimada, K., and Yamamoto, T. 1978. Modification of immunoglobulin G using specific reactivity of sugar moiety. Immunochemistry 15:523-528.
10. O'Shannessy, D. J., and Quarles, R. H. 1985. Specific conjugation reactions of the oligosaccharide moieties of immunoglobulins. J. Appl. Biochem. 7:347-355.
11. O'Sullivan, M. J., Gnemmi, E., Morris, D., Chieregatti, G., Simmonds, A. D., Simmons, M., Bridges, J. W., and Marks, V. 1979. Comparison of two methods of preparing enzyme-antibody conjugates: Application of these conjugates for enzyme immunoassay. Anal. Biochem. 100:100-108.
12. Richards, F. M., and Knowles, J. R. 1968. Glutaraldehyde as a protein cross-linking reagent. J. Mol. Biol. 37:231-233.

Section II.D.

AMPLIFICATION OF SEROLOGICAL REACTIONS

D. P. Durand
Iowa State University
Ames, Iowa 50010

Principle

Nonspecific serological reactions often occur especially with use of high antibody titers for sensitive detection of antigens. Nonspecific reactions can often be diminished by dilution of the antibody. To facilitate the use of dilute antibody preparations amplification of the signal may be required. While reducing nonspecificity, amplification can also dramatically increase sensitivity to render a reaction detectable at extremely low concentrations of antigen. Furthermore, even though additional steps are involved, overall assay times may be reduced because of powerful amplification.

Radioactive isotopes, enzymes, fluorogenic, chemiluminescent compounds, and heavy metals are commonly used serological signals and all have been applied in reaction amplification.

Three main systems of amplification are commonly used to enhance weak signals in immunoassay and immunoblotting. They are indirect antibody, biotin-streptavidin, and biotin-streptavidin complex (see **Section IV.E.2.** and **Section IV.A.2.**). In addition, Protein A-gold and -silver enhancement are employed to magnify weak signals in immunoblots with very impressive results.

Indirect Antibody

One of the early, and still effective, enhancement methods is the use of interspecies antibody. In this approach, primary (detecting) antibody is applied to the system and allowed to react with its immobilized antigen. A secondary polyclonal antibody, generated in a different animal species and conjugated to the signal (available commercially) reacts with multiple epitopes on the primary antibody producing significant signal enhancement (Fig. 1). In addition, the primary antibody is not involved in any coupling reaction and its antibody specificity is better retained.

Biotin-Streptavidin System

For a general review see reference 7.

Biotin

Biotin, a water-soluble vitamin with molecular weight of 244 Da, readily binds to antibodies without reducing their activity. Various antibodies labeled with biotin are available from Sigma, Bio-Rad, and Boehringer-Mannheim. For additional information on biotin see reference 2.

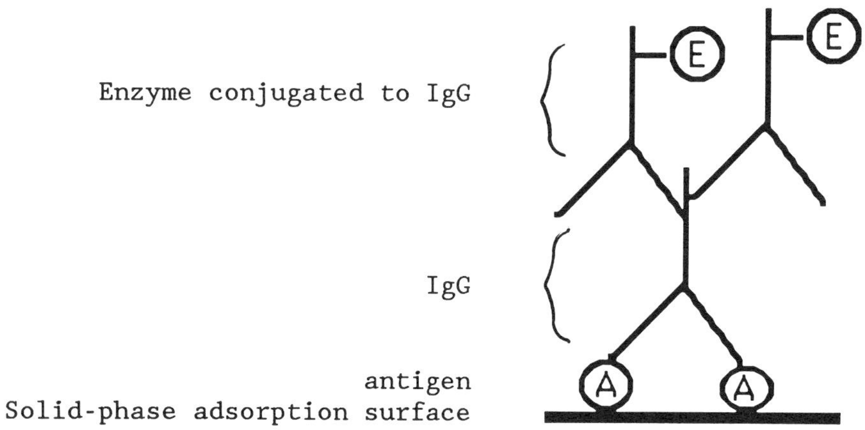

Fig. 1. Indirect ELISA.

Streptavidin

Streptavidin has an isoelectric point near pH 7 and is not glycosolated, two physical properties which make it far superior to avidin obtained from egg white protein. Since streptavidin also has the same binding properties as avidin, our comments will be limited to streptavidin.

Assay systems exploiting the biotin and streptavidin interaction may use antibody to which biotin is linked (either primary or secondary antibody) and signal provided by a streptavidin-bound label. Since the label is not directly linked to antibody but is coupled to streptavidin, the biotin labeled antibody may be used with a number of different streptavidin-bound detection systems. This gives remarkable versatility to the system. (Fig. 2)

The coupling of biotin to macromolecules is tight and steric constraints may prevent labeled streptavidin from normal binding. Unhindered binding may be restored by the use of spacer arms. Many biotinylated proteins prepared with spacer arms are also available commercially, and it is possible to introduce spacer arms into biotinylated proteins in the standard laboratory.

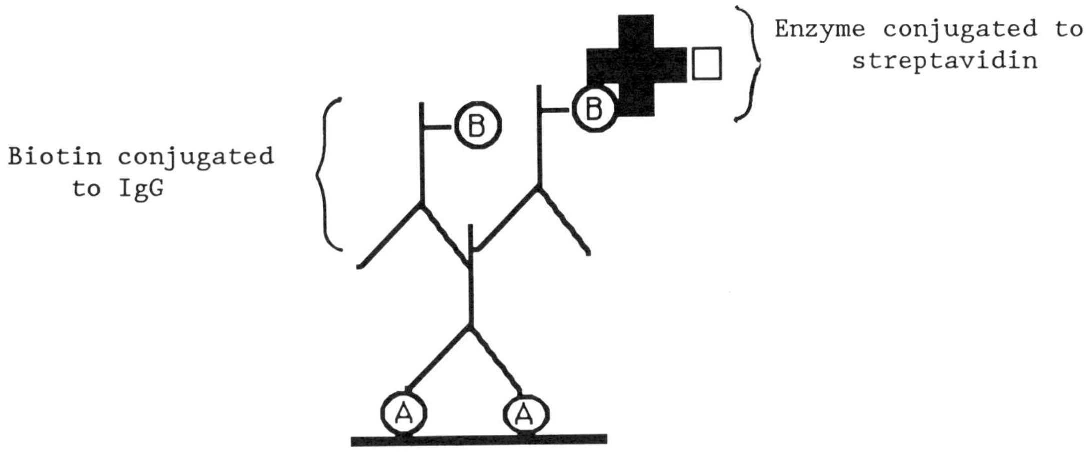

Fig. 2. Antigen detection using labeled streptavidin.

Biotin-Streptavidin Complex

Biotinylated enzymes and streptavidin can be formed into large networks, where many enzyme molecules are linked to many streptavidin molecules. Very significant amplification of a signal results when the preformed complex is added to biotinylated antibody (Fig. 3) (3).

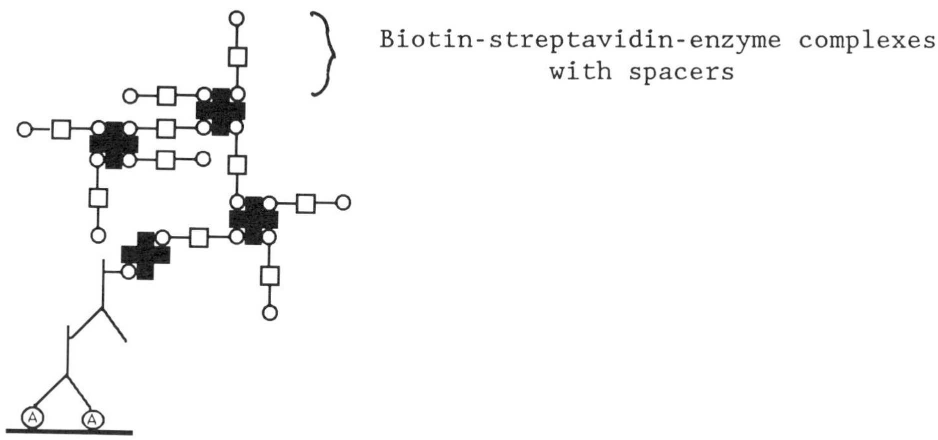

Fig. 3. Amplification using pre-formed complexes.
○ = biotin, □ = enzyme, ✚ = avidin

Other Systems

An interesting, rather innovative ELISA amplification system for alkaline phosphatase-based immunoassays (5) uses enzymatic amplification of the alkaline phosphatase cleavage product. Each molecule of alkaline phosphatase enzyme label generates many molecules of NAD+ which in turn catalytically activates a redox cycle and generates a red color (Fig. 4). Torrance (6) has applied this procedure to the detection of barley yellow dwarf virus in oats and in individual vector aphids. An excellent discussion of the procedure is presented by Johannsson (1).

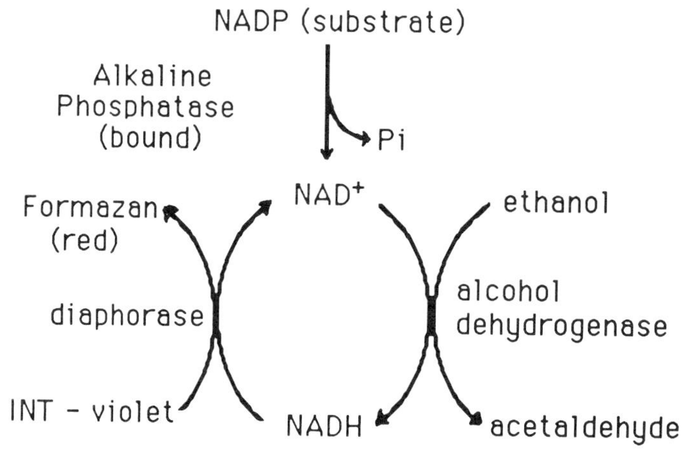

Fig. 4. Alkaline phosphatase amplification scheme. NAD^+ = Nicotinamide-adenine dinucleotide; NADH = Reduced NAD^+; $NADP^+$ = Nicotinamide adenine dinucleotide phosphate; Pi = Phosphate

Another highly sensitive procedure called time-resolved fluoroimmunoassay (TR-FIA) has been used to detect plant viruses in crude plant extracts. TR-FIA was reported to be 5 to 100 times more sensitive than conventional double antibody sandwich ELISA. One major drawback to TR-FIA is the requirement of a special fluorometer (4).

Literature Cited

1. Johannsson, A. 1988. Enzyme amplification of immunoassays: Increased speed and sensitivity. Focus 10:(3)49-53.
2. Kendall, C., Ionescu-Matiu, I., and Dreesman, R. 1983. Utilization of the biotin/avidin system to amplify the sensitivity of the enzyme-linked immunosorbent assay (ELISA). J. Immunol. Methods 56:329-339.
3. Shamsuddin, M. A., and Harris, C. C. 1983. Improved enzyme immunoassay using biotin-avidin-enzyme complex. Arch. Pathol. Lab. Med. 107:514-517.
4. Siitari, H., and Kurppa, A. 1987. Time-resolved fluoroimmunoassay in the detection of plant viruses. J. Gen. Virol. 68:1423-1428.
5. Stanley, C. J., Johannsson, A., and Self, C. H. 1985. Enzyme amplification can enhance both the speed and sensitivity of immunoassays. J. Immunol. Methods 83:89-95.
6. Torrance, L. 1987. The use of enzyme amplification in an ELISA to increase sensitivity of detection of barley yellow dwarf virus in oats and in individual vector aphids. J. Immunol. Methods 85:131-138.
7. Wilchek, M., and Bayer, E. A. 1984. Avidin-biotin complex in immunology. Immunol. Today 5:39-43.

Section III.A.

AGAR DIFFUSION TECHNIQUES, INTRODUCTION

E. M. Ball
University of Nebraska
Lincoln, NE 68583

General concepts

Dilution endpoints determined by liquid-phase serology have great merit in the ease by which they are manipulated and the simplicity of the test procedure. However, they cannot equal the resolving power of agar diffusion techniques. Antigenic components appear in agar as one or more bands formed when antigen and antibody, diffusing toward each other, meet in optimum proportions. Agar diffusion tests establish relationships between antigens (or antibodies) in precisely the same environmental conditions, i.e., side-by-side in the same agar medium (3). Some physical characteristics of antigens that can be determined by agar diffusion techniques are diffusion coefficients, electrophoretic mobilities and relative molecular weights.

There are two kinds of diffusion techniques. Single diffusion occurs when one reactant (antigen) diffuses into a medium containing the other reactant. The antigen is added to a well punched out in agar-serum medium, and a ring(s) of precipitation forms around the antigen well. In contrast, double diffusion produces precipitin bands in the agar medium between wells containing antigen and antibody, respectively. Each procedure has merit and shows distinctive properties of the antigen (1). However, the majority of agar diffusion applications reported by plant pathologists involve double diffusion in plates, the Ouchterlony procedure (2), for purposes of field surveys for pathogens, pathogen diagnosis, relationship studies, strain analyses, and establishment of serogroups.

Composition of the gels

Purified agars have been available for many years. Several of these products used in immunodiffusion are Ionagar No. 2 and Purified Agar, Code L28 (Oxoid) (Colab), Agar-Agar (Baltimore Biological), and Noble Agar (Difco). The clarity of these agars varies, depending upon concentration and the buffers used. Agarose is a highly purified polysaccharide isolated from agar which is found in marine algae. Important characteristics include gelling temperature, gel strength, sulfate content and electroendosmosis. Many companies list highly satisfactory agarose (Miles, Pharmacia, Sigma). Sigma lists 14 types varying in sulfate content (indicator of purity) and gel strength. Properties of five agarose preparations from FMC are presented in Table 1 (from FMC Bioproducts Source Book, 1985).

Table 1. The Gelling Temperature and Electroendosmosis Properties of Five (FMC) Agaroses Recommended for Use in Agar Diffusion Techniques

Seakem agarose	Gelling temp., conc.	EEO (-Mr)	Recommended use
LE	36 C, 1.5%	0.10-0.15	IEP*, Rocket, ADD, RID, cell culture, soft agar cloning
ME	36 C, 1.5%	0.16-0.19	IEP, Crossed IEP, CtEP, Counter-IEP, ADD, RID
HEEO	36 C, 1.5%	0.25-0.26	IEP, CIEP, custom EEO blending, ADD, RID
HGT	42 C	0.10	CIEP, Crossed IEP

*IEP = immunoelectrophoresis, CIEP = counter immunoelectrophoresis, RID = radial immunodiffusion, ADD = agar double diffusion, EEO = electroendosmosis, MR = relative mobility, - = negative migration.

Agarose gels are less fragile than agar gels, can be manipulated more easily, and can be readily fixed, stained and dehydrated. They are optically clean when dried, can be accurately scanned by densitometer and are easily stored in a notebook when dried. Electroendosmosis, in which negatively-charged molecules bound to sulfate esters in the gel migrate toward the cathode when a voltage is applied, is essential for immunoelectrophoresis, rocket-electrophoresis, electrophoresis and isoelectric focusing. Viral and bacterial antigenic components are generally positively charged and thus migrate toward the anode.

Some plant viruses as well as bacteria and/or their components present limitations in analysis by gel diffusion because of their size and shape and because their tendency to aggregate prevents movement in the gel. Improved diffusion patterns have been achieved by the following modifications: degradation of large size particles by detergents (SDS) (**Section III.C.2.c.**) or chemicals (pyridine or pyrrolidine), stabilization by chemicals (formaldehyde), and selecting favorable buffer systems or pH.

Most diffusion assays require addition of preservatives to the gel medium since development of precipitin bands requires several hours and because gels must be incubated in a very moist atmosphere to prevent dehydration (alters gel composition). Sodium azide at 0.01-0.05% or mercurials such as merthiolate or thimerosol at 0.01-0.02% are dispersed in the gel when in the liquid state. These are poisons and must be used with great caution. Mercurials react with some proteins to give false positive precipitin bands. Any product added to the gel should be tested for nonspecific effects.

As a general consideration, buffers for agar diffusion systems should be chosen according to the requirements of the antigen, since antibodies are stable in a wide pH range from pH 3 to pH 10. Phosphate, borate and Tris buffers are used most frequently. Occasionally barbital buffers are used, primarily for immunoelectrophoresis. Physiological saline, 0.15 M NaCl, may also be used. Addition of polyethylene glycol (molecular weight 6,000-8,000), Triton X-100 or a chelating agent such as EDTA to the gel medium may enhance development of precipitation bands, and dyes such as trypan blue increase visibility of bands.

Sources of error

Agar diffusion tests must be done carefully and with precision to obtain reproducible results. Whereas, simple sap extracted antigens may be used in diffusion, clarified or partially purified antigens give better results. Specificity of the antiserum must be carefully evaluated to assure the identity of precipitin bands formed during tests. Environmental factors including temperature can influence the appearance of precipitin bands. Optimal concentrations of each reactant must be determined by test-standardization before antigen relationships are compared. The "immunospecific barrier" of the precipitation zone (a precipitate inhibiting diffusion of similar determinants but permitting dissimilar antigens to pass through, as in intragel specific absorption) is effective only under the operation of a balanced system of antigen and antibody. Appropriate controls are of utmost importance.

Preservation and staining

Agar diffusion precipitin bands can be accentuated by staining. It is easier to stain agarose gels than agar gels, especially if the agarose gel is formed on a polyester film (Gel-Bond, FMC). Gels are soaked in phosphate-buffered saline or Tris-buffered saline, for 2 to 18 hr, to remove excess, unreacted antigen or antibody. They are then soaked for 1-2 hr in demineralized water to leach buffer components. The gel is blotted with filter paper and several layers of paper towels, covered with a glass plate (or, if the gel is in a Petri dish or plastic dish, with an inverted dish of smaller size) and pressed under the weight of a reagent bottle of suitable size. When most of the liquid is removed, final drying can be done under an incandescent lamp. Coomassie stain can be applied for 10-30 min, destained until the background is clear, rinsed in water and dried. [Coomassie stain and destain base: combine 45 ml each of methanol and demineralized water with 10 ml of glacial acetic acid. For stain, add 0.10-0.25% of Brilliant Blue R (C.I. No. 42660).] Other stains that can be used include amido black, azocarmine, bromophenol blue, Ponceau S or red, and light green.

Precipitin bands in agar diffusion plates or slides are usually documented by photography, e.g., polaroid.

Reading and recording

Agar diffusion reactions can be best observed in a dark room over a box with a slit or circular opening over an indirect light source. Changes in the development of precipitin bands can be recorded for experimental reference by drawing on paper with the template well-pattern printed on it. Diffusion plates should be observed for at least one week, the exact conditions and termination of the test being determined by the investigator. Gels containing SDS can be observed for only 24-48 hr.

Literature Cited

1. Crowle, A. J. 1961. Immunodiffusion. Academic Press, New York. 333 pp.
2. Ouchterlony, O. 1968. Handbook of Immunodiffusion and Immunoelectrophoresis. Ann Arbor Science Publishers, Inc., Ann Arbor. 215 pp.
3. Van Regenmortel, M. H. V. 1982. Serology and Immunochemistry of Plant Viruses. Academic Press, New York. 302 pp.

Section III.B.1.

AGAR SINGLE DIFFUSION, TUBES (OUDIN TECHNIQUE)

E. M. Ball
University of Nebraska
Lincoln, NE 68583

Principle

In this procedure, antigen, in a liquid phase, diffuses into an agar-serum medium in small test tubes. A precipitin band(s) appears which migrates toward the bottom of the tube. The rate of migration is dependent on the diffusion coefficient of the antigen, concentration of the antigen and antibody, and properties of the agar or agarose. Multiple zones may develop representing different antigenic components.

The position of the leading edge of the precipitation band in the agar is proportional to the square root of time and a plot of these functions gives a straight line. When antigen is the liquid phase, the relationship between antigens can be determined by Oudin's equation (2).

$$R = D \log/A$$

where R = distance in mm/time in hr)$^{1/2}$, D = diffusion constant of the antigen, and A = concentration of the antigen. Similar calculations can be made with antibody as the external reactant (liquid phase). The test is applicable to any stable antigen in crude sap, clarified sap, seed extract, or purified preparation providing adequate precautions are taken against fungal or bacterial contamination.

Materials

1. Test tubes: 10 x 75 mm or smaller, sterile, plugged with cotton.
2. Agar or agarose, 0.8-1.0%.
3. Preservative: 0.1% thimerosol or 0.25% Naazide.
4. Buffer: phosphate buffered saline or (suitable alternative)
5. Pipets for dilution.
6. Parafilm for sealing the tubes, or a cork soaked in hot wax.
7. Waterbath at 50 C (depending on gel) to maintain the liquid state.

Procedure

1. Prepare agar/agarose in buffer and autoclave for 10 min or microwave carefully just until agar is dissolved.
2. Add preservative to agar to give 1:10 dilution.
3. Mix serum with partially cooled agar. Keep in waterbath to prevent solidification.

4. Transfer a portion of the serum-agar to small tubes. With 10 x 75 mm tubes, 1 ml of the mixture is sufficient. Allow solidification.
5. Add appropriate dilution(s) of antigen. For 10 x 75 mm tubes, use 0.5 ml of antigen (Fig. 1). Antigen should be filtered or preservative added to retard growth of contaminants.
6. Cut cotton plug in the sterile tubes so it is level with the top. Seal tube with waxed cork or several layers of parafilm to prevent evaporation.
7. Set tubes at room temperature and observe for at least one week. Higher temperatures may speed the reaction, but it is important to maintain a constant temperature throughout the incubation period.
8. The migration of the precipitin zone can be measured with reasonable accuracy using calipers and a centimeter rule. Plot distance moved vs. square root of time (1).

Fig. 1. A drawing illustrating the position of antigen and antibody (serum) in Single Diffusion in Tubes, the technique of Oudin.

Comments

1. Avoid excessive heat with the agar/agarose preparations. Waterbath incubation of the agar-serum should be no higher than 50 C, lower if agarose is used, because serum will be denatured and the medium will be opaque from excessive heat.
2. Small tubes can be siliconized or pretreated with hot 1% agar, drained and dried. Such treatment permits the formation of a flatter meniscus between serum-agar and antigen.

Applications (See Fig. 2)

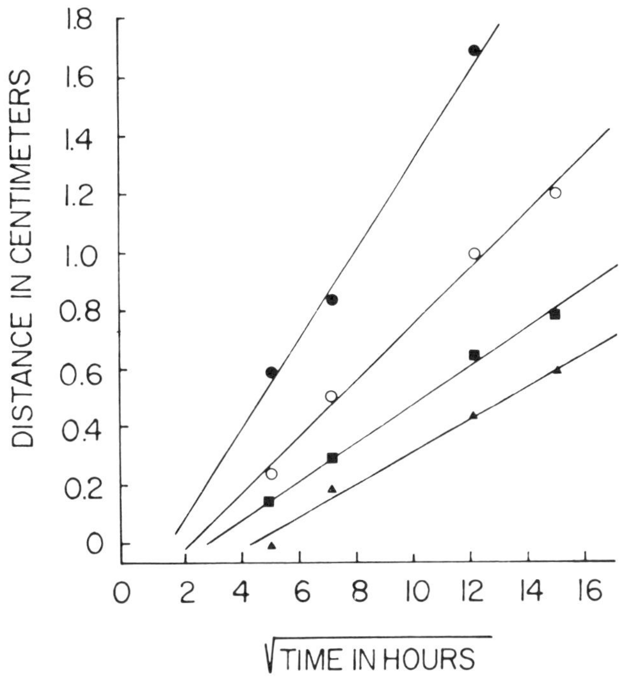

Fig. 2. Use of the Oudin technique in determining the rate of movement of tobacco mosaic virus precipitation zone, relative to time:
● = 1.7 mg/ml, o = 0.34 mg/ml, ■ = 68 μg/ml, ▲ = 14 μg/ml.

Literature Cited

1. Becker, E. L. 1971. Determinations of diffusion coefficient of antigens by simple diffusion in tubes. Pages 174-180 in: Methods of Immunology and Immunochemistry, Vol. III. C. A. Williams and M. W. Chase, eds. Academic Press, New York. 515 pp.
2. Oudin, J. 1952. Specific precipitation in gels and its application to immunochemical analysis. Pages 335-378 in: Methods in Medical Research, Vol. 5. A. C. Corcoran, ed. Year Book Publishers, Inc., Chicago. 394 pp.

Section III.B.2.

AGAR SINGLE DIFFUSION, PLATES

S. A. Slack
Cornell University
Ithaca, NY 14853

E. M. Ball
University of Nebraska
Lincoln, NE 68583

Principle

In this procedure, a single reactant (antigen) diffuses into a medium (agar/agarose) containing the other reactant (antibody) uniformly dispersed in the gel in a dish or plate (Fig. 1) (3). The pattern of diffusion is a ring or halo forming around the antigen well. More than one ring may form depending upon antigenic components, antigen concentration and optimum concentration of both reactants. When the endpoint of diffusion is reached, the ring no longer enlarges. A linear relationship exists between amount of antigen present in the liquid phase and the area of the ring. An inverse linear relationship exists between the area of the ring and antibody concentrations in the gel.

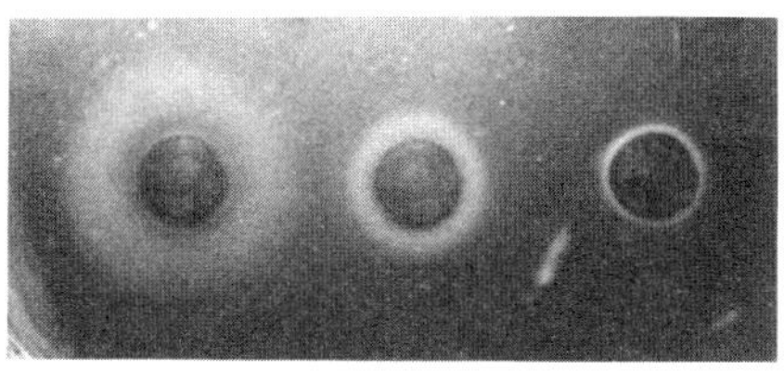

Fig. 1. Agar single diffusion assays of pyrrolidine-degraded potato virus X (i.e., 'protein D') antigen. The gel consisted of 2.7 ml of 1.0% Seakem LE agarose (FMC) in 0.01 M neutral phosphate buffer, mixed with 0.3 ml of crude antiserum to protein-D and poured into a 50-mm-diameter Falcon 'tight-seal' dish. Wells, L to R, received 6 μl of antigen, with concentrations of 1.00, 0.25, and 0.06 mg/ml (i.e., antigen quantities per well were 6, 1.5, and 0.36μg), respectively. Wells were 4 mm diameter, 13 mm apart. Note decreasing size of precipitation rings, with decreasing antigen concentrations. Plate photographed after 24 hr incubation at room temperature.

Materials

1. Agar [Special Noble (Difco)], ionagars (Oxoid) or agarose (Miles, Pharmacia, FMC).
2. Buffer: borate, phosphate, Tris, or one most suitable for antigen. Preservative: sodium azide - 2.5%.
3. Plates: flat-bottom Petri dish, glass or plastic. Tight-seal 50 x 9 mm (Falcon) are convenient since they do not need to be incubated in a moist box.
4. Leveling table to make uniform thickness of gel.
5. Gel punch. Automatic pipet, 20 μl size, or micro-syringe to deliver antigen. Pipets for making dilutions.
6. Humid chamber for incubation--sealable plastic box with moist towels on bottom is sufficient.
7. Waterbath--45-50 C to hold gel in liquid state.

Procedure

1. Dissolve agarose in buffer by autoclaving for 10 min or heating in microwave oven until just melted.
2. Add azide to buffer or to melted agarose to give 1:100 dilution. Dispense into tubes the volume of agarose needed for each plate. Keep in the waterbath.
3. When the agarose has cooled to the temperature of the bath, add the appropriate amount of serum, mix by gently inverting the tubes several times and pour or pipet the gel into a plate which is on a leveling table.
4. When set (30 min or more), punch out antigen wells, removing the plug by light suction. Do not destroy the symmetry of the well or the diffusion pattern will be askew.
5. Fill the well with antigen at appropriate dilutions. Do not fill entirely full to avoid running over the edge and across agar which will destroy the diffusion pattern. Include appropriate positive and negative controls.
6. Incubate plates at a constant temperature (e.g. 22-25 C).
7. Observe the plate in a dark room against a dark background using a slit or circular light source. Measure the diameter of the rings with a caliper and ruler, with an ocular micrometer using low power of a dissecting microscope, or by photographing and measuring on the print.
8. The plate can be washed several times (the number depending on concentrations of reactants) in buffer or saline, then in water, dried and stained (see Agar Diffusion, Introduction, **Section III.A.**).

Comments

1. Optimum concentrations of antigen and antibody can be predetermined by making several small plates, each with a different dilution of serum in the buffer-agarose gel. When solidified and wells cut out, different dilutions of antigen in buffer can be tested. Incubate several days observing ring development and time in which the rings fail to further enlarge, the titration endpoint.
2. The amount of antigen added to the wells is not critical, but it is important to add the same volume to all wells.
3. A template drawn on paper and slipped under the plate is a convenient way to duplicate well distribution between plates. Wells should be spaced adequately to prevent inhibition of halo formation.

4. Duplicate sets of antigens or antigen dilutions can be placed in a plate, but the sets should not be adjacent, so that any non-homogeneity of the agarose-serum gel can be observed.
5. Avoid placing wells near the edge of the gels, since the pattern may be deformed because of lack of uniform diffusion or because of gel dehydration or change in thickness near the side of the dish.
6. Diffusion of rod-shaped viruses or bacterial cells can be facilitated by dissociating the antigen with detergent or chemicals. Precipitation must not take place within the well or at the edge of the antigen well.
7. Commercial kits for radial diffusion are available from Miles, Helena, or Pharmacia. These include cutters, templates, and ruler device for measuring rings.

Applications

This test has been used to detect potato viruses X, S, and M in potato plants (1) and barley stripe mosaic virus in barley seedlings (2). In both applications, procedures are used to dissociate virions into soluble antigens that more readily migrate into the antibody-containing agar medium. Pyrollidone and pyrimidine have been used to dissociate virions in sap extracts of potato prior to adding samples to the wells cut in the agar. Alternatively, detergents and high salt concentrations incorporated directly into the agar medium were sufficient to enhance solubilization of barley stripe mosaic virus. A test modification in which antigen wells are not required is shown in Fig. 2. It can be noted that nonspecific reactions may occur with different dissociation chemicals, viruses, and plant or plant part combinations. Also, the antigenic properties of the dissociated virions may differ from untreated virions. Thus, antibodies prepared against dissociated virions may be required. Each system must be evaluated carefully for reproducibility and specificity. This test can be useful as a rapid serodiagnostic assay for multiple plant samples. Results may be scored in 4-24 hours and sensitivity is about 1 μg antigen/ml.

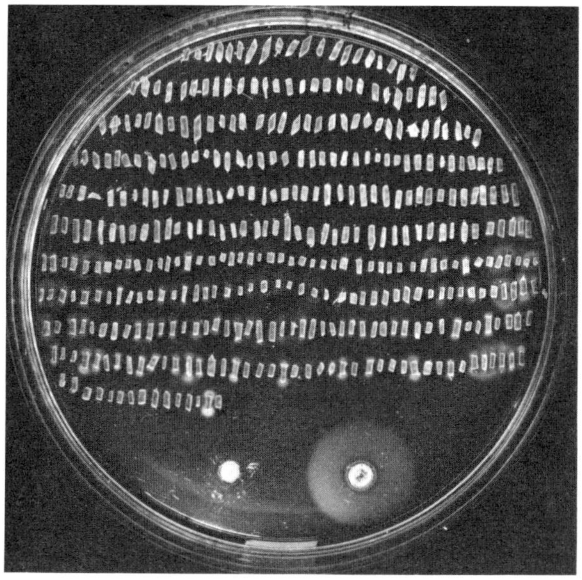

Fig. 2. Application of agar single diffusion serology to detection of barley stripe mosaic virus (BSMV) in barley tissue samples. Plants of the top 200 tissue samples were from a virus-free seed lot. The two wells were filled with sap from a BSMV-infected barley plant (right) and a healthy barley plant (left). Crude BSMV antiserum was incorporated into 0.5% Ionagar No. 2 at a 1:20 (v/v) proportion. Agar was buffered with 0.05 M Tris, pH 7.2, containing 0.85% NaCl and 0.5% Leonil SA (sodium dibutylnaphthalene sulfonate) detergent. Positive reactions were indicated by the presence of opaque precipitates where samples contacted the agar.

Literature Cited

1. Shepard, J. F. 1972. Gel diffusion methods for the serological detection of potato viruses X, S, and M. Montana Agr. Exp. Sta. Bull. 662. 72 pp.
2. Slack, S. A., and Shepherd, R. J. 1975. Serological detection of seed-borne barley stripe mosaic virus by a simplified radial-diffusion technique. Phytopathology 65:948-955.
3. Vaerman, J.-P. 1981. Single radial immunodiffusion. Methods Enzymol. 73: 291-305. Part B.

Section III.C.1.

AGAR DOUBLE DIFFUSION, TUBES (OAKLEY AND FULTHORPE TECHNIQUE)

E. M. Ball
University of Nebraska
Lincoln, NE 68583

Principle

In the tube double diffusion test, three alternate layers containing antiserum, plain agar, and antigen are prepared (Fig. 1). During diffusion the reactants meet in optimum proportions in the blank agar column forming precipitation zones of various densities and positions depending upon the characteristics of the antigen-antibody system and the properties of the medium in buffer and agar composition. With different antigen concentrations, a given precipitin band forms at various distances from the antigen meniscus (Fig. 2), sharpens as diffusion continues, and then decreases. At optimum ratios, the diffusion coefficient of a virus can be estimated by the equation

$$Xg^2/Xb^2 = Dg/Db$$

where Xg = distance between zone and antigen meniscus, Dg = diffusion coefficient of antigen, and Xb and Db = the same characters for antibody (1,3).

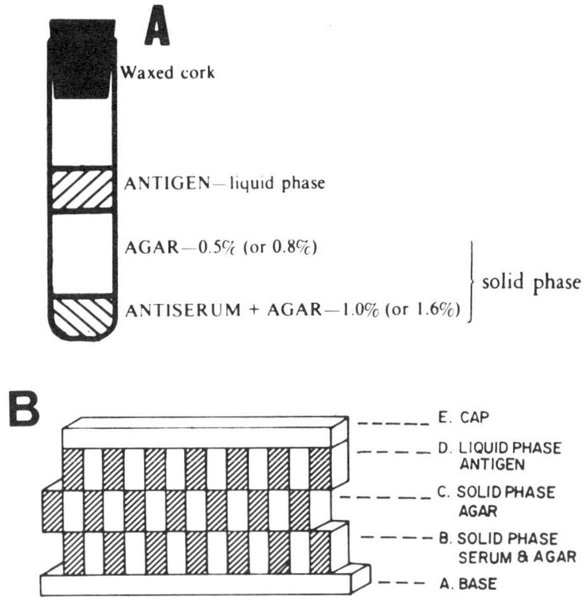

Fig. 1. A drawing illustrating the position of antigen and antibody (serum) in the tubes double diffusion test (A) and in a plastic apparatus (B) designed by Polson (2) to test many samples under exacting conditions for diffusion

coefficient measurement.

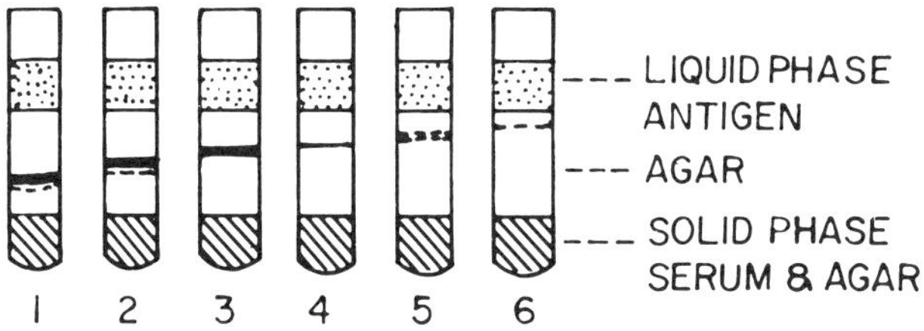

Fig. 2. A drawing of test results of double diffusion reactions in tubes illustrating change in position and width of precipitation zone as a result of variation in antigen concentration. Tube 1 - 0.5 mg/ml, broad heavy zone; Tube 2 - 0.025 mg/ml, broad zone, Tube 3 - 0.12 mg/ml, less heavy zone; Tube 4 - 0.06 mg/ml, thin sharp zone, optimum concentration; Tube 5 - 0.03 mg/ml, light zone; Tube 6 - 0.015 mg/ml, faint zone.

Materials and procedure are similar to those described in "Single Diffusion, Tubes" (**Section III.B.1.**). The serum-agar layer, however, is separated from the external reactant (antigen) by blank agar. The proportions in a 10 x 75 mm tube are 0.5 ml of serum-agar, 1.0 ml of agar at 0.5-0.8%, and 0.5 ml of antigen in the liquid phase. A preservative is included to prevent contamination. Tubes are sealed as described for Single Diffusion (**Section III.B.1.**).

Literature Cited

1. Allen, W. R., and Tremaine, J. H. 1965. Characteristics of antigens and antibodies associated with the Prunus necrotic virus immune system. Virology 25:122-132.
2. Polson, A. 1971. Determination of diffusion coefficients by a method of double diffusion in gels. Pages 180-188 in: Methods in Immunology and Immunochemistry, Vol. III. C. A. Williams, and M. W. Chase, eds. Academic Press, NY. 515 pp.
3. Tremaine, J. H., and Willison, R. S. 1961. Estimation of the size of a stone-fruit virus antigen by the gel-precipitin technique. Can. J. Bot. 39:1843-1845.

Section III.C.2.a.

AGAR DOUBLE DIFFUSION, PLATES (OUCHTERLONY): VIRUSES

E. M. Ball
University of Nebraska
Lincoln, NE 68583

Principle

In this procedure, antigen and antibody are placed in separate wells cut into the agar medium. As they diffuse into the medium, a precipitation zone forms where they meet in optimum proportions. The size, shape, number, and density of the zones depends upon the characteristics of the antigen-antibody system under investigation.

Various patterns of well size and arrangement can be made in the agar using cork borer cutters, preformed squares placed in the plate prior to adding molten agar, or commercially made punches. A pattern arrangement (illustrated in Fig. 1) drawn on a card which is slipped under the plate is a convenient way to cut similar reaction areas in the agar. Templates and cutter systems are available commercially (Miles, Helena, Pharmacia).

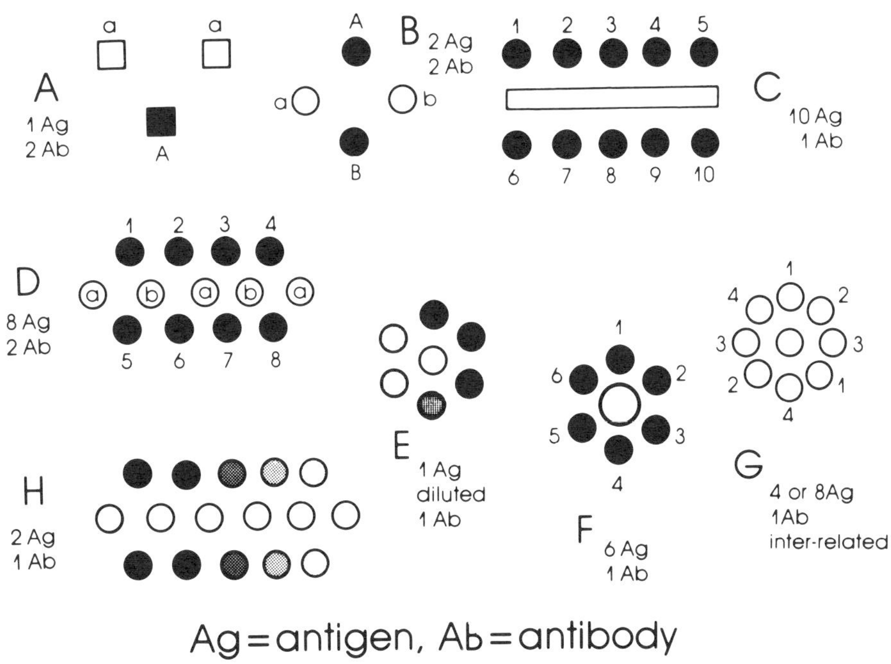

Fig. 1. Agar Diffusion Test Patterns: A drawing illustrating the various arrangements of wells cut, or preformed by a template, in agar to study the relationship between antigen and antiserum samples. The dark or stippled symbols represent antigen and antigen dilutions and the open symbols represent antibody.

Agar diffusion studies are suitable for crude plant sap, clarified extract, and highly purified antigen. Precautions must be taken to avoid contamination. Controls of healthy plant extracts must be included. The advantage of the technique is that many antigen strains, different partially purified antigen preparations or extracts, or several different antisera or gamma globulins can be examined at the same time under identical conditions.

Materials

1. Ionagar (Oxoid), Noble agar (Difco) or a number of kinds of agarose (FMC, Sigma, Pharmacia), 0.5-1.0%.
2. Preservative: sodium azide, 2.5% in water.
3. Plates: flat-bottom glass, or plastic. Tight-seal, 50 x 9 mm (Falcon) are convenient.
4. Buffer: Tris [(Tris-hydroxymethyl-aminomethane)-HCl], phosphate, borate-any suitable for antigen stability. pH 6.5-7.5. 0.85% NaCl, if applicable.
5. Small tubes and pipets for making dilutions and dispensing agar. Automatic or micro-pipets for adding exact amounts of reactants to wells.
6. Cutters: cork borer, commercial punches, knife-scalpel, or removable shapes for preforming in agar.
7. Moist chamber for incubation; tight-seal plastic box with moist towels on bottom.

Procedure

1. Weigh out agar/agarose and dispense in appropriate buffer to give 0.7%.
2. Dissolve agar by heating in autoclave 10 min or in microwave oven just until agar becomes soluble. Add sodium azide, 1:100 w/v, and homogenize just before dispensing agar/agarose.
3. Place plates on leveling table. When agar is cooled to about 60 C, dispense the appropriate amount into plates. The 50 x 9 mm plate requires 2.5-3.0 ml, and the 100 x 13 plate 10-15 ml. Let sit undisturbed until solidified, about 30 min.
4. Slip the drawing of the well design under the plate, punch out wells and remove the plugs by slight suction, being careful to not destroy the clean-cut side of the well. As an example, use pattern E, Fig. 1, of 6 peripheral wells equidistant around a central well.
5. Dilute antigen in small tubes in buffer or in water or saline. Add aliquots of each dilution to the peripheral wells. Add the appropriate dilution of serum to the center well.
6. Incubate the plates at a constant temperature in a moist environment (except for tight-seal plates).
7. Observe plates for at least a week. Read plates in a dark room against a dark background over a slit or ring source of light. Record the change in the development of the precipitation by drawing on paper on which the well pattern has been printed.
8. Precipitin patterns can be preserved by photographing or drying and staining (see Composition of Gels, **Section III.A.**)

Comments

1. Avoid heating the gel for extended periods since properties of the gel may change or be destroyed.
2. Plates may be stored in a tight container or sealed in "freezer" bags at 4 C. They should be used within a month.
3. Optimal concentrations of antigen and antibody can be determined by making the pattern E, Fig. 1, with equal size wells equidistant from each other (or the pattern of choice for the experiment). Add two-fold dilutions of antigen (for example: purified virus at 4.0, 2.0., 1.0, 0.5, 0.2, 0.1 mg/ml) in peripheral wells and a sample of antibody (undiluted) in the central well. Make several identical patterns with antigen using 1/2, 1/4, etc. dilution of antiserum in the center well. The combination of antigen and antiserum giving a dense, sharp line would be considered optimum (Fig. 2).

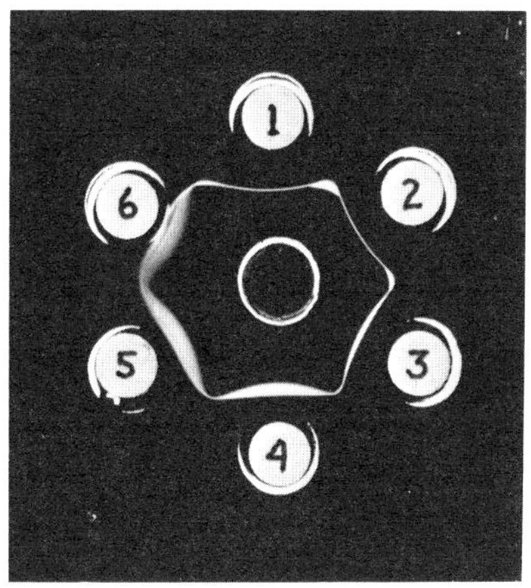

Fig. 2. The determination of the optimal reaction dilution of cowpea mosaic virus in agar double diffusion assay. Antigen concentrations: 4, 2, 1, 0.5, 0.2, 0.1 mg/ml in wells 1-6, respectively. Antiserum in center well, undiluted. Gel: 0.7% agarose (Seakem LE, FMC) in PBS. Photograph taken after 4 days of incubation. Note the change in density and width of zone associated with antigen dilution. Well 3 (1 mg/ml) represents optimum concentration of antigen and antibody as shown by a thin, sharp line of precipitation.

4. Either antiserum or IgG can be used in agar diffusion. Pre-immune serum should be included at some time in the study to assess specificity of the antibody.
5. It is important to report the exact conditions for making the gels and the pattern used in a study. For example, conditions to be described are: 1) DISH - size; 2) GEL - kind, concentration, volume/dish, additives; 3) WELLS - pattern arrangement, diameter, number of wells, distance from center and from each other; 4) VOLUME - in well; 5) INCUBATION - temperature, time experiment terminates (final reading).

Interpretation

The precipitin patterns showing relationships between antigens have been described in detail by Crowle (11) and by Ouchterlony (33). All of the interpretations are based on a balanced system of antigen and antibody concentration. The development of spurs, crossed lines, or blocked lines of precipitation can occur when the concentration of one antigen is out of proportion to the other and therefore can give false representation of the relationships between antigens.

Briefly, four types of reactions can develop: identity, nonidentity, partial identity, and inhibition (Fig. 3). When interference of the lines of precipitation developing between two adjacent antigens is complete (the lines show coalescence, do not deviate, and do not show an alteration in position), the reaction is one of identity. The determinant groups of the antigens are the same. When there is no interference of precipitation lines developing between two adjacent antigens, the lines cross completely without deviating toward an antigen well (bent line) and have independent density, the reaction is one of non-identity. The determinant groups of the two antigens are dissimilar and unrelated. Intermediate reactions occur when precipitin lines formed between adjacent antigens deviate in position and density, even if they coalesce, if they form a spur indicating some determinants are dissimilar, or if they are blocked by the precipitation barrier set up with one antigen-antibody system. Exact interpretation of patterns is difficult sometimes, and it is helpful to study critical relationships in several pattern arrangements and/or in different positions within the same pattern arrangement.

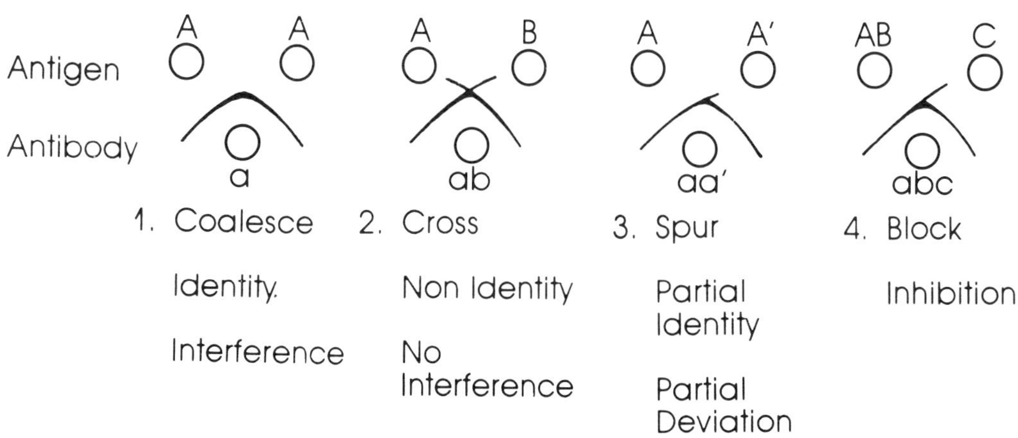

Fig. 3. A drawing illustrating the types of relationships resulting from precipitation reactions between two antigens in agar double diffusion tests [see Ouchterlony (33) for details].

Pre-treatment of Reactants

Regardless of the composition of the gel medium, some antigens or antisera are pretreated before adding them to diffusion plates. Samples containing sonchus yellow net virus were ground for extraction in PBS containing 2% Triton X-100 (also see **Section III.C.2.b.**). Precipitin banding in PBS-agar after antigen pre-treatment was superior to that obtained in SDS-agar medium without antigen pre-treatment (14). Zucchini yellow mosaic virus reacted best when pretreated with 0.3% SDS (32). Maize dwarf mosaic virus was degraded for agar diffusion studies by pre-treatment with an equal volume of 5% pyrrolidine (27). Pyrrolidine and pyrimidine treatment are used effectively in assays of the potato viruses in single radial diffusion (see **Section III.B.2.**).

Use of Agar/Agarose

The type of agar/agarose used and the composition of the medium and use of additives help to determine the appearance of the precipitate forming in agar double diffusion experiments. Variation in types of gel and buffers used with a number of viruses is presented in Table 1. On occasion the percentage of agar used was not mentioned by the author. The range of concentration of gel is 0.4-1.0. Azide concentration varies also (0.02-1.0%). In most instances, the variations depend upon the stability of the antigens used in the studies.

Table 1. Examples of Agars and Buffers Utilized for Agar Double-diffusion Serology, 1984-1988

Types of gel percentage	Virus	Buffer in agar (pH)	Citation no.
A. Noble			
1.0%	alfalfa mosaic, hydrangea ringspot, tobacco ringspot	saline, 0.85-azide or 0.01 M K-PO_4*	9
0.7	prunus necrotic ringspot	0.1 PO_4(7)	30
0.8	sonchus yellow net	PBS	14
-	southern bean mosaic	PBS	28
B. Ionagar			
0.5	barley stripe mosaic-oat	water-azide	8
-	cucumber mosaic-legume	saline or BO_3(9)	40
0.4	cucumber mosaic	PBS	12
1.0	maize dwarf mosaic	0.2 M K-PO_4(7)	27
0.8	prunus tomato ringspot	saline-azide (1%)	20
0.7	tomato aspermy	0.1 M PO_4(7.6)	23
0.75	tomato ringspot	saline-azide (.13%)	6
C. Agarose			
0.5	bean pod mottle	PBS	35
0.9	broad bean wilt	0.07 M PO_4 or saline-0.02 azide	38
1.0	cucumber mosaic	0.01 M $NaPO_4$(7)	41
0.85	cynodon chlorotic streak,	water-azide	26
	cereal chlorotic mottle,	water-azide	25
-	peanut stunt, cucumber mosaic	5 mM BO_3	2
0.9	sowbane mosaic	water	44
0.9	spring beauty latent	0.1 M Na-PO_4(6)	42
0.7	tomato vein yellowing	water-azide	13
D. Other agars			
Bacto 0.8	cowpea chlorotic mottle	0.2 M acetate, saline-azide	5
1.0	lisianthus necrosis	water	21
0.5	maize mosaic, maize stripe	PBS	3
0.6	pepper mild mottle	0.01 M PO_4	43

*PO_4 - phosphate, PBS - phosphate buffered saline, saline - 0.85%, Tris - Tris (hydroxymethyl)aminomethane, BO_3-borate.

Applications

Broad bean wilt virus in 0.03 M neutral phosphate buffer containing 0.02 M 2-mercaptoethanol was added to wells in agar plate 4 hr before testing with antiserum (38). Likewise, cauliflower mosaic virus was added to wells 8 hr before adding serum (37). This was done in order to move the precipitation line away from the edge of the antigen well.

The antiserum may be combined with other materials to increase sensitivity in agar diffusion tests. Lockhart (26) added 5% bovine serum albumin to serum diluted in 0.05 M Tris buffer, pH 7.7, containing 0.85% NaCl. This serum mixture as well as the antigen was added to plates made with 0.8% agarose in water-azide. Nagel (31) titrated antiserum in normal serum before adding to 0.8% Noble agar containing 0.5% SDS.

Additives

An additive usually enhances the movement through the gel, stabilizes the antigen, or enables the researcher to see the reaction more clearly. The chelating agent ethylene diamine tetraacetic acid (EDTA) has been added to agars at 0.05 M to stabilize cucumber mosaic virus (1), at 0.01 M for maize mosaic virus (3), and at 0.005 M for peanut stunt virus and cucumber mosaic virus (2,40). Probably the most complicated medium reported in recent literature is one for cucumber mosaic virus: 0.8% agarose in 0.01 M borate + 0.05 M EDTA + 0.5% Triton X-100 + 0.01 M sodium thioglycolate (1). In addition the stabilization of the virus was improved by treatment with 1% formalin. In many studies, polyvinyl-pyrrolidone, molecular weight 6,000, is recommended as an addition to the extraction buffer, especially for those viruses being used in ELISA tests. However, McLaughlin (29) has reported decreased sensitivity with this compound used on forage legume viruses. Sodium diethyldithiocarbamate performed better as a protective antioxidant.

The dye trypan blue (C.I. No. 23850) can be added to agar. It is a vital stain, but in diffusion media it enhances precipitation zones (17). Casting agar/agarose medium on Gelbond usually gives a thinner layer of gel which can be easily washed and stained. However, it must be handled quickly and carefully while preparing the well pattern and adding the reactant to prevent drying which would change the composition of the gel. Smith (39) examined the reaction of prunus necrotic ringspot virus in a 1.5% agarose gel cast on Gelbond. Wetter (43) used a mini-diffusion procedure by casting a gel on a glass slide pre-coated with a thin film of gel.

Strain studies

Agar diffusion in plates represents an ideal medium for studying strain relationships. Using pattern C, Fig. 1, 10 or more antigens can be related to a single antiserum; in pattern D, 8 antigens or more can be compared with two antisera; and in pattern E or G, 6-8 antigens can be tested with a single serum. Most of the patterns of Fig. 1 are adjusted in well numbers to fit the tight-seal, 50 x 9 mm plate. Additional antigen and antisera combinations can be made on larger plates. However, it is possible to have some interactions between antigenic components in adjacent wells since diffusion from a given well occurs in all directions. For instance, in pattern C antigen 2 interacts with 1 and 3, but not 4 and 5. The positions of antigens in that pattern should be changed in subsequent tests. Also, in pattern E, in which dilutions of antigen are placed in the peripheral wells and with most concentrated at the 12:00 (No. 1) position, undue influence may be placed on the reaction in the well at 10:00 (No. 6) because of the diffusion of the very concentrated antigen in No. 1 position.

Relationship studies have been done in diffusion plates with 10 strains of cucumber mosaic virus and two of peanut stunt virus (2), with 17 isolates of bean yellow mosaic (4) with 34 strains of red clover necrotic mottle virus (34) and with 70 isolates of prunus necrotic ringspot virus (30). Several other relationship studies have been with two isolates, M and S, of watermelon mosaic virus II (7), two strains of Kalanchoe latent virus (19) and serotypes of cowpea severe mosaic virus (24). Koenig (22) reported on a study of tombusviruses in which 35 sera were used to establish relationships in agars made with 0.01 M Tris, pH 8, or with 0.02 M K-Na-PO_4, pH 5.5, buffers. Some 2000 titrations were made.

Survey studies

Agar diffusion tests are used frequently in testing material from field surveys. Some recent studies have been with passion fruit yellow mosaic virus in Brazil (10), asparagus virus I and II (15), asparagus virus II (18), bean pod mottle virus in North Carolina (35), tobacco ringspot virus in squash in South Carolina (36), cucumber mosaic and peanut stunt viruses in Arkansas (2), and viruses in burley tobacco in North Carolina (16).

Literature Cited

1. Ahmad, I. B., and Scott, H. A. 1984. An improved immunodiffusion test for the detection of intact cucumber mosaic virus in crude tobacco sap. Phytopathology 74:1097-1100.
2. Ahmad, I. B., and Scott, H. A. 1985. Identification and serotyping of cucumber mosaic and peanut stunt viruses in Arkansas. Plant Dis. 69:891-893.
3. Ammar, E. D., Gingery, R. E., and Nault, L. R. 1987. Interactions between maize mosaic and maize stripe viruses in their insect vector, *Peregrinus maidis*, and in maize. Phytopathology 77:1051-1056.
4. Barnett, O. W., Randles, J. W., and Burrows, P. M. 1987. Relationships among Australian and North American isolates of the bean yellow mosaic potyvirus subgroup. Phytopathology 77:791-799.
5. Bijaisoradat, M., and Kuhn, C. W. 1985. Nature of resistance in soybean to cowpea chlorotic mottle virus. Phytopathology 75:351-355.
6. Bitterlin, M. W., and Gonsalves, D. 1988. Serological grouping of tomato ringspot virus isolates: Implications for diagnosis and cross-protection. Phytopathology 78:278-285.
7. Chala, V. H., Harrison, C. W., and Halliwell, R. S. 1987. Identification of two distinct strains of watermelon mosaic virus-2 affecting cucurbits in Texas. Plant Dis. 71:750-752.
8. Chiko, A. W. 1984. Increased virulence of barley stripe mosaic virus for wild oats: Evidence of strain selection by host range. Phytopathology 74:595-599.
9. Chiko, A. W., and, Godkin, S. E. 1986. Occurrence of alfalfa mosaic, hydrangea ringspot and tobacco ringspot viruses in *Hydrangea* spp. in British Columbia. Plant Dis. 70:541-544.
10. Crestani, O. A., Kitajima, E. W., Lin, M. T., and Marinho, V. L. A. 1986. Passion fruit yellow mosaic virus, a new Tymovirus found in Brazil. Phytopathology 76:951-955.
11. Crowle, A. J. 1961. Immunodiffusion. Academic Press, New York. 333 pp.

12. Davis, R. F., and, Hampton, R. O. 1986. Cucumber mosaic virus isolates seedborne in Phaseolis vulgaris: Serology, host-pathogen relationships, and seed transmission. Phytopathology 76:999-1004.
13. El Maataoui, M., Lockhart, B. E. L., and Lesemann, D. E. 1985. Biological, serological and cytopathological properties of tomato vein-yellowing virus, a rhabdovirus occurring in tomato in Morocco. Phytopathology 75:109-115.
14. Falk, B. W., Purcifull, D. E., and Christie, S. R. 1986. Natural occurrence of sonchus yellow net virus in Florida lettuce. Plant Dis. 70:591-593.
15. Falloon, P. G., Falloon, L. M., and Grogan, R. G. 1986. Survey of California asparagus for asparagus virus I, asparagus virus II, and tobacco streak virus. Plant Dis. 70:103-105.
16. Gooding, G. V., Jr., and Ruffy, R. C. 1987. Distribution, incidence, and strains of viruses in Burley tobacco in North Carolina. Plant Dis. 71:38-40.
17. Granett, A. L., and Shalla, T. A. 1970. The relation of tobacco mosaic virus X-protein to amorphous cellular inclusions (X-bodies). Phytopathology 60:426-430.
18. Hartung, A. C., Evans, T. A., and Stephens, C. T. 1985. Occurrence of asparagus virus II in commercial asparagus fields in Michigan. Plant Dis. 69:501-504.
19. Hearon, S. S. 1984. Comparison of two strains of Kalanchoe latent virus, carlavirus group. Phytopathology 74:670-676.
20. Hoy, J. W., Mircetich, S. W., and Lownsbery, B. F. 1984. Differential transmission of prunus tomato ringspot virus strains by Xiphinema californicum. Phytopathology 74:332-335.
21. Iwaki, M., Hanada, K., Maria, E. R. A., and Onogi, S. 1987. Lisianthus necrosis virus, a new Necrovirus from Eustoma russellianum. Phytopathology 77:867-870.
22. Koenig, R., and Gibbs, A. 1986. Serological relationships among tombus-viruses. J. Gen. Virol. 67:75-82.
23. Kuti, J. O., and Moline, H. E. 1984. A mild strain of tomato aspermy virus isolated from tomato in Maryland. Plant Dis. 68:959-961.
24. Lin, M. T., Hill, J. H., Kitajima, E. W., and Costa, C. L. 1984. Two new serotypes of cowpea severe mosaic virus. Phytopathology 74:581-585.
25. Lockhart, B. E. L. 1986. Occurrence of cereal chlorotic mottle virus in North Africa. Plant Dis. 70:912-915.
26. Lockhart, B. E. L., Khaless, N., El Maataoui, M., and Lasfra, R. 1985. Cynodon chlorotic streak virus, a previously undescribed plant rhabdovirus infecting Bermuda grass and maize in the Mediterranean area. Phytopathology 75:1094-1098.
27. McDaniel, L. L., and Gordon, D. T. 1985. Identification of a new strain of maize dwarf mosaic virus. Plant Dis. 69:602-607.
28. McGovern, M. H., and Kuhn, C. W. 1984. A new strain of southern bean mosaic virus derived at low temperatures. Phytopathology 74:95-99.
29. McLaughlin, M. R., Barnett, O. W., Gibson, P. B., and Burrows, P. M. 1984. Enzyme linked immunosorbent assay of viruses infecting forage legumes. Phytopathology 74:965-969.
30. Mink, G. I., Howell, W. E., Cole, A., and Reger, S. 1987. Three serotypes of prunus necrotic ringspot virus isolated from rugose mosaic-diseased sweet cherry trees in Washington. Plant Dis. 71:91-93.
31. Nagel, J., Zettler, F. W., and Hiebert, E. 1983. Strains of bean yellow vein virus compared to clover yellow vein virus in relation to gladiolus production in Florida. Phytopathology 73:449-454.

32. Nameth, S. T., Dodds, J. A., Paulus, A. O., and Kishaba, A. 1985. Zucchini yellow mosaic virus associated with severe disease of melon and watermelon in southern California desert valleys. Plant Dis. 69:785-788.
33. Ouchterlony, O. 1968. Handbook of Immunodiffusion and Immunoelectrophoresis. Ann Arbor Science Publishers, Inc., Ann Arbor. 215 pp.
34. Rao, A. L. N., Lakshman, D. K., Ohki, S. T., and Hiruki, C. 1987. Identification of new serotype and antigenic relationships among six strains of red clover necrotic mosaic virus. Phytopathology 77:995-999.
35. Ross, J. B., and Butler, A. K. 1985. Distribution of bean pod mottle virus in soybeans in North Carolina. Plant Dis. 69:101-103.
36. Sammons, B., and Barnett, O. W. 1987. Tobacco ringspot virus from squash grain in South Carolina and transmission of the virus through seed of smooth pigweed. Plant Dis. 71:530-532.
37. Schoetz, J. E., Shepherd, R. J., and Richins, R. D. 1986. Properties of an unusual strain of cauliflower mosaic virus. Phytopathology 76:451-454.
38. Scott, S. W., and Barnett, O. W. 1984. Some properties of an isolate of broad bean wilt virus from dogwood (*Cornus florida*). Plant Dis. 68:983-985.
39. Smith, D. R., and Skotland, C. B. 1986. Host range and serology of prunus necrotic ringspot virus serotypes isolated from Hops (*Humulus lupulus*) in Washington. Plant Dis. 70:1019-1023.
40. Sommerville, P. A., Campbell, R. N., Hall, D. H. and Rowhani, A. 1987. Natural infection of potatoes (*Solanum tuberosum*) by a legume strain of cucumber mosaic virus. Plant Dis. 71:18-20.
41. Valverde, R. A. 1984. Unusual strain of cucumber mosaic virus causing flower-breaking symptoms in wild violets. Plant Dis. 68:913-915.
42. Valverde, R. A. 1985. Spring beauty latent virus: A new member of the bromovirus group. Phytopathology 75:395-398.
43. Wetter, C., Conti, M., Altschuh, D., Tabillion, R. and van Regenmortel, M. H. V. 1984. Pepper mild mottle virus, a tabamovirus infecting pepper cultivars in Sicily. Phytopathology 74:405-410.
44. Zebzami, M., Lockhart, B. E. L., and Hill, J. H. 1987. Identification of three serotypes of sowbane mosaic virus. Phytopathology 77:571-574.

Section III.C.2.b.

OUCHTERLONY DOUBLE-DIFFUSION TESTS IN THE PRESENCE OF SODIUM DODECYL SULFATE FOR DETECTION OF VIRION PROTEINS AND VIRUS-INDUCED INCLUSION BODY PROTEINS

D. E. Purcifull
University of Florida
Gainesville, FL 32611

Principle

Sodium dodecyl sulfate (SDS) is an anionic detergent which may be used to dissociate virions and virus-induced inclusion bodies into antigenically active, diffusible fragments. The SDS is included as a component of the agar gel medium and also is often mixed with the test antigen to facilitate dissociation. As with other double diffusion on plate tests, the antigens and antibodies diffuse radially from reactant wells and form immunoprecipitin bands between the wells under suitable conditions. The chief advantage of the test is its simplicity for routine virus detection, virus identification, and assessment of the relationship between virus-specified antigens. The technique was developed by Gooding and Bing (11) for use with flexuous, elongated viruses, but it has also been used for viruses with other morphologies as well as for inclusion body proteins.

Materials

1. Plastic disposable Petri dishes (90 x 13mm) with clear, flat bottoms
2. Chemicals
   Noble agar (Difco)
   sodium dodecyl sulfate (e.g., Sigma Cat. #L-5750)
   sodium azide
3. Gel cutting device
4. Vacuum pump or vacuum line
5. Disposable Pasteur pipets (5 3/4") and rubber bulbs, or adjustable pipettor with plastic tips.
6. Moist chamber (plastic box with moist towels on bottom)
7. Mortars and pestles
8. Cheesecloth
9. Antigens, which may consist of freshly prepared or preserved (e.g., freeze-dried or frozen) tissue extracts from infected and noninfected plants, and purified viruses or virus-induced inclusions
10. Antisera
11. Normal serum
12. Black box with indirect lighting for reading plates

Procedure

1. Preparation of medium. (0.8% Noble agar - 1.0% NaN_3 - 0.5% SDS). For a batch of approximately 40 plates, add 4 g agar to 300 ml of distilled water

in a beaker and autoclave for 5 min at 121 C or melt the agar in a microwave oven. Add 5 g sodium azide to the molten agar, and mix well. Add 2.5 g SDS to 100 ml of hot distilled water and add to the agar-azide mixture. Bring up to a volume of 500 ml with hot distilled water and mix. Place Petri dishes on a level surface and add 12 ml/dish with the aid of a graduated centrifuge tube. Bubbles that may occur can be moved to the edge of the dish by gentle swirling. Allow medium to gel and store plates in closed plastic bags at 4 C. Upon cooling, the plates will become cloudy due to the SDS but they will clear when warmed to room temperature. Use plates within one month.

2. Antigens. For routine testing, use a mortar and pestle to prepare tissue extracts. For each gram of tissue, add 1 ml of distilled water, triturate, add 1 ml of 3% SDS (in water), triturate again, and express sap through cheesecloth. The expression step can sometimes be eliminated if wide-mouth pipets are used to avoid clogging the tips with tissue debris. Purified virions or inclusion bodies and their corresponding dissociated proteins may also be used. For some virus-host combinations, leaf disks, freeze-dried sap, frozen sap, and dried tissue have been used as reliable sources of antigen. Depending on the virus, it is sometimes useful to boil the antigen for 4 min in SDS to facilitate dissociation. It is important that for each antiserum being tested, antigen known to react homologously with that antiserum be included in all tests to serve as a reference for antigenic relationships and to verify that test conditions are appropriate. Extracts from healthy plants are also included as controls to help verify the specificity of precipitin reactions.

3. Antiserum. Undiluted antiserum with known reactivity to the SDS-dissociated antigen is normally used. Preparation of such antisera may require immunization with SDS-treated antigens (15, 27). In some cases, however, antiserum to untreated virus reacts well in SDS-immunodiffusion (e.g. with many potyviruses). If serum dilutions are made for titering or other purposes, the serum should be diluted with normal serum or with 0.05 M Tris-HCl, pH 7.2, containing 0.85% NaCl and 5% bovine serum albumin. The use of antiserum diluted with glycerol should be avoided. Remove the glycerol by dialysis, freeze dry the antiserum, and resuspend it in a volume of distilled water equivalent to the volume of the undiluted serum. Normal (pre-immune) serum should be included as a control to aid in detection of nonspecific precipitation or precipitates due to the presence of antibodies specific for host antigens. With some host-virus-antiserum systems, excess nonspecific precipitations occur. Sometimes these can be eliminated by adding bovine serum albumin to the gel medium (27).

4. Preparation of gel patterns and addition of reactants. Gel patterns are cut in the agar and the agar disks are removed by aspiration. A convenient pattern consists of 6 peripheral wells and one central well, each 7 mm in diameter. Each peripheral well is 4-5 mm from the central well at the closest point. Three such patterns can be used in nonpartitioned dishes. Reactants are pipetted into the wells until they are nearly full (i.e., over-flow tendency is increased by SDS). Usually, antigens are added to the peripheral wells and antisera to the central wells.

5. Incubation of plates and recording results. The Petri dishes are placed in covered plastic boxes to maintain humidity and incubated for 2-4 da at 25 C. Precipitin lines usually develop within 6-24 hr, and they usually start to disappear within 2-5 da after the tests are initiated. The precipitin lines can be preserved to some extent if the reactant wells are emptied 24-48 hr after the tests are initiated and re-filled with a slurry

of Norit A charcoal (15%, w/v, in distilled water). Extracts from some hosts occasionally interact to form granular, nonspecific precipitates. These can often be removed by submersing the plates in a solution containing 0.85% NaCl in 0.05 M Tris-HCl, pH 7.2, with gentle stirring for 1-2 da. The immunoprecipitin bands remain whereas the granular precipitates disappear.

Results of the tests may be recorded photographically or by drawing. The use of a hand lens aids visualization of precipitin bands. Viewing and photography of the precipitin lines are enhanced by use of a specially designed, black box[a] which has indirect, fluorescent ring lighting. The plates should fit snugly in a circular hole cut in the top of the box. For photography, the lids of the plates should be removed. A light coating of mineral oil applied to the outer surface of the bottom of the plates diminishes the appearance of scratches on the plastic.

Comments

The methods described above have been used for detecting virion proteins of flexuous, filamentous viruses (Fig. 1) (1,5-8,16-18,20-22,27,28,33-38,40), rigid-rod-shaped viruses (2,41), isometric viruses (23,29,32) and enveloped bacilliform and spherical viruses (9,17). They have also been used for detecting the proteins associated with potyviral-induced cylindrical (Fig. 2), nuclear, and amorphous inclusion bodies (3,4,15,19,25,26,27,30,31,39). The SDS-immuno-diffusion tests have proven extremely valuable as a standard diagnostic tool for viruses which have moderate to high virus concentrations (e.g., potyviruses, comoviruses, potexviruses, cucumoviruses, tobamoviruses, etc.). The tests are relatively simple to conduct and interpret. With tests set up in late afternoon, results are usually available the next morning. Some of the many applications include detection of multiple infections and diagnosis of viruses in epidemi-ological studies, studies of antigenic relationships, assay of various fractions for loss of virions during purification, assay for virus infections in host range studies, and detection of viruses in seedlings.

---

[a] See "Dark field illuminator". Pages 258-260 in: Crowle, A. J. 1961. Immunodiffusion. Academic Press, New York, N.Y. 333 pp.

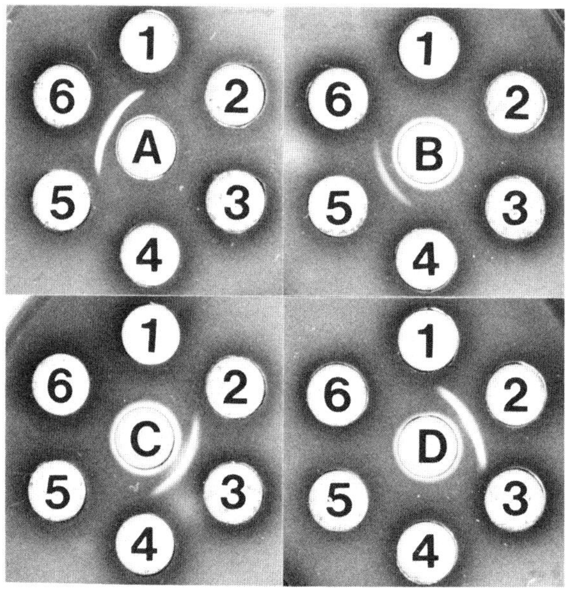

Fig. 1 Serological detection and distinction of filamentous and isometric viruses by SDS-immunodiffusion tests. Central wells contain antisera to: A = clover yellow mosaic virus (potexvirus) protein derived by SDS-treatment; B = SDS-treated cucumber mosaic virus (cucumovirus); C = SDS-treated cowpea mosaic virus (comovirus); and D = untreated watermelon mosaic virus-2 (potyvirus). Peripheral wells contain SDS-treated leaf extracts from _Nicotiana benthamiana_: 1 & 4 = noninfected plants; 2 = watermelon mosaic virus-2; 3 = cowpea mosaic virus; 5 = cucumber mosaic virus, and 6 = clover yellow mosaic virus. (adapted from data in reference no. 29)

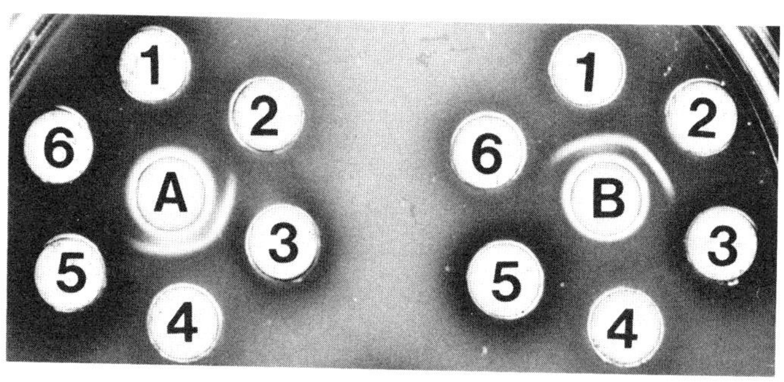

Fig. 2. Serological detection and distinction of two potyviruses based on the antigenic properties of their cylindrical inclusion proteins by SDS-immuno-diffusion. A = antiserum to tobacco etch virus cylindrical inclusions and B = antiserum to potato virus Y cylindrical inclusions. Peripheral wells contain SDS-treated extracts from: 1 = potato virus Y-infected _Datura metel_; 2 = potato virus Y-infected tobacco; 3 = tobacco etch virus-infected tobacco; 4 = tobacco etch virus-infected _D. metel_; 5 = healthy _D. metel_; 6 = healthy tobacco.

It is clear from the literature that various modifications of the SDS-immunodiffusion tests can optimize results in some cases. Factors relating to the gel medium that have been modified include choice of agar or agar concentration (2,18,34,37,39), SDS concentration (2,10,18,20,34,36,37), sodium azide concentration (2,18,34,36), and addition of various salts and buffers (18,37,38). The most suitable concentrations of antigen and antiserum can be determined by testing different dilutions of each in a particular system. The ways in which antigens are prepared also vary considerably (1,8,10,21,22,27, 28,29,35,38). If samples are to be used for either electron microscopy of particles or infectivity, samples for such purposes should be removed before adding SDS. In some cases, it is possible to use the same extract for both serological and biological assays (21).

Relationships between virion or inclusion body proteins can be evaluated by studying the intersection of precipitin lines and by intragel absorption tests (6,30,31) (see **Section III.C.2.e.**). As with other plate double diffusion tests, spur formation can occur with antigenically related but different antigens (Fig. 3) (23,27,30). Intragel absorption tests can also be useful for removal of antibodies to host proteins (40).

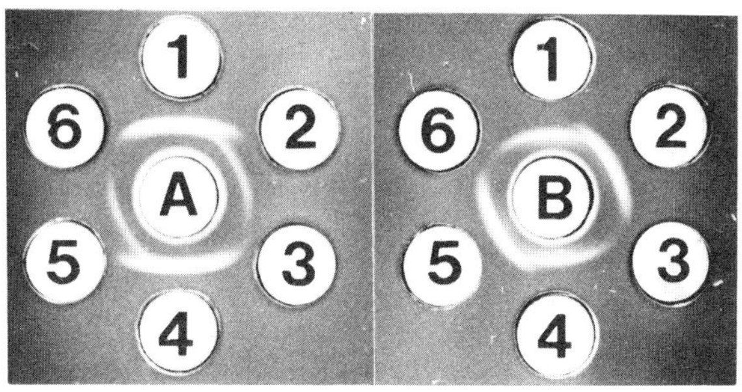

Fig. 3. Distinction of two serologically related, but antigenically different, strains of a potyvirus by SDS-immunodiffusion. Note reciprocal spur formation. The central wells contain antisera to: A = papaya ringspot virus type W capsid protein and B = antiserum to papaya ringspot virus type T. Peripheral wells contain SDS-treated leaf extracts from Cucurbita pepo: 1 and 4 = infected with papaya ringspot virus type W; 2 and 5 = infected with papaya ringspot virus type T; 3 and 6 = noninfected plants.

Literature Cited

1. Barnett, O. W., Randles, J. W., and Burrows, P. M. 1987. Relationships among Australian and North American isolates of the bean yellow mosaic potyvirus subgroup. Phytopathology 77:791-799.
2. Carroll, T. W., Gossel, P. L., and Batchelor, D. L. 1979. Use of sodium dodecyl sulfate in serodiagnosis of barley stripe mosaic virus in embryos and leaves. Phytopathology 69:12-14.
3. Chang, C.-A., Hiebert, E., and Purcifull, D. E. 1988. Purification, characterization, and immunological analysis of nuclear inclusions induced by bean yellow mosaic and clover yellow mosaic potyviruses. Phytopathology 78:1266-1275.
4. DeMejia, M. V. G., Hiebert, E., and Purcifull, D. E. 1985. Isolation and partial characterization of the amorphous cytoplasmic inclusions associated with infections caused by two potyviruses. Virology 142:24-33.

5. Dijkstra, J., Bos, L., Bouwmeester, H. J., Hadiastono, T., and Lohuis, H. 1987. Identification of blackeye cowpea mosaic virus from germplasm of yard-long bean and from soybean, and the relationships between blackeye cowpea mosaic virus and cowpea aphid-borne mosaic virus. Neth. J. Plant Pathol. 93:115-133.
6. Dodds, J. A., Lee, J. G., Nameth, S. T., and Laemmlen, F. F. 1984. Aphid- and whitefly-transmitted cucurbit viruses in Imperial County, California. Phytopathology 74:221-225.
7. Garnsey, S. M., Gonsalves, D., and Purcifull, D. E. 1979. Rapid diagnosis of citrus tristeza virus infections by sodium dodecyl sulfate-immunodiffusion procedures. Phytopathology 69:88-95.
8. Giorda, L. M., Toler, R. W., and Miller, F. R. 1986. Identification of sugarcane mosaic virus strain H isolate in commercial grain sorghum. Plant Dis. 70:624-628.
9. Gonsalves, D., and Trujillo, E. E. 1986. Tomato spotted wilt virus in papaya and detection of the virus by ELISA. Plant Dis. 70:501-506.
10. Gooding, G. V., Jr. 1975. Serological identification of tobacco viruses. Tobacco Sci. 29:104-109.
11. Gooding, G. V., Jr., and Bing, W. W. 1970. Serological identification of potato Y virus and tobacco etch virus using immunodiffusion plates containing sodium dodecyl sulfate. Phytopathology 60:1293.
12. Hamilton, R. E., and Nichols, C. 1978. Serological methods for detection of pea seed-borne mosaic virus in leaves and seeds of Pisum sativum. Phytopathology 68:539-543.
13. Hearon, S. S. 1984. Comparison of two strains of Kalanchoe latent virus, Carlavirus group. Phytopathology 74:670-676.
14. Hibino, H., Usugi, T., Omura, T., Tsuchizaki, T., Shohara, K., and Iwasaki, M. 1985. Rice grassy stunt virus: a planthopper-borne circular filament. Phytopathology 75:894-899.
15. Hiebert, E., Purcifull, D. E., and Christie, R. G. 1984. Purification and immunological analyses of plant viral inclusion bodies. Pages 225-280 in: Methods in Virology, Vol. 8. K. Maramorosch and H. Koprowski, eds. Academic Press, Inc., Orlando, FL. 396 pp.
16. Hiebert, E., Tremaine, J. H., and Ronald, W. P. 1984. The effect of limited proteolysis on the amino acid composition of five potyviruses and on the serological reaction and peptide map of the tobacco etch virus capsid protein. Phytopathology 74:411-416.
17. Jackson, A. O., and Christie, S. R. 1977. Purification and some physicochemical properties of sonchus yellow net virus. Virology 77:344-355.
18. Jones, R. T., and Diachun, S. 1977. Serologically and biologically distinct bean yellow mosaic virus strains. Phytopathology 67:831-838.
19. Knuhtsen, H., Hiebert, E., and Purcifull, D. E. 1974. Partial purification and some properties of tobacco etch virus induced intranuclear inclusions. Virology 61:200-209.
20. Lawson, R. H., Brannigan, M. D., and Foster, J. 1985. Clover yellow vein virus in Limonium sinatum. Phytopathology 75:899-906.
21. Lecoq, H., and Lot, H. 1982. Intérêt de l'immunodiffusion en milieu gélosé contenant du SDS pour la détection du virus de la mosaique de la pasteque type 2 (WMV2) chez les Cucurbitacées. Use of SDS immunodiffusion for the detection of watermelon mosaic virus type 2 (WMVZ) in Cucurbits. 1982. Agronomie 2:786-787.

22. Lima, J. A. A., and Purcifull, D. E. 1980. Immunochemical and microscopical techniques for detecting blackeye cowpea mosaic and soybean mosaic viruses in hypocotyls of germinated seeds. Phytopathology 70:142-147.

23. Lot, H., and Lecoq, H. 1986. Interest of SDS-immunodiffusion for cucumber mosaic virus serotyping in field samples. Phytopathology 76:1131.

24. Nagel, J., and Hiebert, E. 1985. Complementary DNA cloning and expression of the papaya ringspot potyvirus sequences encoding capsid protein and a nuclear inclusion-like protein in Escherichia coli. Virology 143:435-441.

25. McDonald, J. G., and Hiebert, E. 1975. Characterization of the capsid and cylindrical inclusion proteins of three strains of turnip mosaic virus. Virology 63:295-303.

26. Provvidenti, R., Gonsalves, D., and Humaydan, H. S. 1984. Occurrence of zucchini yellow mosaic virus in cucurbits from Connecticut, New York, Florida and California. Plant Dis. 68:443-446.

27. Purcifull, D. E., and Batchelor, D. L. 1977. Immunodiffusion tests with sodium dodecyl sulfate (SDS)-treated plant viruses and plant viral inclusions. Fla. Agric. Exp. Stn. Tech. Bull. 788. 39 p.

28. Purcifull, D. E., Christie, S. R., and Batchelor, D. L. 1975. Preservation of plant virus antigens by freeze-drying. Phytopathology 65:1202-1205.

29. Purcifull, D. E., Christie, S. R., and Lima, J. A. A. 1981. Detection of four isometric plant viruses in sodium dodecyl sulfate immunodiffusion tests. Phytopathology 71:1221-1224.

30. Purcifull, D. E., Hiebert, E., and McDonald, J. G. 1973. Immunochemical specificity of cytoplasmic inclusions induced by viruses in the potato Y group. Virology 55:275-279.

31. Quiot-Douine, L., Purcifull, D. E., Hiebert, E., and de Mejia, M. V. G. 1986. Serological relationships and in vitro translation of an antigenically distinct strain of papaya ringspot virus. Phytopathology 76:346-351.

32. Rana, G. L., Gallitelli, D., Kyriakopoulou, P. E., Russo, M., and Martelli, G. P. 1980. Host range and properties of artichoke yellow ringspot virus. Ann. Appl. Biol. 96:177-185.

33. Richter, J., Haack, I., and Schuster, K. 1979. Untersuchungen uber die serologische Verwandschaft zwischen den kartoffelviren S and M. Arch. Phytopathol. und Pflanzenschutz 15:161-168.

34. Simmonds, D. H., and Cumming, B. G. 1979. Detection of lily symptomless virus by immunodiffusion. Phytopathology 69:1212-1215.

35. Timmer, L. W., Scorza, R., and Lee, R. F. 1981. Incidence of tristeza and other citrus diseases in Bolivia. Plant Dis. 65:515-517.

36. Tolin, S. A. 1977. Identification of legume viruses in the field by serology. Fitopatologia Brasileira 2:1-7.

37. Tolin, S. A., and Ford, R. H. 1983. Purification and serology of peanut mottle virus. Phytopathology 73:899-903.

38. Wisler, G. C., and Zettler, F. W. 1987. Virus infections of Vanilla and other orchids in French Polynesia. Plant Dis. 71:1125-1129.

39. Yeh, S.-D., and Gonsalves, D. 1984. Purification and immunological analyses of cylindrical-inclusion protein induced by papaya ringspot virus and watermelon mosaic virus 1. Phytopathology 74:1273-1278.

40. Zee, F., Gonsalves, D., Goheen, A., Kim, K. S., Pool, R., and Lee, R. F. 1987. Cytopathology of leafroll-diseased grapevines and the purification and serology of associated closteroviruslike particles. Phytopathology 77:1427-1434.

41. Zettler, F. W., and Nagel, J. 1983. Infection of cultivated gesneriads by two strains of tobacco mosaic virus. Plant Dis. 67:1123-1125.

Section III.C.2.c.

OUCHTERLONY DOUBLE DIFFUSION, PLATES: BACTERIA

H. Bouzar
Université Blida
Blida, Algeria

L. W. Moore
Oregon State University
Corvallis, Oregon 97331

Principle

Ouchterlony double diffusion tests (ODD) consist of agar gels wherein preparations of bacterial antigens and specific antibodies are placed in adjacent sample wells (12,22,30,31) (Fig. 1). ODD is a qualitative test that has been used successfully with all the major plant pathogenic bacterial genera to show antigenic relatedness, to confirm identity (2,5,6,37), and to clarify taxonomic position (37). Bacterial-taxonomic relationships are more readily assessed by ODD than any other serological method, with the exception of immunoelectrophoresis, because the number of precipitin bands produced between opposing wells of reactants corresponds to the number of antigenic components in a given gel-antigen-antibody system. Conversely, ODD is one of the least sensitive of the common serological methods because it requires high and equivalent concentrations of antigens and specific antibodies for precipitin bands to develop. This low sensitivity generally limits its usefulness for identification of unknown bacteria directly from environmental samples (16). Exceptions occur with high bacterial populations, concentrated bacterial antigens or high-titer antiserum (41).

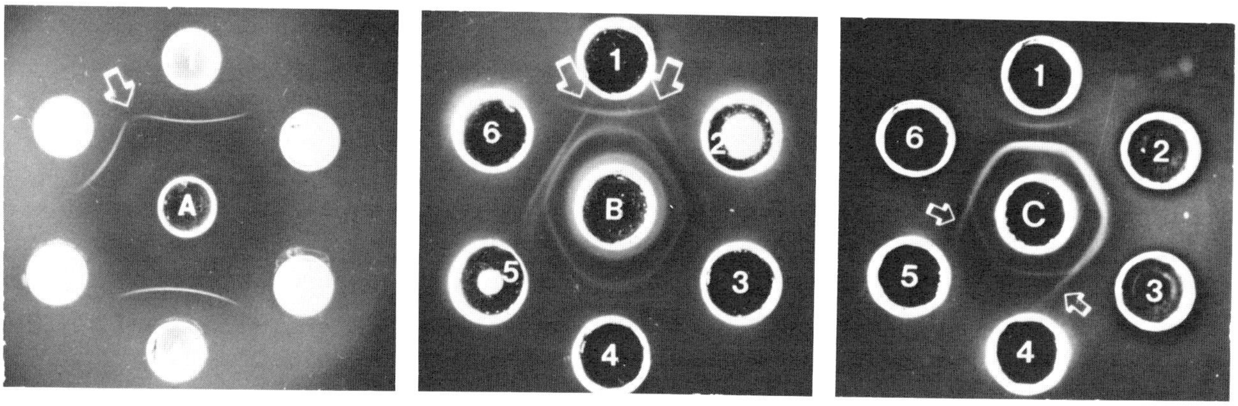

Fig. 1 Reaction types I, II, and III of double diffusion. Center wells A, B, and C contain 15 μl antiserum to 50 S ribosomal subunits of Agrobacterium. Outer well 1 contains the homologous strain. Pattern A shows a reaction of complete fusion or type I (reaction of identity) (arrow). Reaction type II (reaction of non-identity), where precipitin bands cross each other, is shown in pattern B (arrows). Spur formation or reaction type III (reaction of partial identity) is shown in pattern C (arrow).

Materials

(See Materials, **Section III.C.2.a.**).

1. Plastic Petri dishes, to hold the gel. If glass Petri dishes are used, coat bottoms with 2% Formvar in ethylene chloride or use "Gel-Bond" (FMC BioProducts), cut to fit in plates, before pouring the gel to provide a hydrophobic surface which prevents reactants from running under the gel.
2. Stain (optional), to color the gel for enhanced contrast between gel and precipitin bands (e.g. trypan blue, 2%, stains the gel blue). Alternatively, if the precipitin bands are not sufficiently clear, stain them with Coomassie Brilliant Blue for 10 minutes, then destain (24). It is recommended that plates be lined with Gel-Bond if the gel is to be dried for permanent storage.
3. Waterbath, 50-60 C, to maintain the gel in a liquid state before pouring into plate.

4. Lighted tube[a] to allow visualization of precipitin bands, especially unstained bands.
5. Drying oven (optional), 37 C, to dry gel for permanent storage.
6. Camera (e.g. Polaroid) and film to photograph the bands and provide a permanent record (optional).
7. Template to cut wells in gel.

Procedure

(See Procedures, **Section III.C.2.a.**)

1. Weigh purified agar or agarose and dispense in a salt solution containing, per liter: 2 g $MgCl_2.6H_2O$ and 8.5 g NaCl adjusted to pH 7.2. Add thimerosal or sodium azide to a final concentration of 0.02 to 0.19% as a preservative. Optional: add 1 ml of an aqueous 2% trypan blue solution to 100 ml of gel.
2. Select well diameter (e.g. 4-9 mm) according to quantity of reactants available.
3. Select distance between antigen and antibody wells according to well diameter, antigen concentration, and antiserum titer (Fig. 1, 2).

---

[a] A simple lighted tube can be constructed from a round cardboard or metal cylinder. (We use a coffee can about 18 cm high and 15 cm in diam.) Center the tube vertically around a simple light bulb socket mounted on a wooden base. To optimize visualization of the precipitate, light must bounce off the walls of the tube before illuminating the gel; for that reason, the inner surface of the tube should be either naturally reflective, painted or covered with aluminum foil. A circular black disk, larger than the bulb and with a diameter about 50% smaller than that of the tube, is suspended a few centimeters above the bulb by wires attached to the sides of the tube to block the light from shining directly on the gel. A lid is positioned over the top of the tube within a few centimeters (determined by the experimenter) of the black disk. Cut a hole (4-8 cm diam.) in the center of the lid and position the gel over the hole such that indirect light can illuminate the precipitin lines. Simply holding the ODD plate at an angle in the light beam of a microscope illuminator also works well to observe precipitin bands.

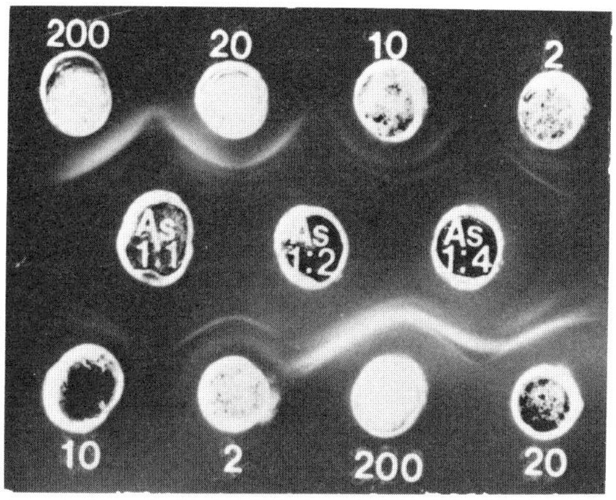

Fig. 2. Piazzi test. The center row of wells contains 20 μl of three different concentrations of antiserum (As) to 50 S ribosomal subunits of *Agrobacterium tumefaciens* B6: undiluted (As, 1:1), diluted by half with buffered saline (As, 1:2), and diluted to one fourth its strength with buffered saline (As, 1:4). The top and bottom rows of wells contain 2, 10, 20, or 200 μg of *A. tumefaciens* B6 cell lysate antigen. When the antigen and antibody concentrations are high, the immunoprecipitin lines are thick and diffused. At low antigen and/or antibody concentrations, fewer bands are visible. At the optimum antigen-antibody ratio, sharp immunoprecipitin lines develop (as shown by the reaction between 10 μg of antigen and the undiluted antiserum).

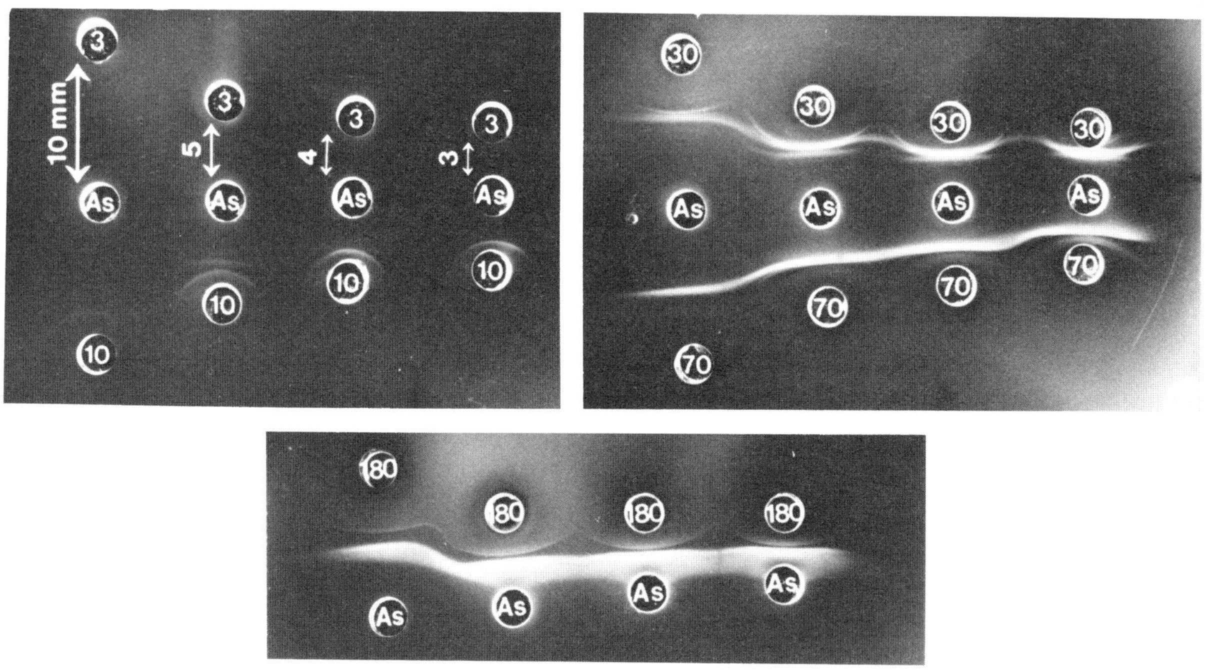

Fig. 3. Effects of distance between antigen and antiserum wells at different antigen concentrations. The distance between the antiserum (As) and antigen wells is 3, 4, 5, or 10 mm. Antiserum wells contain 15 μl of antiserum to 50 S ribosomal subunits of *Agrobacterium tumefaciens* B6. Antigen wells contain 3, 10, 30, 70, or 180 μg of ribosomal antigens from *A. tumefaciens* B6. At this As concentration, the best resolution of immunoprecipitin lines is when the wells are 5 mm apart and when the antigen concentration is 10 μg per well. The patterns may be varied with different antisera and antigens.

4. Avoid overfilling the wells; reagent overflow tends to destroy diffusion patterns.
5. Balance concentration of reactants by the double immunodiffusion method of Piazzi (34) (see **Section III.C.2.a.**). Sharp precipitin lines form at the optimum antigen antibody ratio (i.e. equivalence) (Fig. 2, 3).
6. Include appropriate positive and negative controls.
7. Incubate the gel at a constant temperature (e.g. 20-25 C) for 24-48 hr.
8. Examine precipitin bands with the aid of a light box illuminated by indirect light. Record and measure the pattern of bands or photograph the patterns.
9. Dry (optional) developed plates at 37 C for a permanent record. Cover gel with several sheets of moist filter paper before placing in the oven, preferably in the path of moving air. The agar gel dries to a thin film.

Comments

(See Comments, **Section III.C.2.a.**)

1. Concentrations of antigen should be kept constant for reproducible results. However, it is time consuming to standardize all the antigenic determinants associated with whole bacterial cells. Refer to **Section III.C.2.a.**, Fig. 2, for further information regarding the principle of reacting serially diluted test-antigens against a constant concentration of antibody.
2. When antiserum and antigen react to form a precipitin band, the band acts as a barrier against further diffusion of the same reactants. However, other antigens may diffuse through the barrier and react with the antiserum, the end result being formation of multiple precipitin bands (Figs. 1-3). This capacity to visualize multiple precipitin bands simultaneously in the gel (Fig. 3) is one of the major strengths of the ODD test. The pattern of precipitin bands allows one to distinguish more readily a cross-reaction between strains which are not identical antigenically than can be achieved with the whole-cell agglutination technique. Based on ODD patterns of cross-reaction, a cross-reacting antibody can often be recognized and removed by cross-absorption to obtain more specific precipitin bands.
3. When antigen and antibody are in wells of identical size and shape, antigens having a molecular weight greater than 200,000 d will form precipitin bands close to the antigen well (13). The curvature of the precipitin bands depends in part upon the relative molecular weights of antigen and antibody. If both reagents have similar molecular weights, the line is usually straight; otherwise, the line curves toward the reagent of highest molecular weight (29).
4. Although precipitin bands from adjacent antigen wells may fuse in a reaction of identity (Fig. 3A), the two antigen sources may differ in other antigenic determinants to which there are no corresponding antibodies in the antiserum employed.
5. Absence of fusion between two precipitin bands from adjacent antigen wells (Fig. 3B) indicates that the antiserum contained antibodies which reacted with antigenic determinants on the antigen diffusing from each well, but these determinants were unique to each antigen source.
6. Spur formation (Fig. 1C) is attributed to different antigenic determinants associated with the homologous antigen (immunogen) and the heterologous antigen. False spurs may develop when antigens in adjacent wells differ greatly in concentration. A false spur may also be due to a precipitin band which overlaps another precipitin band common to the two test-antigens. It

is a true spur if the "spur" cannot be removed by (i) cross-absorption to eliminate one of the two precipitin bands or by (ii) reacting the antiserum against different antigen concentrations to separate the precipitin bands.

7. Unstable antigens which degrade under ODD conditions may produce false precipitates because serological specificity of the antigen changes during antigen denaturation. Extra precipitin bands may appear if antibodies to the denatured antigen are also present in the antiserum. Similarly, certain antigens may degrade, releasing fragments with distinct molecular weights and antigenic specificities, as was the case with the multiple precipitin bands formed against bacteria containing proteolytic enzymes (29).
8. Serotyping of ribosome antigens by ODD should be done with caution. Extreme care is needed to insure that the ribosomal preparations are free of contaminating bacterial surface antigens (7,8). Multiple precipitin bands were observed with antiserum from rabbits immunized to partially purified ribosomal preparations, and 18 serovars were tentatively described (3,6). In contrast, only one major precipitin band developed (one serovar) when ribosomes were reacted with antiserum from rabbits that had been immunized with pure ribosomal preparations.
9. A disadvantage of ODD is that only the major soluble antigens can be studied and minor components are typically ignored or undetected. For example, whole bacterial cells can be used in ODD tests, but their associated antigens may be released in low concentration or migrate poorly through the gel matrix and form equivocal precipitin bands. Cell concentrations of 10^{10} CFU/ml may be required to detect somatic antigens directly (25). More reproducible ODD results were reported when whole cells, bacterial membranes, or samples of plant sap which contained bacteria were: (i) sonicated (26,28,42), (ii) shaken with glass beads (41), (iii) extracted with hot phenol (15), phenol-water (7,8,21,26,35), or trichloroacetic acid (26), or (iv) solubilized with detergents such as sodium deoxycholate, Triton X-100 or sodium dodecyl sulfate (26,38). (See **Section III.C.2.b.**)
10. The following factors are important to obtain reproducible ODD results:
    a. Critical serotyping of bacterial antigens requires use of standardized antiserum (reference population of antibodies). When changes in antisera are required (e.g. from subsequent bleedings), new antisera should be standardized against a set of selected antigens.
    b. The formation of visible precipitin bands is influenced not only by relative and total reactant concentrations but also by distances between reactant wells and well sizes. Visible precipitin bands may develop between closely set wells (e.g. 3-5 mm), but may fail to develop with greater well distances. When the antibody concentration is very low, precipitin bands can form if larger wells are employed and the wells are refilled to provide a continuous diffusion of antibody (8).
11. In general, the sensitivity of ODD is lower than immunofluorescence (1,2,16,17,18,19,41), ELISA (1,19,39,40), and microagglutination (28) for detection of bacterial antigens. The specificity of the method, however, is generally superior to these other serological methods because various antigens and antisera can be compared directly with one another in the same gel to observe antigenic differences and relationships among bacteria with cross-reacting antigens (2,5,42).

Applications

Serotyping

ODD has a long history of use for routine serotype recognition in the rhizobia (43). Similarly, De Boer et al (17) proposed a serotyping scheme for isolates of *Erwinia carotovora* subsp. *carotovora* from potatoes that was based on diffusible somatic antigens. The original scheme described 18 serogroups and was expanded to 40 by 1984 (35); it now contains over 50 serogroups (M. L. Powelson, personal communication). In contrast to isolates of *E. c.* subsp. *carotovora*, the serology of *E. c.* subsp. *atroseptica* isolates had been rather homogeneous until recently with only two serogroups reported (a major and a minor group) (33). However, De Boer et al (17) reported that authentic *E. c.* subsp. *atroseptica* strains typed into serogroups I, XVIII, XX, or XXII, indicating increased complexity of the serotyping scheme for this group.

Twenty-five strains of *Xanthomonas campestris* pv. *campestris* were typed into four serovars using antiserum raised against a ribosomal extract from *X. c.* pv. *campestris* (42). Of thirty-two other bacteria only three strains of *X. vesicatoria* and one strain of *X. campestris* pv. *transluscens* cross-reacted with antisera specific for the four serovars. These bacteria included 16 strains of five other pathovars of *X. campestris*, 12 strains of five other genera, and four unidentified bacteria from crucifer seeds. In contrast, none of the antisera was sufficiently specific in immunofluorescence tests (IF) to differentiate *X. c.* pv. *campestris* from other xanthomonads.

In a later study (2), the serological relationships among 26 strains of *X. campestris* pv. *transluscens* from a variety of plants were determined with antisera to membrane proteins. Based on the major ODD band and cross-absorption tests, 25 of the 26 strains were grouped into two serovars and two subserovars. There was no correlation between host of origin and serovar. In contrast to the high specificity observed with ODD, cross-reactions occurred between the strains in immunofluorescence tests with the same antisera.

Lucas and Grogan's (25) ODD analysis of antigen preparations from nearly 300 isolates of *Pseudomonas lachrymans* indicated there were three distinct serovars. Although all the plant pathogenic *Pseudomonas* nomenspecies had several antigens in common, each one also had at least one specific antigen. It was necessary, however, to determine the range of serological variation before the method could be used with confidence. The specific antigens were similar in physical, chemical, and serological properties to the O antigens (the lipopolysaccharides) of members of the Enterobacteriaceae (26).

Immunodiffusion studies with *Streptomyces* have revealed differences in precipitin band patterns (27). To obtain bands useful for identification, it was necessary to cross-absorb the antisera numerous times with heterologous lyophilized antigen. Manzer et al (27) reported that the serological tests provided an additional basis for separation of the acid scab pathogen (*Streptomyces* spp.) from *S. scabies*, the usual incitant of common scab of potato. The serological differences between the two organisms were consistent with other morphological and physiological traits used to differentiate the two microbes. In an earlier serological study of *Streptomyces* spp. (9), however, there appeared to be little correlation between the precipitin patterns of various isolates and their pathogenicity. Studies of other *Streptomyces* species by the ODD technique resulted in a complex pattern of cross-reactions which was of little taxonomic value (23). At the genus level, Ridell and Williams (36) showed that *Nocardia mediterranea* should be separated from *Streptomyces* and *Streptoverticillium* even though they could not be separated by numerical phenetic tests. The latter two genera shared a limited number of antigens and were not clearly distinguished serologically, but the authors concluded that the serological analysis supported the phenetic classification and provided further evidence that many species in the genus Streptomyces should be combined.

Detection and Identification

To detect plants infected with bacterial wilt caused by Pseudomonas solanacearum, Morton et al (28) sampled infected stem pieces using two methods: a) stem pieces were placed in physiological saline for several hours and the exudates collected, or b) stem pieces were squeezed with pliers to express sap into physiological saline. Samples were sonicated for 1 min before adding to peripheral wells of an ODD plate. Positive precipitin lines usually developed within 2 days. There was complete agreement between results from ODD and microagglutination tests with samples extracted from tomato, tobacco, pepper, and eggplant, but the latter test was completed in less than 20 min in contrast to 2 days for the ODD. Morton et al (28) concluded that both techniques were useful for identification of plant infections and suggested that they would be useful for detecting the bacterium in other habitats such as soil or wild plants.

Sinha (38) reported the first use of ODD to analyze the antigens of nonculturable plant mycoplasmas. Lengthy purification steps were required to obtain mycoplasmalike organisms (MLOs) from plants to use as immunogens, but once the antiserum was obtained, partially purified preparations were satisfactory for analysis of antigenic determinants of MLOs. Partially purified preparations were less time consuming to obtain, and they contained about 90% more MLO than the purified preparation. The minimum concentration of detectable MLO was 20 mg/ml; increasing the concentration to 160 mg/ml did not affect the number of bands produced.

A means of diagnosing the causal agent of phony disease of peach was also provided by ODD tests. Although rickettsia-like bacteria (RLB) were isolated and cultured from diseased grapevines, almonds, plum, and alfalfa, all such attempts at isolation and culture failed with the putative RLB associated with phony disease of peach. Nevertheless, precipitin bands formed between sonicated preparations of tissues from peach trees with phony disease and antiserum to RLB, suggesting that an agent was present in the diseased peach tissue which was antigenically related to the RLB causing Pierce's disease and almond leaf scorch (15).

Cappaert et al (11) utilized 21 different antisera in ODD tests to serotype 3,556 isolates of Erwinia carotovora subsp. carotovora obtained from irrigation water in Oregon and Colorado to determine whether contaminated irrigation water was a potential source of inoculum of soft rot Erwinias for aerial stem rot of field grown potatoes. In Oregon, seven different serogroups were detected, first in irrigation water during the season, then on leaflets, and later in diseased stems, indicating that contaminated water sources were a potential source of inoculum for aerial stem rot. Similarly, ODD was utilized to determine the relative importance of isolates from soil and seed tubers to serve as inoculum for infection of field-grown potatoes (35). Interestingly, some of the same serogroups of E. carotovora, found in North America, were associated with water, soil, tuber, and stems of potato plants in Western Australia (32), indicating that these bacteria from widely separated geographic regions share common epitopes. However, these authors also reported that significant proportions of isolates from all environments belonged to none of the known serogroups as was also the case in the North American studies.

Dickey et al (21) investigated the use of ODD for the diagnosis and identification of strains of E. chrysanthemi isolated from corn and other hosts. Five serovars were identified. They could find no definite relationship between formation of secondary precipitin bands and the original host, geographical location, and phenotypic characteristics of the strains. Their results also emphasized the variability in the antigenic properties of the strains tested. For example, antigenic heterogeniety was common among strains of E. chrysanthemi isolated from a specific host and among strains from different hosts.

To determine the reliability of serological detection of C. sepedonicum in infected plant parts, Slack et al (41) obtained a specific polyclonal antiserum against glutaraldehyde-fixed cells of C. sepedonicum and evaluated its sensitivity and specificity in agglutination, ODD, and indirect fluorescent antibody stain tests. The sensitivity was reportedly much higher with immunofluorescence, 10^1 to 10^2 cells/ml, than with agglutination and ODD whose sensitivities were about 2 X 10^7 cells/ml. Cross-reactivity occurred with agglutination but not with ODD or immunofluorescence. The authors concluded that the confidence level in diagnoses of ring rot could be increased significantly by serological assays, particularly immunofluorescence because of its greater sensitivity. However, they pointed out that the higher level of sensitivity of immunofluorescence may not be required for confirmation of ring rot because the ODD test was sensitive enough to detect the bacterium in visibly infected plant parts, probably because the bacterium was present in high populations.

In contrast to the specificity reported by Slack et al, other reports show that antisera to C. sepedonicum cross-reacted in ODD tests with other Corynebacterium species and with a nonpathogenic coryneform bacterium isolated from healthy potato stems (14). When each cross-reacting bacterium was cross-absorbed individually against the antiserum, the antiserum titers to C. sepedonicum were reduced significantly. Since both research groups used polyclonal antisera, the contradiction in specificity may have been due to factors such as the method of preparation of the immunogen or a differential response by the rabbits to the immunizing antigens. or because different strains were used to test antisera specificity. Obviously, cross-reaction limits the usefulness of the serological technique for detecting the presence of C. sepedonicum in potato tissue.

To avoid the problems of cross-reactivity associated with polyclonal antisera, De Boer et al (16,18) selected a monoclonal antibody to a soluble, extracellular antigen from C. sepedonicum that was extracted from the culture filtrate. This monoclonal antibody was highly specific to the extracellular antigen in ODD, and it produced a precipitin band that fused completely with a precipitin band formed by the reaction of the same antigen against the reference polyclonal antiserum that had been developed earlier.

Definitive applications of ODD in our laboratory, using antiserum to 50 S ribosomal subunits, are outlined below.

Ribosomal subunits (50 S) were isolated from five Agrobacterium strains and injected into rabbits. When the harvested antisera were reacted with purified ribosomes from 34 Agrobacterium and 4 Rhizobium strains in ODD, all members of the Rhizobiaceae formed a single precipitin band of identity. Although these strains were of diverse origin and taxonomic affiliation, their 50 S ribosomal subunits were serologically identical, suggesting a high degree of conservation among ribosomes from members of the Rhizobiaceae. In an ecological study, these antisera proved useful to confirm the identity of putative agrobacteria isolated from a natural oak savanna and tallgrass prairie which had never been disturbed agriculturally (4).

When crude ribosome preparations, in contrast to purified, were tested against these antisera, heterogenous precipitation patterns with multiple precipitin lines were produced. Fifteen serovars were tentatively identified. However, further attempts to designate serovars were discontinued because the ODD precipitin patterns were not always reproducible. It was subsequently shown that a somatic antigen contaminant was usually associated with the unwashed ribosomes (7). Interestingly, the somatic antigenic contaminant was responsible for a strain-specific precipitin band observed in ODD tests (5). This strain-specificity was exploited successfully to monitor the survival of *A. radiobacter* K84, biological control agent of crown gall, under field conditions. Colonies of K84 could be identified *directly* on the selective medium onto which the sample had been initially streaked, using the technique described in **Section III.C.2.d.**

Subsequent research demonstrated that the antigenic ribosomal contaminant responsible for inducing the strain-specific antiserum could be extracted from whole cells with phenol-water (8). When this phenol-water extract was placed in an ODD well adjacent to a well with purified lipopolysaccharide (LPS) from *Agrobacterium*, a single precipitin line of identity formed between the wells.

The antigenic specificity of phenol-water extracts of whole bacterial cells observed here with *Agrobacterium* has also been reported for members of the genus *Corynebacterium* (16), *Pseudomonas* (20,26), and *Erwinia* (17,21,35). With the exception of the corynebacteria, it is probable that a LPS moiety (10), representing a unique constituent in each of these taxons, is extracted in the phenol-water treatment. The LPS constituents appear to be good antigens for typing bacteria by ODD.

In summary, ODD has been used for detection, identification, and serotyping of members of the phytopathogenic genera *Agrobacterium*, *Corynebacterium* (*Clavibacter*), *Erwinia*, *Pseudomonas*, *Xanthomonas*, *Streptomyces*, and of the mycoplasma-like and fastidious procaryotes, thus illustrating the utility of gel immunodiffusion.

Literature Cited

1. Alvarez, A. M., Benedict, A. A., and Mizumoto, C. Y. 1985. Identification of Xanthomonads and grouping of strains of *Xanthomonas campestris* pv. *campestris* with monoclonal antibodies. Phytopathology 75:722-728.
2. Azad, H., and Schaad, N. W. 1988. Serological relationships among membrane proteins of strains of *Xanthomonas campestris* pv. *transluscens*. Phytopathology 78:272-277.
3. Bouzar, H. 1982. A survey of *Agrobacterium* strains associated with Georgia pecan trees and an immunological study of the bacterium. M.S. thesis, Oregon State University, Corvallis. 67 pp.
4. Bouzar, H., and Moore, L. W. 1987. Isolation of different *Agrobacterium* biovars from a natural oak savanna and tallgrass prairie. Appl. Environ. Microbiol. 53:717-721.
5. Bouzar, H., and Moore, L. W. 1987. Complementary methodologies to identify specific *Agrobacterium* strains. Appl. Environ. Microbiol. 53:2660-2665.
6. Bouzar, H., Moore, L. W., and Schaad, N. W. 1986. Serological relationship between 50 S ribosomal subunits from strains of *Agrobacterium* and *Rhizobium*. Phytopathology 76:1265-1269.
7. Bouzar, H., Moore, L. W., and Schaup, H. W. 1986. Importance of ribosome purity in ribosomal serology. Phytopathology 76:1323-1325.
8. Bouzar, H., Moore, L. W., and Schaup, H. W. 1988. Lipopolysaccharide from *Agrobacterium tumefaciens* B6 induces the production of strain-specific antibodies. Phytopathology 78:1237-1241.

9. Bowman, T., and Weinhold, A. R. 1963. Serological relationships of the potato scab organism and other species of _Streptomyces_. Nature 200:599-600.
10. Bradshaw-Rouse, J. J., Sequeira, L., Kelman, A., and Dickey, R. S. 1988. Partial characterization and serological specificity of the lipopolysaccharide of _Erwinia chrysanthemi_. Phytopathology 78:996-999.
11. Cappaert, M., Powelson, M. L., Franc, G. D., and Harrison, M. D. 1989. Irrigation water as a source of inoculum of soft rot Erwinias for aerial stem rot of potatoes. Phytopathology 78:1668-1672.
12. Chaparas, S. D., Lind, A., Ouchterlony, O., and Ridell, M. 1983. Terminology guidelines for serotaxonomic studies using immunodiffusion and immunoelectrophoresis. Int. J. Syst. Bacteriol. 33:414-416.
13. Clausen, J. 1981. Immunochemical techniques for the identification and estimation of macromolecules, 2nd edition. Elsevier/North-Holland Biomedical Press, Amsterdam, 387 pp.
14. Crowley, C. F., and De Boer, S. H. 1982. Nonpathogenic bacteria associated with healthy potato stems cross-react with _Corynebacterium sepedonicum_ antisera in immunofluorescence. Am. Pot. J. 59:1-8
15. Davis, M. J., Stassi, D. L., French, W. J., and Thomson, S. V. 1978. Antigenic relationship of several rickettsia-like bacteria involved in plant diseases. Proc. 4th Int. Conf. Plant Path. Bact. 1:311-315.
16. De Boer, S. H. 1982. Cross reaction of _Corynebacterium sepedonicum_ antisera with _C. insidiosum_, _C. michiganense_, and an unidentified coryneform bacterium. Phytopathology 72:1474-1478.
17. De Boer, S. H., and Wieczorek, A. 1984. Production of monoclonal antibodies to _Corynebacterium sepedonicum_. Phytopathology 74:1431-1434.
18. De Boer, S. H., Copeman, R. J., and Vruggink, H. 1979. Serogroups of _Erwinia carotovora_ potato strains determined with diffusible somatic antigens. Phytopathology 69:316-319.
19. De Boer, S. H., Wieczorek, A., and Kummer, A. 1988. An ELISA test for bacterial ring rot of potato with a new monoclonal antibody. Plant Dis. 72:874-878.
20. DeWeger, L., Jann, B., Jann, K., and Lugtenberg, B. 1987. Lipopolysaccharides of _Pseudomonas_ spp. that stimulate plant growth: composition and use for strain identification. J. Bacteriol. 169:1441-1446.
21. Dickey, R. S., Claflin, L. E., and Zumoff, C. H. 1987. _Erwinia chrysanthemi_: serological comparisons of strains from Zea mays and other hosts. Phytopathology 77:426-430.
22. Elek, S. D. 1948. The recognition of toxicogenic bacterial strains _in vitro_. Br. Med. J. 1:493.
23. Kwapinski, J. B. G. 1970. Serological taxonomy and relationships of Actinomycetales. Pages 345-369 in: The Actinomycetales. H. Prauser, ed. Jena Gustav Fischer Verlag. 439 pp.
24. Leinonen, M. 1985. Serological methods for the study of bacterial surface antigens. Pages 184-185 in: Enterobacterial Surface Antigens. T. K. Korhonen et al, eds. Elsevier, Amsterdam. 364 pp.
25. Lucas, L. T., and Grogan, R. G. 1969a. Serological variation and identification of _Pseudomonas lachrymans_ and other phytopathogenic _Pseudomonas_ nomenspecies. Phytopathology 59:1908-1912.
26. Lucas, L. T., and Grogan, R., G. 1969b. Some properties of specific antigens of _Pseudomonas lachrymans_ and other _Pseudomonas_ nomenspecies. Phytopathology 59:1913-1917.
27. Manzer, F. E., McIntyre, G. A., and Merriam, D. C. 1977. A new potato scab problem in Maine. Univ. of Maine at Orono, LSA Exp. Sta. Tech. Bull. 85.

28. Morton, D. J., Dukes, P. D., and Jenkins, S. F., Jr. 1965. Serological identification of Pseudomonas solanacearum in four solanaceous hosts. Phytopathology 55:1191-1193.
29. Munoz, J. 1971. Double diffusion in plates. Pages 146-160 in: Methods in Immunology and Immunochemistry, Vol. 3. C. A. Williams and M. W. Chase, eds. Academic Press, New York. 515 pp.
30. Ouchterlony, O. 1948. Antigen-antibody reactions in gels. Ark. Kem. Min. Geol. 26B:1-9.
31. Oudin, J. 1980. Immunochemical analysis by antigen-antibody precipitation in gels. Pages 166-197 in: Methods in Enzymology, Vol. 70. H. Van Vunakis and J. J. Langone, eds. Academic Press, New York. 525 pp.
32. Peltzer, S., and Sivasithamparam, K. 1988. Sero-groups of Erwinia carotovora associated with water, soil, tuber, and stems of potato plants in Western Australia. New Zealand J. Exp. Agricul. 16:265-270.
33. Perombelon, M. C. M., and Kelman, A. 1980. Ecology of the soft rot Erwinias. Annu. Rev. Phytopathol. 18:361-387.
34. Piazzi, S. E. 1969. A simple method for preliminary immunodiffusion test of antigen-antibody systems having unknown ratios of reaction. Anal. Biochem. 27:281-284.
35. Powelson, M. L., and Apple, J. D. 1984. Soil and seed tubers as sources of inoculum of Erwinia carotovora pv. carotovora for stem soft rot of potatoes. Phytopathology 74:429-432.
36. Ridell, M., and Williams, S. T. 1983. Serotaxonomical analyses of some Streptomyces and related organisms. J. Gen. Microbiol. 129:2857-2861.
37. Schaad, N. W. 1979. Serological identification of plant pathogenic bacteria. Annu. Rev. Phytopathology 17:123-147.
38. Sinha, R. C. 1979. Purification and serology of mycoplasma-like organisms from aster yellows-infected plants. Can. J. Plant Pathol. 1:65-70.
39. Sinha, R. C., and Benhamou, N. 1983. Detection of mycoplasmalike organism antigens from aster yellows-diseased plants by two serological procedures. Phytopathology 73:1199-1202.
40. Sinha, R. C., and Chiykowski, L. N. 1984. Purification and serological detection of mycoplasmalike organisms from plants affected by peach eastern X-disease. Can. J. Plant Pathol. 6:200-205.
41. Slack, S. A., Kelman, A., and Perry, J. B. 1978. Comparison of three serodiagnostic assays for detection of Corynebacterium sepedonicum. Phytopathology 69:186-189.
42. Thaveechai, N., and Schaad, N. W. 1984. Comparison of different immunogen preparations for serological identification of Xanthomonas campestris pv. campestris. Phytopathology 74:1065-1070.
43. Vincent, J. M. 1970. A Manual for the Practical Study of Root-nodule Bacteria. IBP Handbook No. 15. Blackwell Scientific Publications, Oxford. 164 pp.

Section III.C.2.d.

DIRECT IMMUNODIFFUSION FOR BACTERIAL COLONIES ON AGAR PLATES

J. W. L. van Vuurde and Ph. M. de Vries
Research Institute for Plant Protection
Wageningen, The Netherlands

Principle

Direct immunodiffusion (2) was developed to serologically identify bacterial colonies directly in the original isolation plates. Alternatively, the method can be used as a faster and less laborious modification of the Ouchterlony double diffusion procedure for routine identification of bacterial strains in pure culture and colonies on dilution-plates.

Materials

For materials see **Section III.C.2.a.**

Procedure

1. Prepare dilution series of the test sample on a semi-selective or differential agar medium for the target bacterium (1). Prepare similar dilution series from known infected samples as controls.
2. Incubate as short as possible under appropriate conditions to enable recognition of the colony type of the target bacterium.
3. Cut wells 7 to 15 mm away from a suspected colony or in a suitable area of the plate to test various colonies simultaneously (e.g., in case of poor visible colony differentiation or search for cross-reacting microorganisms).
4. Fill the wells with antiserum at proper dilution as described for agar double diffusion.
5. Incubate as in step 2 until the precipitin bands in the positive control are visible at similar relative distances from the wells as in the test plates. If increasing colony size causes problems, incubation should be continued at a lower temperature.
6. Read plates with darkfield illumination for precipitin bands near colonies that are similar in appearance to the target bacterium.
7. Isolate bacteria from serologically positive colonies for further confirmation and/or pathogenicity testing if needed. Isolation from other serologically positive colonies can be used to identify cross-reacting organisms or non-typical colonies of the target organism.

Comments

1. Direct immunodiffusion also can be used as a rapid agar double-diffusion procedure by preparing wells in nutrient agar plates suitable for growth of the target bacterium, (i.e., not in water agar). For this purpose, fill the central well with diluted antiserum and inoculate the bottom inner side of a

peripheral well with a thick smear from the colony to be tested. Inoculate an adjacent well in a similar way with a known culture of the target bacterium. Incubate for one or two days at optimal temperature for the target bacterium and read the plate as in agar double diffusion. Cultures of serologically positive bacteria can be isolated from positive wells.

2. Serological detection of colonies of the target bacterium directly in a nutrient agar is possible by mixing the sample and antiserum with the medium at 45 C (2), followed by incubation. In a positive serological reaction, a precipitation halo is formed around the colony. In experiments with *Erwinia carotovora* subsp. *atroseptica*, a positive reaction was related to the size of the colony; small colonies did not react. Because large quantities (1 to 2 ml per 6 cm plate) of sterile undiluted antisera are required for this test, immunofluorescence colony staining (see **Section IV.G.3.**) is recommended for direct immuno-colony identification in agar.
3. The composition of the growth medium may affect the production of soluble antigens involved in direct immunodiffusion. This should be investigated for each pathogen/growth medium combination.

Applications

The direct immunodiffusion method proved useful for identification of *Xanthomonas campestris* pv. *begoniae*, (Fig. 1), *Erwinia chrysanthemi*, and *E. carotovora* subsp. *atroseptica* on dilution-plates among saprophytic bacteria (2).

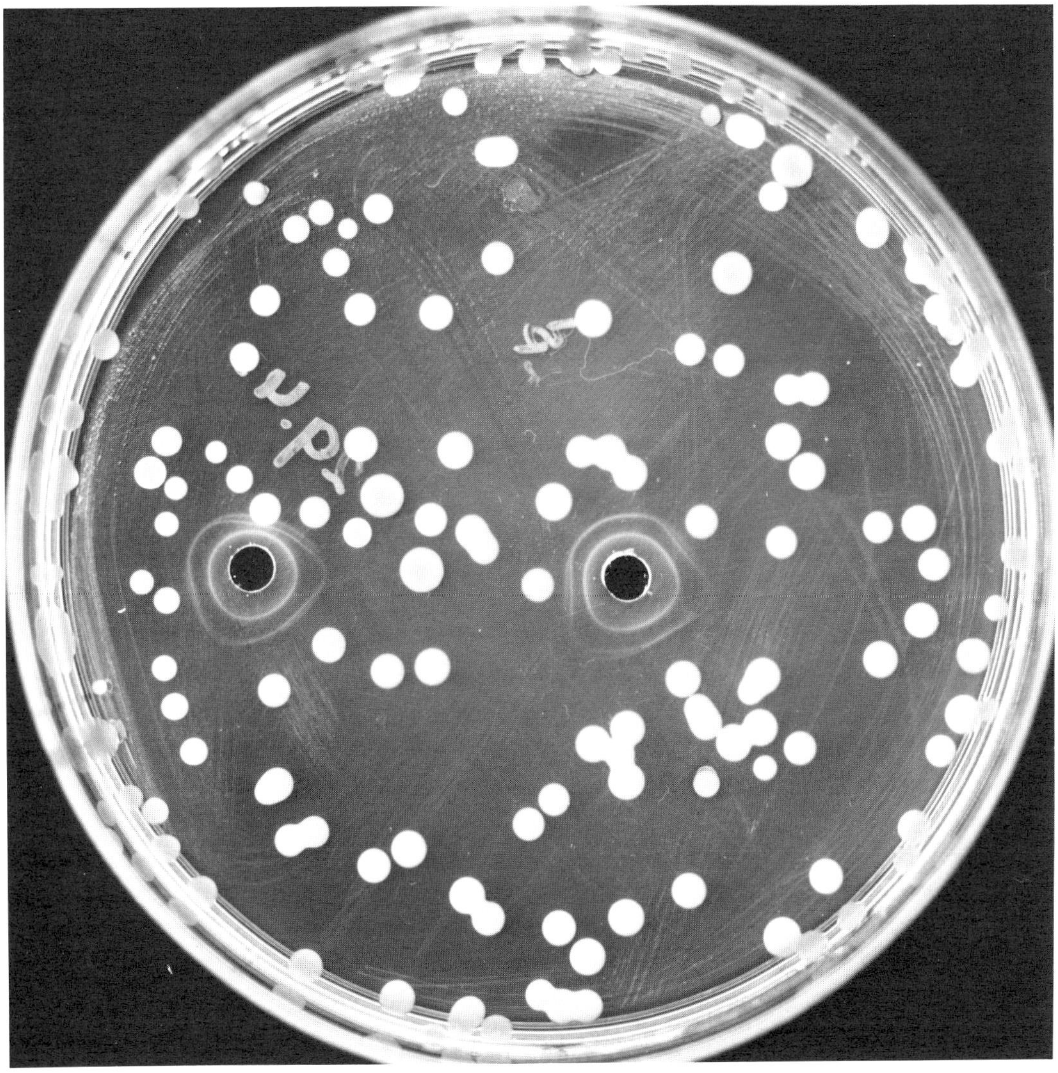

Fig. 1. Direct immunodiffusion for colonies of *Xanthomonas campestris* pv. *begoniae* after plating the soaking solution of a naturally infected begonia leaf on Tripticase soy agar. The antiserum was added undiluted to fresh cut wells at 5 to 1.5 mm distance from the colonies to be tested. Precipitin bands were visible after incubation of the plate for 24 hr at room temperature.

Literature Cited

1. Fahy, P. C., and Hayward, A. C. 1983. Media and methods for isolation and diagnostic tests. Pages 337-378 in: Plant Bacterial Diseases: A Diagnostic Guide. P. C. Fahy and G. J. Persley, eds. Academic Press, Sidney. 393 pp.
2. van Vuurde, J. W. L. 1987. New approach in detecting phytopathogenic bacteria by combined immunoisolation and immunoidentification assays. EPPO Bulletin 17:139-148.

Section III.C.2.e.

INTRAGEL SPECIFIC ABSORPTION

E. M. Ball
University of Nebraska
Lincoln, NE 68583

Principle

Intragel techniques were first described by Feinberg (3) and applied to plant viruses in studies with TMV by van Regenmortel (11). Cross-absorption of antigens aids in establishing relationships among viral strains. Removal of similar antigenic groups by precipitation and analyses of remaining antigenic determinants indicates degrees of serological relationships. In intragel specific absorption procedures, the absorption is performed in the agar at the same time the relationships between antigens are determined.

Materials

Materials required for intragel specific absorption are the same as those for agar double diffusion (see **Section III.C.2.a.**).

Procedure

1. Add antigen to the central well of matrix and allow to diffuse into the agar, e.g., as illustrated in Agar Double Diffusion (see Figure 1-E, **Section III.C.2.a.**). For diffusion time, see Comments.
2. Place antiserum in the same well to which antigen has been added, and add antigens to be compared in peripheral wells.

Comments

1. Time allotted for diffusion of the absorbing antigen, before adding antiserum, is determined by its characteristics. Hibiscus chlorotic ringspot virus was allowed to diffuse "several" hours before antiserum was added (5). Kalanchoe latent virus was degraded with 0.1% SDS, then added to the central well 4 hr before adding serum (4).
2. Upon diffusion, antibodies specific for determinant groups of the absorbing antigen precipitate, forming a ring around the central well. Unabsorbed antibodies pass through the immune barrier and diffuse into the agar where they form precipitates with reactive antigens from peripheral wells.
3. As with other serological tests, optimum proportions and conditions must be established. The precipitin barrier around the central well, formed by the absorbing antigen, can be overridden by adding excess antiserum. Preliminary tests must therefore be performed to determine the concentration of antigen required to precipitate all specific antibody, i.e., leaving none to produce precipitin bands with the same antigen from peripheral wells.

Applications

Intragel tests are applied frequently in determining relationships of viral strains. The relationship between TMV type strain and an isolate causing mosaic disease of Rhoeo discolor was determined in agar diffusion assays (9). Antiserum to TMV type strain intragel-absorbed by the Rhoeo isolate produced a precipitin band with type-strain antigen but not with Rhoeo isolate antigen. Reciprocal absorption (i.e., Rhoeo strain antiserum by type strain antigen) verified the serological distinction between the TMV isolates, establishing them as separate serotypes. A new isolate of Andean potato mottle virus from Brazil was compared to two known strains of the virus from the CIP (International Potato Center, Peru) collection (2). Precipitin band spurs resulted when the isolates were compared by agar double diffusion. Intragel absorption between antiserum to the new isolate and homologous antigen left fractional amounts of antibody reactive to the CIP-C strain of APMV. Two strains of carnation ringspot virus, R and N, were studied because of their unique property of reversible aggregation with respect to temperature and pH. In intragel tests, R antigen was added to the central well containing N antiserum and N and R antigens placed in adjacent wells. A precipitation band formed between N serum and N antigen but not with R antigen (10). All of these studies demonstrated relatedness but not identity among strains. The conclusions were reinforced by the intragel technique.

Two new serotypes of cowpea severe mosaic virus, III and IV were established by agar diffusion and the relationship reinforced by intragel tests. A common determinant - (A) was shared by all strains studied. An additional determinant - (B) was shared by serogroups I, II and IV and determinant - (C) by I and III. In addition, each serotype had specific antigenic determinants - D, E, F, G for serogroups I, II, III, and IV, respectively (7). Seven unique antigenic determinants were demonstrated in agar diffusion and intragel studies with 3 serogroups of sowbane mosaic virus (12). In a study of six strains of red clover necrotic mosaic virus, an English and an Australian isolate formed a new serogroup based on agar diffusion tests (8). Intragel studies demonstrated specific antigens for each serotype.

A mutant of cucumber mosaic virus, isolated by single lesion transfer, formed large lesions of Chenopodium quinoa and became systemic whereas the parent was not. Agar diffusion serology demonstrated that the lesion was caused by CMV. In intragel tests, 0, 10, 15, 20, 25 μg/ml of purified large-lesion virus were placed in peripheral wells. After 18 hr, the wells were refilled with 30 μl of CMV antiserum and the center well filled with extract from infected leaves. No zones formed, confirming the identity of CMV (6).

Another use of intragel specific absorption was demonstrated in a study of 10 isolates of cucumber mosaic virus and two of peanut stunt virus. Twenty-four hours prior to establishing relationships among the strains by agar diffusion, healthy plant sap was added to the central serum well to remove plant material that cause a non-specific precipitin band to form in the main test (1).

Literature Cited

1. Ahmad, I. B., and Scott, H. A. 1985. Identification and serotyping of cucumber mosaic and peanut stunt viruses from Arkansas. Plant Dis. 69:891-893.

2. Avila, A. C., Salazar, L. F., Ortega, M., and Daniels, J. 1984. A new strain of Andean potato mottle virus from Brazil. Plant Dis. 68:997-998.

3. Feinberg, J. G. 1957. Identification, discrimination and quantification in Ouchterlony gel plates. Int. Arch. Allergy Appl. Immunol. 11:129-152.

4. Hearon, S. S. 1984. Comparison of two strains of Kalanchoe latent virus, carlavirus group. Phytopathology 74:670-676.
5. Hurt, S. S. 1987. Detection and comparison of electrophorotypes of hibiscus chlorotic ring spot virus. Phytopathology 77:845-850.
6. Lakshman, D. K., Gonsalves, D., and Fulton, R. W. 1985. Role of _Vigna_ species in the appearance of pathogenic variants of cucumber mosaic virus. Phytopathology 75:751-757.
7. Liu, M. T., Hill, J. H., Kitajima, E. W., and Cosat, C. L. 1984. Two new serotypes of cowpea severe mosaic virus. Phytopathology 74:581-585.
8. Rao, A. L. N., Lakshman, D. K., Ohki, S. T., and Hiruki, C. 1987. Identification of a new serotype and antigenic relationships among six strains of red clover necrotic mosaic virus. Phytopathology 77:995-999.
9. Thompson, S. M., and Corbett, M. K. 1985. Mosaic disease of _Rhoeo discolor_ caused by a strain of tobacco mosaic virus. Plant Dis. 69:356-359.
10. Tremaine, J. H., Roland, W. P., and McGavley, E. M. 1984. Temperature-reversible aggregation of two strains of carnation ringspot virus. Phytopathology 74:161-165.
11. van Regenmortel, M. H. V. 1967. Serological studies on naturally occurring strains and chemically induced mutants of (TMV). Virology 31:467-480.
12. Zebzami, M., Lockhart, E. L., and Hill, J. H. 1987. Identification of three serotypes of sowbane mosaic virus. Phytopathology 77:571-574.

Section III.D.

RING INTERFACE TEST

E. M. Ball
University of Nebraska
Lincoln, Nebraska 68583

Introduction

Optimal concentrations of antigen and antibody are less important in ring interface tests than in microprecipitin tests since precipitate forms in the tubes where optimal proportions occur. Antiserum must be diluted to 1:10 or greater to give a relatively clear solution; antigen must be clean, free of insoluble particles, and diluted so it also gives a relatively clear solution.

Materials

1. Small glass tubes, 0.5 x 5 cm or 0.7 x 7.0 cm, or microtiter plates for making dilutions.
2. Pipets: Pasteur, hand-drawn, or long-tip 1 ml pipets.
3. Buffer: most suitable (i.e. phosphate - saline, Tris, etc.)
4. Glycerin.

Procedure

1. Make appropriate dilutions of antigen in small tubes (i.e., 10 x 75 mm) or in a microtiter plate. (See **Section III.E.**).
2. Make an appropriate dilution of antiserum in buffer. If dilution is greater than 1:10, glycerin must be added to buffer to give 10% concentration.
3. Place a small volume of serum or serum-glycerin in the bottom of each tube. The smaller size tubes listed above can contain 0.2 ml of serum. The top of the serum solution must be above the curvature of the bottom of the tube to avoid light reflections.
4. Holding the tube at approximately a 30° angle, gently add the antigen dilution letting it slide down the side of the tube and contact the serum/glycerin in an overlayer, not a mix (Fig. 1). Repeat overlayering of antigen dilutions onto antiserum/glycerin.
5. Set the tubes upright in a small rack, in styrofoam holder, or in plasticene.
6. Observe the tubes after 15-20 min. The precipitate at the interface of antigen and antibody is best seen in a dark room with the tube being illuminated by a slit light source.

Comments

1. Antisera of high titer can be diluted for use; however, the density of the solution must be maintained (i.e., glycerin may be added) to a concentration of 10-20% to support the weight of the antigen and to prevent mixing when

the antigen (or antibody) is added as the overlayer (i.e., by use of 10-20% glycerin).

2. Occasionally the glass tubes will have scratches on the outside that refract light. Coating the tube with glycerin can help eliminate the problem. Wipe off excess glycerin with lens tissue or Kimwipes.
3. The amount of precipitate forming in a given time can be recorded on a scale of 4 to 0 (Fig. 2). The time of appearance of the visible ring can be used also as a test reading. When time is plotted against concentration on a semilog scale, a linear relationship exists within certain ranges of concentration of antigen and is directly related to antibody concentration.
4. This test is sensitive and relatively inexpensive. Results can be obtained in a short period of time, 1 hr. Once the overlayer technique is learned, the test can be set up easily. Specific antigen can be identified throughout purification procedures when antiserum is the underlayer. However, the method may also be applied to assays of antisera using partially purified antigen in glycerin-buffer as the underlayer. The test is more sensitive than microprecipitin techniques in detecting low concentrations of antibody. Recently, ring interface assays of antiserum have been applied to studies of soybean chlorotic mottle virus (7), a carlavirus from eggplant (8), rice ragged stunt virus (1), rice grassy stunt virus (2,3), sweet clover necrotic mosaic virus (6), Indian peanut clump virus (9), and asparagus virus II (11). Rice tungro disease was shown to be a composite of a bacilliform virus and a spherical virus, the latter being required for insect transmission by leaf hoppers (4,5). In a study of ilarviruses, ring interface and agar double diffusion tests were used with eleven sera to establish interrelationships that resulted in a revision of subgroup A (12). Two new viruses isolated from hops were shown to be serotypes of Prunus necrotic ringspot virus using ring interface, agar double diffusion and ELISA procedures (10).

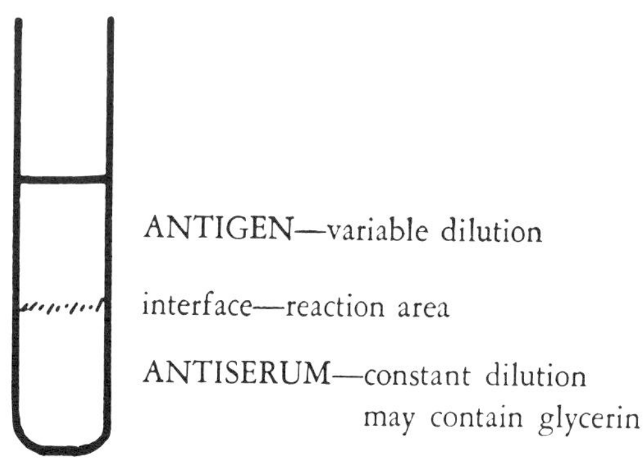

Fig. 1. A drawing illustrating the distribution of reactants for the ring interface test. Depending on the size of the tubes used, as little as 0.1 ml of each reactant can be tested.

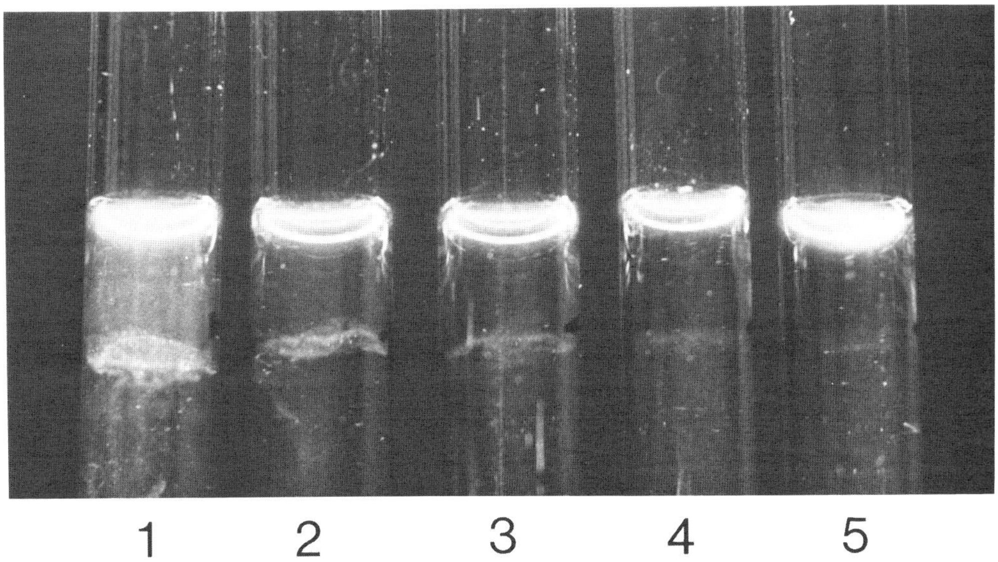

Fig. 2. Photograph of precipitates of tobacco mosaic virus and anti-TMV antibodies produced at the antigen-antiserum interface. The antiserum dilution was constant at 1:40 in 0.01 M neutral phosphate containing 10% glycerin. Type strain antigen concentration in µg/ml: a., 1000; b., 250; c., 62.5; d., 15.6; e., 3.9; Reading 1 hr: numerical scores of 4, 4, 3, 2, and trace, respectively. Pellicle precipitate of antigen at 1 mg/ml formed below original meniscus, was very dense and fell out of solution upon additional incubation.

Literature Cited

1. Hibino, H., and Kimura, I. 1982. Detection of rice ragged stunt virus in insect vectors by enzyme-linked immunosorbent assay. Phytopathology 72:656-695.
2. Hibino, H., Cabavatan, P. Q., Omura, T., and Tsuchizaki, T. 1985. Rue grassy stunt virus strain causing tungrolike symptoms in the Philippines. Plant Dis. 69:538-541.
3. Hibino, H., Usugi, T., Omura, T., Tsuchizaki, T., and Shohava, K. 1985. Rice grassy stunt virus: a plant hopper-borne circular filament. Phytopathology 75:894-899.
4. Hibino, H., and Cabavatan, P. Q. 1987. Infectivity neutralization of rice tungro-associated viruses acquired by vector leaf hoppers. Phytopathology 77:473-476.
5. Hibino, H., Tionglo, E. R., Cabunagan, R. C., and Flores, Z. M. 1987. Resistance to rice tungro-associated viruses in rice under experimental and natural conditions. Phytopathology 77:871-875.

6. Hiruki, C., Rao, D. V., Chem, M. H., Okuno, T., and Figueriredo, G. 1984. Characterization of sweet clover necrotic mosaic virus. Phytopathology 74:482-486.
7. Iwaki, M., Isogawa, Y., Tsuzuki, H., and Honda, Y. 1984. Soybean chlorotic mottle, a new caulimovirus on soybean. Plant Dis. 68:1009-1011.
8. Khalil, J. A., Nelson, M. R., and Wheeler, R. E. 1982. Host range, purification, serology, and properties of a carlavirus from eggplant. Phytopathology 72:1064-1068.
9. Nolt, B. L., Rajeshwari, R., Reddy, D. V. R., Bharatham, N., and Manohar, S. K. 1988. Indian peanut clump virus isolates: host range, symptomology, serological relationships, and some physical properties. Phytopathology 78:310-313.
10. Smith, D. R., and Skotland, C. B. 1986. Host range and serology of prunus necrotic ringspot virus serotypes isolated from hops (Humulus lupulus) in Washington. Plant Dis. 70:1019-1023.
11. Uyeda, I., and Mink, G. I. 1981. Properties of asparagus virus II, a new member of the Ilarvirus Group. Phytopathology 71:1264-1269.
12. Uyeda, I., and Mink, G. I. 1983. Relationships among some ilarviruses: Proposed revision of subgroup A. Phytopathology 73:47-50.

Section III.E.

MICROPRECIPITATION (VIRUS) AND MICRO-AGGLUTINATION (BACTERIA)

E. M. Ball
University of Nebraska
Lincoln, Nebraska 68583

Principle

The simplest application of serology to diagnosis of plant pathogens is the chloroplast agglutination test wherein a drop of plant extract - leaf, stem, flower, fruit - is placed on a slide or in a dish and a drop of antiserum is added. Within 10-20 min, agglutination of chloroplasts to which virions are adsorbed is observable by low magnification of a dissecting microscope. The test is rapid and inexpensive but requires higher antigen concentration in host tissue than more refined tests.

Sensitivity of precipitation tests for soluble antigens (viral) or agglutination tests for insoluble antigens (bacterial) is improved when adapted as tube tests (1,10,11). Crude plant extract is clarified by physical or chemical means to remove spurious precipitates. Partially purified antigens react well in the tube precipitation or tube agglutination tests. Antigens and antibodies are mixed in 0.5-1.0 ml volumes in tubes. During incubation, they form a complex which is soluble at first but which increases in size and gradually settles out of solution. The ratios at which antigen and antibody react are critical since an excess of one reactant compared to another may produce soluble complexes and thus "false-negative" tests.

The disadvantage of the tube test is the requirement of large volumes of reactants. The microprecipitin or micro-agglutination test provides a procedure that conserves reactants. The test is equally suitable to viral precipitins and bacterial agglutinins as well as to whole serum or immunoglobulin G (IgG) antibody preparations. The technique provides a rapid and sensitive assay system to monitor antibody production throughout the immunization schedule and subsequently to detect antigen in plant extracts. Critical ratios of reactants apply to this test as to the tube test.

Reviews of various modifications of the technique are available in Matthews (13) and Van Regenmortel (19).

Materials

1. Plastic Petri dishes: 100 x 13-15 mm or tight-seal 50 x 9 mm (Falcon) dish.
2. Buffer: 0.01 M neutral phosphate, PBS (phosphate + 0.85% NaCl), 0.01-0.05 M Tris (hydroxymethyl) aminomethane, pH 7.0, TBS (Tris + 0.85% NaCl), or any appropriate buffer.
3. Small test tubes or a micro-titer plate for making dilutions.

Table 1. Illustration of the microprecipitation technique to determine the optimal reacting dilution of a viral antigen-antibody system and the application of that information to establishing relationships among strains of the virus. A. Readings, in a scale of 4, 3, 2, 1, tr (trace), and - (negative) of the precipitation reaction between type strain-tobacco mosaic virus and antiserum. [ ] = optimal reacting dilution, typically typically selected at highest corresponding dilutions of antigen and antiserum producing a scale-3 reaction. B. Readings of the precipitate in determination of the antiserum dilution endpoints of 6 strains of tobacco mosaic virus. Antigens at optimum proportions: 0.1 mg/ml of type strain and 0.25 mg/ml of Canary Island strain. Conditions: buffer = 0.01 M neutral phosphate + 0.85% NaCl, uniform drops of reactants delivered from a 27 gauge needle, reading after 2 hr incubation at room temperature and rechecked in 24 hr.

A.

Antigen		Antiserum Tube No. and Dilution (Reciprocal)							
		1	2	3	4	5	6	7	8
Tube No.	Dilution	32	64	128	256	512	1024	2048	-
1	2	4	4	4	4	2	1	tr	-
2	4	4	4	4	4	3	2	tr	-
3	8	4	4	4	4	[3]	2	tr	-
4	16	4	4	4	3	2	1	-	-
5	32	4	3	3	2	1	tr	-	-
6	64	4	3	2	1	tr	-	-	-
7	128	2	1	1	tr	-	-	-	-
8	-	-	-	-	-	-	-	-	-

B.

Antiserum Titration

Antiserum to	TMV Strain (Antigen)	Tube No.	Antiserum Tube No. and Dilution (Reciprocal) 1 32	2 64	3 128	4 256	5 512	6 1024	7 2048	Dilution endpoint (reciprocal)
Type Strain	Type	1	4	4	4	4	3	2	tr	1024
	South Carolina	2	4	4	4	3	1	-	-	512
	Canary Island	3	4	4	4	3	1	-	-	512
	Latent	4	4	4	4	4	2	1	-	1024
	Fenne	5	4	4	4	3	1	-	-	512
	#2	6	4	4	4	3	1	-	-	512
Canary Island Strain	Type	1	3	1	-	-	-	-	-	64
	South Carolina	2	4	4	4	4	3	1	-	1024
	Canary Island	3	4	4	4	4	3	1	-	1024
	Latent	4	4	4	4	4	2	1	-	1024
	Fenne	5	3	1	-	-	-	-	-	64
	#2	6	3	1	-	-	-	-	-	64

4. Pipets: 1 ml for diluting antigen or serum in tubes, 100 or 200 μl automatic pipet, any instrument suitable for making dilutions in volumes of 200 μl or less. Micropipets that will deliver microliter drops of dilutions to assay plate.
5. Wax pencil.

Procedure

1. With wax pencil draw a square pattern of eight rows and eight columns on the inside, bottom of the petri dish. The size of the squares, e.g. 5-8 mm, can be determined by the size of the drops (30-60/ml) to be used in the test. Rows and columns should be numbered, for accurate recording of results (Table 1-A).
2. For a grid titration of antigen and antiserum, make an appropriate dilution series of each reactant in buffer.
   a. Arrange 2 rows of 7 or 8 small test tubes, one row for antigen, one for antiserum.
   b. Add 0.2 ml of buffer to all tubes in each row.
   c. Prepare antigen stock solution containing 1 mg/ml of purified virus.
   d. To the first tube in the antigen dilution row add 0.2 ml of stock antigen. Using a 1 ml pipet, mix and remove 0.2 ml to the second tube in the row, mix gently without making bubbles, remove 0.2 ml to the third tube, mix and continue transferring 0.2 ml until the last tube of the series of diluted antigen. This tube will contain 0.4 ml after mixing. The completed series contains antigen dilutions of 1/2 to 1/128, relative to stock antigen.
   e. Prepare a stock solution of antiserum. To a small tube, add 1.5 ml of buffer containing 0.025% sodium azide as a preservative. Mix 0.1 ml of undiluted antiserum in the buffer, providing a 1:16 dilution of antiserum (stock).
   f. Prepare antiserum dilution series. Add 0.2 ml of stock-antiserum to the first tube of the second row (already containing 0.2 ml buffer), mix gently. Remove 0.2 ml and mix with the buffer in the second tube, etc. to complete the 7-tube series. The last tube will have 0.4 ml. The completed series contains antiserum dilutions of 1/32 to 1/2048.

3. Add reactants to the plate.
   a. Using a micropipet and beginning with the highest dilution of serum (1:2048), add a microdrop (i.e., 6-8 μl) to each square of row 7 in the plate, replace excess back in the tube. Repeat with dilution 1:1024 adding a drop to each square of row 6. Repeat until microdrops of all serum dilutions are placed on rows 1-7 of the plate. A reliable pipetor dispenses droplets of known volume; however, a micropipet fitted with a 27-gauge needle, delivers droplets of similar size and provides adequate test reproducibility.
   b. Using another micropipet and beginning with the most dilute solution of antigen (1:128), add a microdrop to each square in column 7. Make sure it is a full drop and that the pipet does not touch the serum drop already on the plate. Repeat as described in the serum dilution series, adding drops of each dilution of antigen to columns 1-7, ending with column 1 which is the 1/2 dilution. The edge of the plate may be gently tapped (be careful to maintain droplet integrity) to encourage mixing of reagents.
   c. To all the squares of row 8 and of column 8 add 1 microdrop of buffer. These represent antigen and serum controls, respectively. They must be negative (clear) for valid test results.
4. Incubation, reading and interpretation.
   a. Incubate the Petri dish in a moist environment at room temperature (e.g., sealed plastic container containing wet paper towels). The test can be read after 2 hr using a binocular dissecting microscope in a dark room with the light source set obliquely so as to give a dark background. Incubation should be continued and the test read again the next day. By this time convection currents would have made the precipitation pattern clearly visible. Tapping the plate gently may enhance precipitate aggregation. With bacterial systems gentle vibration of the plate for ~ 30 sec improves the agglutination pattern.
   b. The squares representing heavy precipitation can be read as 4 with downward gradations to 1. Trace reactions (+) may be observable also. Negative reactions must be equivalent to controls (i.e., clear).
   c. Optimal proportions of antigen and antibody can be designated by readings of 3 representing the greatest dilution of antigen and the least amount of serum producing precipitation. If a reading of 2 can be duplicated in subsequent tests, it may be selected as a practical endpoint, i.e., conveniently observable.

Comments

1. The technique requires some practice in manipulating the small drops. However, it is possible to use 10 x 10 grids in a 100 mm plastic dish. The wax-pencil grid serves 2 purposes: clarity in reading the results and preventing mergence of drops during incubation.
2. Different types of precipitation are produced in this test. For example, TMV or PVX form amorphous, cloudlike precipitates whereas southern bean mosaic virus produces a granular, sand-like precipitate.
3. Plant-extract antigen preparations must be clarified to eliminate anomalous precipitates of plant products that could confuse the reading. Also, optimal proportions of antigen and antiserum obtained with purified virus differ from those of plant extract and with IgG as compared to whole serum. Therefore, optimal proportions must be determined with each combination of reactants by grid-titration.

4. Using the optimal endpoint concentration of antigen, titrations were made of antisera taken during the weeks of immunization as illustrated in Fig. 1, with barley stripe mosaic virus antiserum prepared in chickens and in studies with American plum line pattern virus (8) and potato virus V (7). The characterization of antisera by microprecipitin titers prior to subsequent studies in ELISA and agar double diffusion tests has been reported recently with soybean mosaic virus (2), bean pod mottle virus (3), cucumber mosaic virus (4), clover yellow vein virus (5), maize dwarf mosaic virus (12), watermelon mosaic virus II (16), peanut mottle virus (18) and a new tobamovirus from *Passiflora edulis* (6)

   A frequent application of microprecipitin tests, using optimal concentrations of antigen or antibody, is for viral strain comparisons in which one reactant (serum, for example) is diluted and the other (antigen) is held constant to compare endpoint reactions of homologous and heterologous antibodies. This is illustrated with strains of tobacco mosaic virus (Table 1, B) and also in strain studies with tomato ringspot virus (17) and with maize dwarf mosaic virus (9,15).
5. Precipitin or agglutination tests are frequently used as a second test, enzyme-amplified assays being more sensitive. However, precipitin tests are especially useful in laboratories lacking expensive equipment and supplies necessary for newer tests.
6. The tube precipitation (or agglutination) test can be made highly sensitive and informative regarding antigen-antibody interactions using the quantitative procedure (14). A row of tubes is set-up containing a constant volume and concentration of antiserum (or IgG) and an equal volume of a series of concentrations of antigen. After incubation and centrifugation, the amount of precipitate is measured by weight, protein analysis or Kjeldahl nitrogen determination. Tubes having the greatest amount of precipitate represent *optimal proportions*. Supernatant fractions are analyzed for excess antigen or antibody by placing aliquots from each tube into 2 rows of small tubes. Antigen is added to one set and antiserum to the other. After incubation three zones of reactivity are observed: precipitation indicating *antigen excess* in the first few tubes of the row tested with antiserum, *equilibrium* in those tubes in each row in which no precipitate is present, and precipitation indicating *antibody excess* in the last few tubes of the row tested with antigen. Equilibrium is a reference value for each antigen-antibody system. If a plot is made of antigen added vs amount of precipitate, a curve results with optimal proportions at the peak of the curve, and equilibrium on the antigen excess side of the curve.

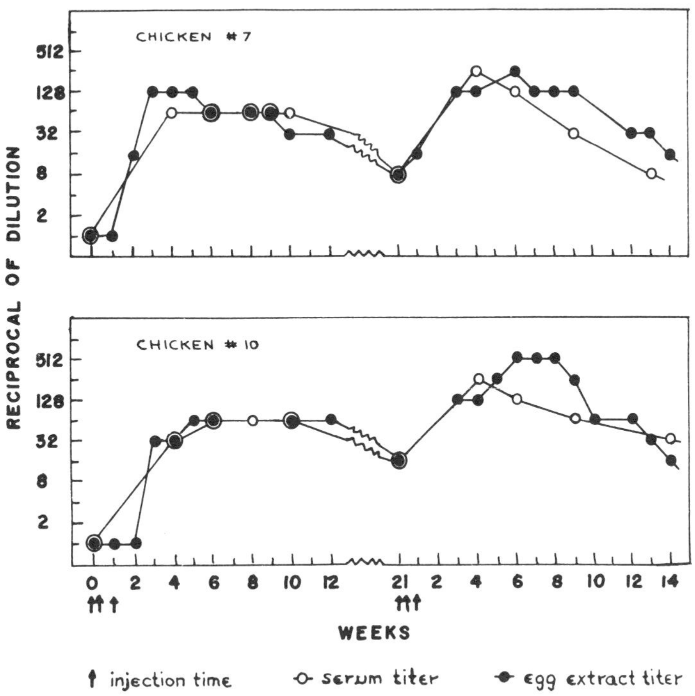

Fig. 1. Application of the microprecipitin test to the determination of antibody response in two chickens immunized with barley stripe mosaic virus. A series of two intramuscular injections at 9 day intervals followed by one intravenous injection 4 days after the last intramuscular injection were given in both the primary and booster response period. A total of 7.0 mg and 13.6 mg were injected in the two response periods, respectively. Serum samples and eggs were collected at 2 week intervals for assay. The dilution endpoint of the yolk extract (see **Section II.B.1.** for extraction method) was plotted as 5 times the actual titer because of the initial dilution during extraction. The ordinate scale includes the yolk dilution factor, i.e. 1:8 serum = 1:10 yolk titer.

Literature Cited

1. Ben-Sin, C. and Po, T. 1982. Infection of barley protoplasts with barley stripe mosaic virus detected and assayed by immunoperoxidase. J. Gen. Virol. 58:323-327.
2. Bryant, G. R., Durand, D. P., and Hill, J. H. 1983. Development of a solid-phase radioimmunoassay for detection of soybean mosaic virus. Phytopathology 73:623-629.
3. Calvert, L. A., and Ghabrial, S. A. 1983. Enhancement by soybean mosaic virus of bean pod mottle virus titer in doubly infected soybean. Phytopathology 73:992-997.
4. Davis, R. F., and Hampton, R. O. 1986. Cucumber mosaic virus isolates seedborne in Phaseolis vulgaris: Serology, host-pathogen relationships and seed transmission. Phytopathology 76:999-1004.
5. Fox, M., and Corbett, M. K. 1985. Winged bean mosaic caused by clover yellow vein virus. Plant Dis. 69:352-354.
6. Fribourg, C. E., Koenig, R., and Lesemann, D. E. 1987. A new tobamovirus from Passiflora edulis in Peru. Phytopathology 77:486-491.
7. Fribourg, C. E., and Nakashima, J. 1984. Characterization of a new potyvirus from potato. Phytopathology 74:1363-1369.
8. Fulton, R. W. 1982. Ilar-like characteristics of American plumline pattern virus and its serological detection in Prunus. Phytopathology 72:1345-1348.
9. Jarjees, M. M. and Uyemoto, J. K. 1984. Serological relatedness of strains of maize dwarf mosaic and sugarcane mosaic viruses as determined by microprecipitin and enzyme-linked immunosorbent assays. Ann. Appl. Biol. 104:497-501.
10. Johns, L. J. 1982. Purification and partial characterization of a carlavirus from Taraxacum officinale. Phytopathology 72:1239-1242.
11. Kuti, J. O., and Moline, H. E. 1984. A mild strain of tomato aspermy virus isolated from tomato in Maryland. Plant Dis. 68:959-961.
12. Lei, J. D., and Agrios, G. N. 1986. Detection of virus infection and spread by immunofluorescent staining of enzyme treated leaves. Phytopathology 76:1031-1033.
13. Matthews, R. E. F. 1970. The precipitation reaction. Pages 480-484 in: Plant Virology. Academic Press, New York. 778 pp.
14. Maurer, P. H. 1971. The quantitative precipitin reaction. Chapter 13A, pages 1-58 in: Methods in Immunology and Immunochemistry. Vol. III. C. A. Williams and M. W. Chase, eds. Academic Press, New York. 515 pp.
15. McDaniel, L. L., and Gordon, D. T. 1985. Identification of a new strain of maize dwarf mosaic virus. Plant Dis. 69:602-607.
16. Moyer, J. W., Kennedy, G. G., and Ramanow, L. R. 1985. Resistance to watermelon mosaic virus II multiplicaton in Cucumis melo. Phytopathology 75:201-205.
17. Puffinberger, C. W., and Corbett, M. K. 1985. Euonymus chlorotic ringspot disease caused by tomato ringspot virus. Phytopathology 75:423-428.
18. Tolin, S. A., and Ford, R. H. 1983. Purification and serology of peanut mottle virus. Phytopathology 73:899-903.
19. van Regenmortel, M. H. V. 1982. Precipitation reaction and microprecipitin tests. Pages 74-81 in: Serology and Immunochemistry of Plant Viruses. Academic Press, New York. 302 pp.

Section III.F.

DENSITY GRADIENT PRECIPITATION

J. E. Duffus
USDA ARS
Salinas, California 93905

Principle

The assay is based on the reduction or elimination of a virus antigen in the scanning patterns of sucrose density gradient columns resulting from centrifugation of antigen-antibody mixtures (1,5).

Materials

1. Small test tubes.
2. PBS.
3. Incubator or water bath at 37 C.
4. Dropping pipets.
5. Density-gradient columns.
6. Density-gradient fractionator.
7. Planimeter.

Procedure

1. Place 0.5 ml of virus at 5.0 to 50 μg/ml concentration in each of several test tubes.
2. Add and mix 0.5 ml of serial twofold dilutions of antiserum in each tube; incubate at 37 C for 30 min.
3. Layer each virus-antiserum mixture on a gradient column.
4. Centrifuge for usual time, depending on virus, in a density gradient rotor.
5. Scan gradient columns photometrically with a density-gradient fractionator.
6. Determine area under the peaks by a planimeter.

Comments

This system is a very sensitive test that is similar in many respects to neutralization (see **Section V.B.**). The technique is especially useful with low titer sera or sera containing antibodies to normal plant constituents, or when other viruses are present in the test solutions (2,3,4).

The test is more sensitive than visual range precipitation tests. The sensitivity can be increased by using smaller amounts of virus and by using infectivity rather than ultraviolet absorption to detect virus (see **Section V.B.**). As with infectivity neutralization, low antiserum dilutions show group relationships and higher dilutions show distinctions between members of the group (Fig. 1).

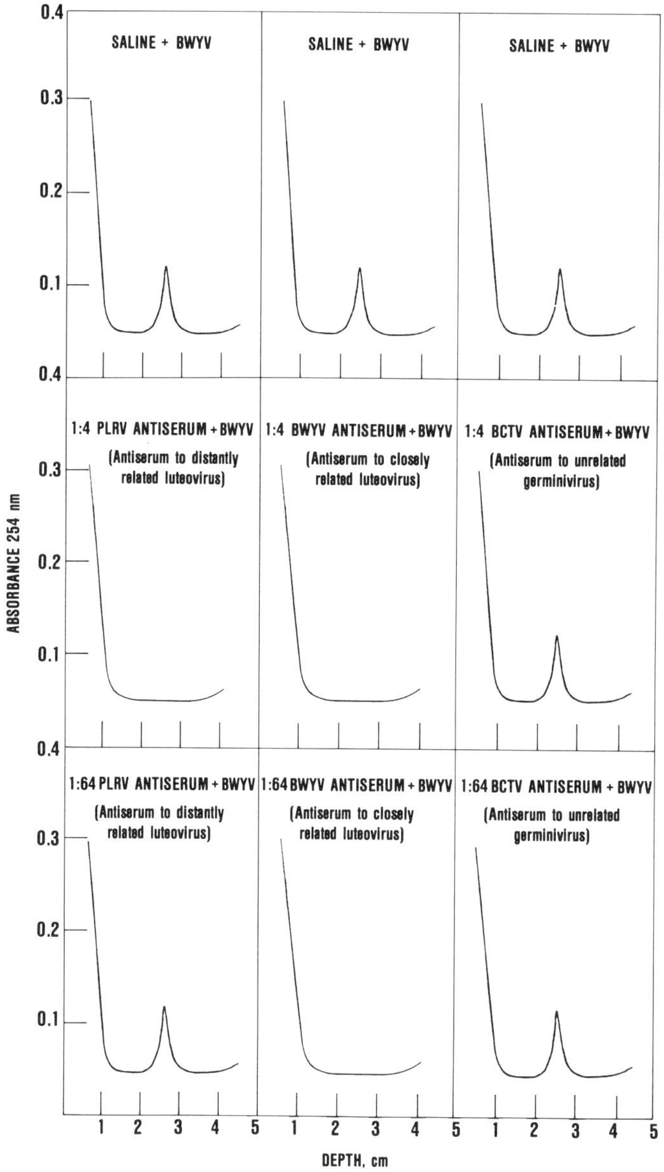

Fig. 1. Photometric scanning patterns of mixtures of beet western yellows virus (BWYV) with antiserum to potato leaf roll virus (PLRV), antiserum to BWYV, and antiserum to beet curly top virus (BCTV) and controls. Virus samples were mixed with two dilutions of antiserum, incubated 0.5 hr at 37 C and subjected to density-gradient centrifugation. Low antiserum dilutions show group relationships between BWYV and PLRV and higher dilutions show distinctions between the two viruses. An unrelated geminivirus (BCTV) shows no relationship by this test.

Literature Cited

1. Ball, E. M., and Brakke, M. K. 1969. Analysis of antigen-antibody reactions of two plant viruses by density-gradient centrifugation and electron microscopy. Virology 39:746-758.
2. Duffus, James, E. 1977. Serological relationships among beet western yellows, barley yellow dwarf, and soybean dwarf viruses. Phytopathology 67:1197-1201.
3. Hamilton, R. I., and Dodds, J. A. 1970. Infection of barley by tobacco mosaic virus in single and mixed infection. Virology 42:266-268.
4. Rochow, W. F., and Ball, E. M. 1967. Serological blocking of aphid transmission of barley yellow dwarf virus. Virology 33:359-362.
5. Whitcomb, R. F., and Spendlove, R. S. 1966. Density gradient centrifugation of virus-antibody complexes: A sensitive serological method. Virology 30:752-754.

Section III.G.

IMMUNOELECTROPHORESIS

J. H. Tremaine
Agriculture Canada Research Station
Vancouver, B.C., Canada

Principle

Complex mixtures of proteins can be resolved by electrophoresis in agar gels. Immunoelectrophoresis combines this resolution with the sensitivity and specificity of serological analysis by double diffusion (8,9,11) (Fig. 1). A comparison of relative electrophoretic mobilities may be useful in distinguishing between serologically related viruses and mutants.

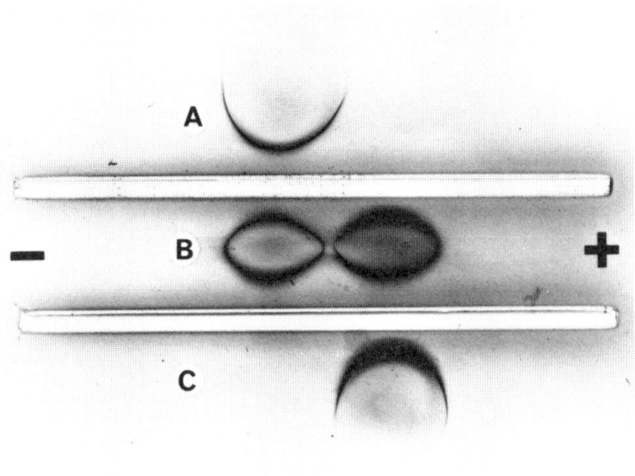

Fig. 1. Immunoelectrophoresis of purified preparations of two strains of southern bean mosaic virus in 0.02 M Tris phosphate buffer, pH 7.0, at 133 volts (30 ma) for 1 hr at 4 C. Sample volume was 3 µl of preparations containing 3 µg/µl. Well A contained the bean strain; well C contained the NCP cowpea strain of McGovern and Kuhn (7); and well B contained both strains. The antiserum troughs were loaded with 400 µl of bean strain antiserum at a 1/5 dilution in 0.15 M NaCl and 0.2% sodium azide.

Materials

1. Electrophoresis apparatus. A typical apparatus consists of a direct current power supply with regulated voltage or amperage and an electrophoresis cell. The cell may use paper wicks to conduct the current to the agar bed or may be of the submarine type commonly used for electrophoresis of nucleic acid and used in the methodology described here. The submarine gel system

has two buffer tanks with electrodes and a connecting support bed holding a removable ultraviolet transparent plastic tray.

2. Supporting medium. Electrophoresis of virus particles is most suitably conducted in 0.7% agarose in tank buffer. Agarose with a low content of sulfate provides a low electroendosmotic flow of buffer through the gel. Gels can be prepared and stored in suitable aliquots for later use. Gel Bond (FMC Corporation, Marine Colloids Division) provides a firm plastic surface which bonds to the agar gel, and wells for the antigen or troughs for the antiserum can be cut from the gel without breaking the seal between the gel and the Gel Bond.

3. Buffers. Immunoelectrophoresis is usually conducted at pH 7 to 8 and the tank buffer should be of sufficiently low molarity to produce a voltage drop of 5 to 10 volts per cm across the agarose bed without producing excessive heating. The virus antigens must be stable in the chosen buffer. Buffers which lack neutral salts are preferable. Tris-phosphate buffer in the pH range 6.2 to 8.4 can be prepared by adjusting the pH of 0.02 M Tris with 0.02 M NaH_2PO_4. Extended periods of electrophoresis may be required to move the virus a sufficient distance and the pH in the anode and cathode compartments may change markedly. Small submarine gel electrophoresis cells (Bio-Rad Mini-Sub cell) can be operated in a refrigerator and buffer can be circulated using the ports on the electrode compartments and a two channel peristaltic pump. As an alternative the power could be turned off and the spent buffer exchanged with fresh cold buffer.

4. Antigen. A crude or purified virus preparation is suspended in the tank buffer and glycerol is added to 30% - 50% to facilitate application. Bromophenol blue (0.001%) added as a tracker dye also facilitates application. The method is sufficiently sensitive to readily detect less than 1 μg of virus. If dilute samples are used, the volume of sample applied can be increased to 50 μl by using larger antigen wells.

5. Antiserum. After completion of electrophoresis the virus-specific antiserum diluted in 0.15 M NaCl containing 0.2% sodium azide is placed in the trough. The ratio of antiserum to antigen is important to the development of well defined precipitin bands, but good results can be obtained with fairly wide ratios of reactant concentrations.

Procedure

1. Preparation of electrophoresis bed. Gel Bond, hydrophilic side up, is attached to the gel tray with water and the open ends of the gel tray are sealed securely with tape. The tray is placed on a level surface and melted agarose gel at 65 C is poured into the tray. The volume of gel used should provide a bed of 1.5 to 3 mm depth (10 to 20 ml for the Bio-Rad Mini-Sub cell). Troughs for antiserum can be formed by placing glass microscope slides (75 x 25 x 1 mm) on edge in the hot agar and supporting them by a plastic comb. After the agarose has solidified for up to one hr, buffer is added on top of the agarose and the slot formers are extracted. Wells for the antigen are cut with a suitably sized cork borer or hypodermic needle. The antiserum troughs alternatively can be cut with a razor blade.

2. Sample application and electrophoresis. After removal of the tape, the tray is placed in the electrophoresis cell and buffer is added to cover the agarose bed to a depth of 2 to 4 mm. Samples are applied with a micropipetor by underlayering into the antigen wells. Conditions of electrophoresis, the voltage, amperage and duration, will vary depending upon the ionic strength of the buffer, and the size of the electrophoresis bed and properties of the antigen. The voltage can be kept as high as possible without denaturing the virus or other antigens under test. Electrophoresis should be done in a cold room if current exceeds 2 ma. Since many factors affect the electrophoretic mobility of an antigen, the conditions appropriate for a specific virus or virus system have to be determined by trial and error. Many spherical plant viruses can be detected by adding 2 μl of ethidium bromide (10 mg/ml) to the melted agarose gel (6). Ethidium bromide is a mutagen and must be used with caution (plastic gloves). It penetrates some virus particles and stains the RNA. The electrophoresis can be stopped, and the position of the ethidium bromide-stained virus can be observed under ultraviolet light. The electrophoresis may be continued until the virus or viruses have migrated to suitable positions. Detection by antiserum is not affected by ethidium bromide staining. With this method Gel Bond NF (non-fluorescent) must be used. The position of migration in trial runs can also be determined by staining overnight with filtered stain solution containing 0.2 g Coomassie Blue R250, 450 ml methanol, 90 ml acetic acid and 460 ml of water. Destaining is done in 15% methanol and 5% acetic acid in water. The denatured stained bands usually do not react with virus antiserum, but the method can be used to compare mobilities of purified viruses at any pH without subsequent development with antisera.

3. Development. Antiserum can be applied in the troughs formed or cut as described above or in filter paper strips soaked in the antiserum and laid on the agarose gel in appropriate positions. Sodium azide in the antiserum will prevent bacterial growth. The gels are placed in plastic petri dishes containing paper soaked in water to provide a humid atmosphere. The dishes are taped shut and incubated for 24-72 hr at a constant temperature of 20 C. The gels can be photographed or, if staining is required, they are washed in 0.15 M NaCl overnight to remove unreacted antiserum and antigen components. Staining and destaining can be done as described above or by the Crowle and Cline procedure (2). A permanent record of the experiment can be obtained by washing the stained gel in 1% glycerol in water and drying the gel on plastic Gel Bond surface. Dried unstained gels may show precipitin bands or they may be stained and redried when required.

Applications

Immunoelectrophoresis has been used to study the electrophoretic mobility of isometric viruses but rod-shaped viruses may yield broader zones of precipitation because of particle size variability. Valverde and Fulton (10) were unable to distinguish mobility differences among five distinct bean strains of southern bean mosaic virus by immunoelectrophoresis. Gallo and Musil (3) studied the mobilities of three serotypes of red clover necrotic mosaic virus in four buffers at three pH values and different ionic strengths. Although there were no differences in mobility of each serotype at pH values of 7.2, 8.0 and 8.6, the mobilities determined in the different buffers and at different ionic strengths differed. They were able to distinguish between the serotypes even in serotype mixtures in 0.01 or 0.1 M HEPES. Hollings and Stone (4) noted that immunoelectrophoresis offered a simple means of typing members of the tombusvirus group. Koenig and Gibbs (5) determined electrophoretic mobilities of purified tombusviruses by staining rather than immunoelectrophoresis. The mobilities they determined differed from those of Hollings and Stone, probably because they used an agar with low electroendosmosis. They noted that some serologically indistinguishable tombusviruses had distinct differences in electrophoretic migration and suggested that differences in migration of some tombusviruses may be caused by size differences induced by swelling rather than virus particle charge differences.

Bercks and Querfurth (1) used immunoelectrophoresis in some novel experiments. They found serological cross reactions between some cocksfoot mild mosaic virus (CMMV) antisera and some tymoviruses, between some tymovirus antisera and CMMV, and between some carnation ringspot virus antisera and some tymoviruses. To demonstrate the validity of these serological cross reactions between unrelated virus groups, they showed these serological reactions occurred with the viruses in immunoelectrophoresis.

Literature Cited

1. Bercks, R., and Querfurth, G. 1972. Serologische Beziehungen von cocksfoot mild mosaic virus und carnation ringspot virus zu Viren der turnip yellow mosaic virus group. Phytopath. Z. 75:215-222.
2. Crowle, A. J., and Cline, L. J. 1977. An improved stain for immunodiffusion tests. J. Immunol. Methods 17:379-381.
3. Gallo, J., and Musil, M. 1984. Serotypes of red clover necrotic mosaic virus. III. Immunoelectrophoresis under different conditions. Acta Virol. 28:78-81.
4. Hollings, M., and Stone, O. M. 1975. Serological and immunoelectrophoretic relationships among viruses in the tombusvirus group. Ann. Appl. Biol. 80:37-48.
5. Koenig, R., and Gibbs, A. 1986. Serological relationships among tombusviruses. J. Gen. Virol. 67:75-82.
6. Lane, L. C., and Joshi, B. 1982. A simple system for gel electrophoresis of intact viruses. Phytopathology 72:937.
7. McGovern, M. H., and Kuhn, C. W. 1982. Southern bean mosaic virus strain derived by preferential systemic movement. Phytopathology 72:937-938.
8. Osterman, L. A. 1984. Methods of Protein and Nucleic Acid Research, Vol. 2.. Immunoelectrophoresis; Application of Radioisotopes. Springer Verlag, Berlin. 204 pp.
9. Smith, I. 1968. Chromatographic and Electrophoretic Techniques. Vol. II. Zone Electrophoresis. W. Heinemann Medical Books, Ltd. Bath, England. 524 pp.
10. Valverde, R. A., and Fulton, J. P. 1982. Characterization and variability among strains of southern bean mosaic virus. Phytopathology 72:1265-1268.
11. Williams, C. A. 1971. Immunoelectrophoretic analysis in agar gels. Pages 237-273 in: Methods in Immunology and Immunochemistry. Vol. III. Reactions of Antibodies with Soluble Antigens. C. A. Williams and M. W. Chase, eds. Academic Press, New York. 515 pp.

Section III.H.

ROCKET IMMUNOELECTROPHORESIS

E. M. Ball
University of Nebraska
Lincoln, Nebraska 68583

Principle

The technique of rocket immunoelectrophoresis is attributed to the early work of Laurell (5). It is similar to immunoelectrophoresis except that antigens are drawn into a medium containing antibody. Where the two reactants meet in optimum proportions, precipitation occurs in the form of a cone or rocket (Fig. 1A). The area of the rocket is directly proportional to the concentration of the antigen. Electrophoretic separation is run under constant current to generate as little heat as possible thereby eliminating heat gradients and/or a change in medium viscosity. Optionally, the reactant positions can be reversed and many other procedural adaptations are possible, several of which are discussed.

Materials

1. Agarose: low electroendosmosis and low gelling temperature are required as well as good gel strength to prevent breakage during manipulation. See **Section III.A.** for a description of properties of agaroses. For this procedure, 1% concentration, Seakem LE (FMC Corporation) is suggested.
2. Polyester film to support a thin gel: Gelbond (FMC Corporation).
3. Suitable buffer depending on antigen characteristics: e.g., 0.01M Na_2HPO_4-KH_2PO_4, pH 8.5.
4. Antiserum: 1% in initial studies, adjust antiserum concentration to yield rockets of 10 mm or more in height.
5. Antigen: at least 50-100 ng of protein are required for rocket visualization: (Tobacco mosaic virus strains at 4, 2, and 1 mg/ml in electrophoresis buffer illustrate a relationship that may be established by this technique.)
6. Agar punch: cork borer or commercial punch.
7. Automatic pipet to deliver microliter volumes.
8. Coomassie stain, destain: destain base = 45 ml methanol, 45 ml demineralized water, 10 ml glacial acetic acid. Stain = 0.25% (w/v) dye in base solution.

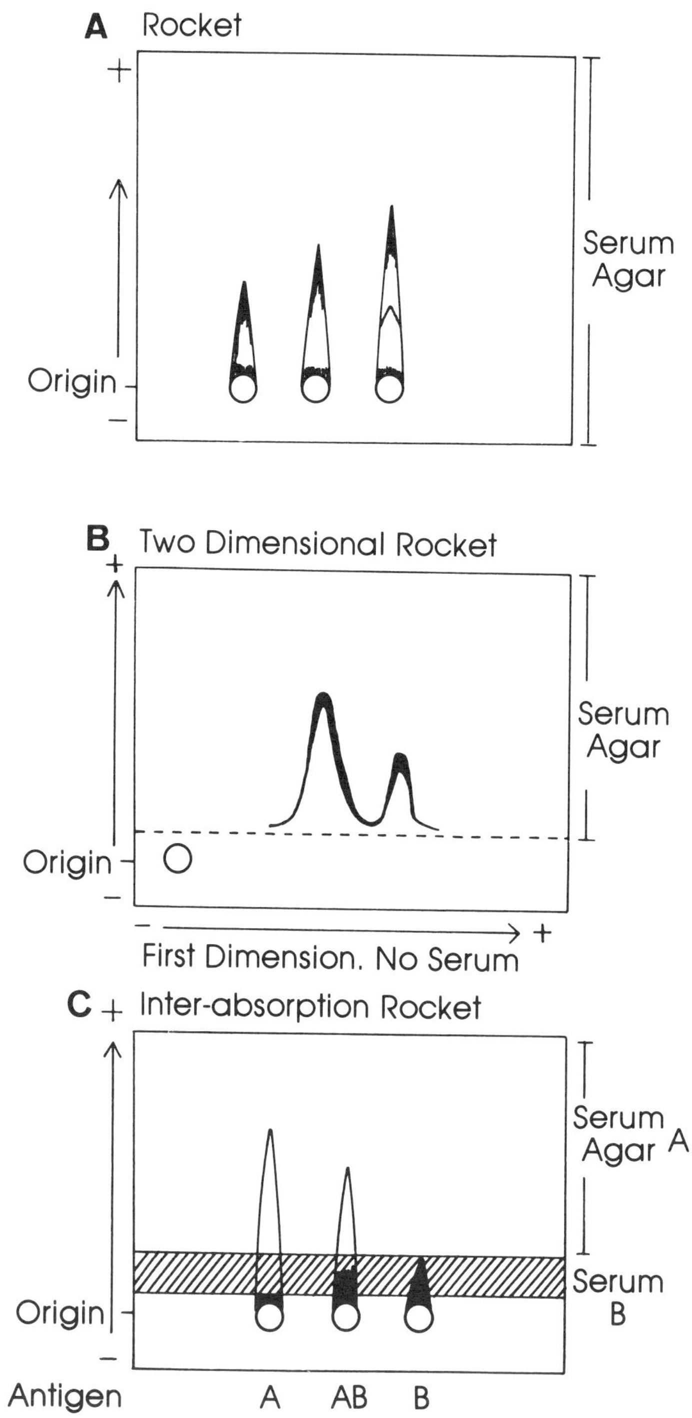

Fig. 1. Drawings illustrating three adaptations of rocket immunoelectrophoresis. A. One dimensional electrophoresis in which antigen from wells has migrated into a serum-agarose medium. B. Two dimensional electrophoresis in which antigenic components are separated in agarose by conventional electrophoresis, then immunoprecipitated by electrophoresing them into serum-agarose by a current perpendicular to the first dimension. C. One dimensional electrophoresis in which an immuno-absorbing barrier is implanted next to antigen wells for cross reaction studies of antigenic components.

Procedure

1. Prepare 1 liter of buffer, or volume required for slab-gel electrophoresis apparatus. Fill apparatus reservoirs. Place a 6 x 9 cm strip of filter paper in each reservoir to be used as wicks.
2. Prepare 10 ml of agarose in a 20-25 ml beaker at 1% in the same buffer as above. Dissolve in autoclave - 10 min or in microwave - 2 min. Avoid excessive heat.
3. Place a 8.5 x 10 cm glass slide on a leveling table. Position Gelbond (8.5 x 10 cm) on top of slide, hydrophilic side up (test by adding a drop of water to film, drop spreads on hydrophilic side).
4. When agarose has cooled to 50-55 C, add 0.1 ml of antiserum, mix by stirring gently, and pour immediately onto Gelbond using the lip of the beaker to evenly spread the gel to the edges of the film. Allow to set 10-15 min; cover to prevent evaporation.
5. Cut 12 wells, 2 mm apart (6 mm center-to-center), in the agarose using a 4 mm punch. Position wells along the long axis of the gel about 5.0 cm from one edge. Wells are positioned centrally in initial experiments to observe any endosmotic migration of antigenic components and can be located 2.0-3.0 cm from the edge in subsequent tests designed for maximum migration of antigens. Remove agarose plugs from the wells by gentle suction, being careful not to disturb the smooth sides of the wells.
6. Using the automatic pipet, quickly add antigens in a sequence of 1, 2, 4, mg/ml to the wells, filling each with 7 μl/dilution of antigen. Thus, wells will contain 7, 14, and 28 μg of antigen.
7. Place slide in an electrophoresis apparatus so that antigen migrates toward the anode. Place a filter paper wick in contact with each edge of the gel, assuring that it extends into the buffer reservoir.
8. Use a constant current of 6 mA/gel, 200 V for 2 hr at room temperature.
9. Remove slide from apparatus. Place slide on a paper towel, cover the gel with a 8.5 x 10 cm filter paper, then add 1-2 folded paper towels and a glass plate. Place a suitable weight on the plate to gently press buffer from the gel. Finish drying the gel under a lamp.
10. Place dried gel in a dish of a size larger than the Gelbond. Add sufficient phosphate buffer (pH 7.0) or saline to cover the gel. Rotate gently for 2 hr. Replace the buffer with demineralized water and continue rotating for 1 hr.
11. Stain by adding sufficient Coomassie dye to cover the gel. Rotate 30 min. Remove dye and destain for 10 min. If background is not clear, repeat destaining process. Rinse in distilled water, blot, and let dry.
12. It is evident in the illustration with strains of tobacco mosaic virus (Fig. 2) that one strain, 'mild dark green', is different from type strain, latent strain or rosette strain which are serologically similar to the type strain. The dark green strain as well as the virion coat protein forms only a light extraneous smear.

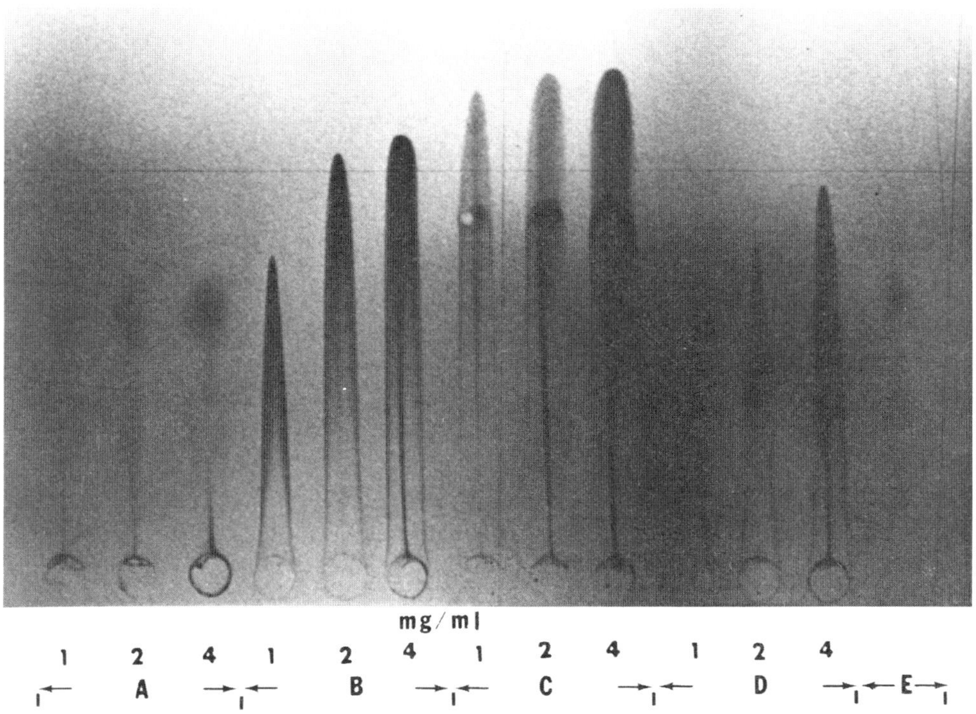

Fig. 2. Rocket immunoelectrophoresis of strains of tobacco mosaic virus. Strains: A. Mild dark green; B. Type; C. Rosette; D. Latent; E. Type-coat protein. Agarose: Seakem LE, 1% in 0.01 M phosphate buffer, pH 8.5; 0.025 ml type strain serum in 10 ml agarose/8 x 10 cm Gelbond support. Seven μl of antigen/well. Constant current, 8 mA, 235-192 volts, 2 hr. Stain: 0.25% Coomassie. NOTE: Flat tops on rockets B4, C2, C4 due to an imbalance of antigen concentration compared to antiserum dilution. Mild dark green strain virions and coat protein are serologically distinct from type strain. Latent and rosette strains are closely related to type strain.

Comments

1. If problems are encountered in obtaining a crystal clear background in the stained gels, purified IgG may be substituted for antiserum. Generally, the least amount of antigen and antibody which produces a good stain should be used in the gel because reactants are conserved, heat during electrophoresis is reduced, and a cleaner stained product results.
2. A thin agarose slab on Gelbond can be cleared readily of unreacted antigen and antiserum. However, to prevent evaporation and change in viscosity the gel must be prepared quickly and placed in a covered electrophoresis apparatus as soon as possible.
3. Coomassie dye or silver stain gives satisfactory results with viruses. Bacterial extracts may require other dyes, such as Amido Black, Ponceau S or a dye system specific for the component being studied.
4. The choice of a buffer system depends upon the characteristics of the antigen. Barbituric acid-sodium barbital, pH 8.6, or barbital-glycine/Tris,

pH 8.6, buffer systems have been used with cauliflower mosaic (4) and panicum mosaic (2) viruses, respectively. Whereas, barbital buffer may be the "classical" system in most literature citations, especially medical references, tris, borate and phosphate buffers work equally well. In studies of relationships among three bromoviruses: brome mosaic, cowpea chlorotic mottle and broad bean mottle viruses, a 0.01 M potassium phosphate buffer at pH 6 and 7 was used. Heterologous relationships were more defined in rocket tests than was observed by agar double diffusion tests (7). To study globulins specific for human IgG and IgM present in hen yolk (1), a low pH of 5.5-6.0 MES buffer [2-(N-morpholine) ethane sulfonic acid] was used for rocket tests. The ionic strength of any buffer is kept as low as possible to reduce heat buildup in the medium during the electrophoretic run.

5. The pH of agarose is usually adjusted to 8.5 - 8.8 to keep the antibody relatively stationary compared to the migration of the antigen. If the charge of the antigen is close to that of the antibody, either charge can be modified by carbamilation of the reactant (6).
6. Reactions in agarose can be enhanced by additives. Divalent cations such as Mg^{++} were used to stabilize the bromoviruses (7) and Ca^{++} improved the resolution of rocket tests with human serum (3). Chelating agents, such as EDTA, may enhance rocket precipitates. Although SDS must be removed because of its denaturation of antigens, Triton X-100 (4) or nonidet P-40 detergents may intensify rocket patterns. Monoclonal antibodies may not precipitate the antigen themselves, but they can contribute to specificity and increase the intensity of precipitates when added to the serum (IgG) medium (3).
7. One dimensional rocket immunoelectrophoresis can be used to resolve differences between strains of viruses (Fig. 1A). In studies with 7 isolates of St. Augustine decline virus (2) the titer of each strain in a common host could be quantitated. The slope of the linear regression line of the log of the peak area of the rockets and the log of the relative concentration of plant sap was a measure of differences between strains.
8. An adaptation of the test can reveal differences/similarities among antigenic components. The two dimensional test (Fig. 1B) involves electrophoresis under conventional conditions in agarose. A strip representing separated antigens, is cut out of the gel and transferred to a slide with Gelbond. Serum-agarose is then poured on the remainder of the Gelbond. Electrophoresis is continued with the current flowing at a 90° angle to the first dimension. Antigenic components or determinant groups separated in the first dimension can be visualized clearly in the second dimension by specific antibody. This is referred to as crossed immunoelectrophoresis and may resolve components not possible by other methods.
9. Another adaptation is illustrated in Fig. 1C. A section of the serum-agarose slab is cut out, or blocked out during pouring, across the Gelbond and close to the position of the antigen wells. A different serum in agarose is poured in this area (8). Cross-absorption occurs within this band similar, in principle, to intragel specific absorption as described in **Section III.C.2.e.**

   Two monospecific antibodies, can be added directly to the agarose medium. Precise measurements of comparative ratios of antigenic components can be calculated (3).
10. The resolution of rocket components resulting from the use of polyspecific serum depends upon the antibody titers present for individual antigenic components. Antisera from 2 rabbits immunized with the same antigen may reveal different results in rocket formation. A mix of different sera may

enhance precipitate formation of one component but not the other.

11. The advantages of using rocket immunoelectrophoresis have been summarized in the studies of an assay for cauliflower mosaic virus (4). The test is efficient, accommodates large numbers of samples, is quantitative over a wide range of virus concentrations (typically, 25 ng to 10 μg) requires less than 8 hr to complete and can detect virus in 25 mg samples of tissue.

Literature Cited

1. Altschuh, D., Hennache, G., and van Regenmortel, M. H. V. 1984. Determination of IgG and IgM levels in serum by rocket immunoelectrophoresis using antibodies from immunized chickens. J. Immunol. Methods 69:1-7.
2. Berger, P. H., and Toler, R. W. 1983. Quantitative immunoelectrophoresis of panicum mosaic virus and strains of St. Augustine decline. Phytopathology 73:185-189.
3. Gross, W., and Marz, W. 1988. Immunoelectrophoretic techniques in protein analysis and quantitation. American Biotech. Laboratory 6:6-19.
4. Hagen, T. J., Taylor, T. B., and Meagher, R. B. 1982. Rocket immunoelectrophoresis for cauliflower mosaic virus. Phytopathology 72:239-242.
5. Laurell, C. B. and McKay, E. J. 1981. Electroimmunoassay. Pages 339-369 in: Methods in Enzymology. Part B. Immunochemical Techniques, Vol. 73. J. J. Langone and H. van Vunakis, eds. Academic Press. 739 pp.
6. Polson, A., von Wechmar, M. B., and Fazakerley, G. 1980. Antibodies to proteins from yolk of immunized hens. Immunological Communications 9:495-514.
7. Rybicki, E. P., and von Wechmar, M. B. 1981. The serology of the bromoviruses. 1. Serological interrelationships of the bromoviruses. Virology 109:391-402.
8. Weeke, B. 1973. Rocket immunoelectrophoresis and crossed immunoelectrophoresis. Pages 37-56 in: A Manual of Quantitative Immunoelectrophoresis. N. H. Axelsen, J. Kroll, and B. Weeke, eds. Scand. J. Immunol. 2., Suppl. 1. 169 pp.

Section III.I.

IMMUNO-OSMOPHORESIS

J. H. Tremaine
Agriculture Canada Research Station
Vancouver, B.C., Canada

Principle

This technique combines electrophoresis and immuno-precipitation in agar (1,2). Many viral antigens migrate electrophoretically toward the anode at pH 7 while the immunoglobulins in antiserum migrate toward the cathode in the electro-endosmotic flow of buffer through the agar containing sulfate groups. Precipitation of virus and antibody occur in the agar where they meet in optimal proportions. The technique is more rapid and sensitive than immunodiffusion in gels. The increase in sensitivity is attributed to the unidirectional migrations of antigen and antibody as opposed to the multidirectional diffusion occurring in immunodiffusion techniques. The method can be used with crude or purified virus preparations.

Materials

The materials used in immuno-osmophoresis are identical to those used in immunoelectrophoresis (**Section III.G.**) including: the Bio-Rad Mini-Sub cell or other submarine gel electrophoresis apparatus; the agarose with moderate electroendosmotic flow; Gel Bond or Gel Bond NF; buffer; and antigen preparation. The antiserum must also be prepared in 30 to 50% glycerol for application by underlaying. Various dilutions of antiserum should be tested to determine optimum proportions for maximum sensitivity.

Procedure

1. The bed preparation is also similar to immunoelectrophoresis (**Section III.G.**). However, wells for the antigen and antiserum are cut in pairs, 1.4 cm apart and aligned in the direction of current flow as in Fig. 1. Larger antigen wells can be used to apply larger volumes of dilute samples. Elipsoidal wells with major axis in direction of current flow are best for dilute samples and the wells can be made with a cork borer using multiple cuts.

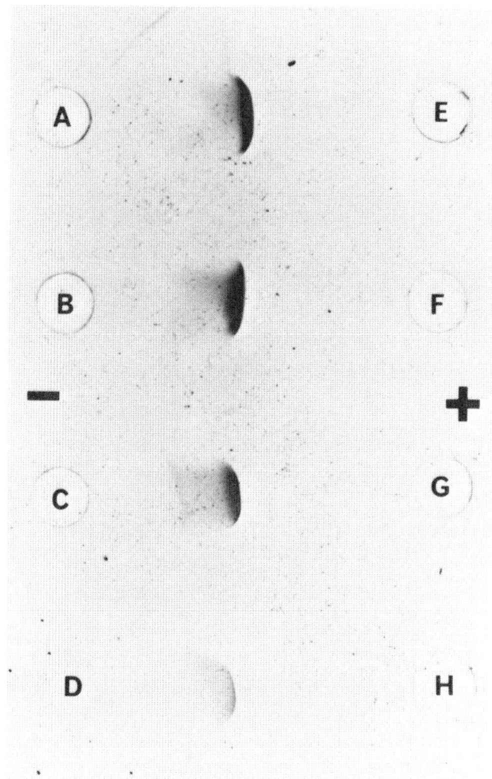

Fig. 1. Immuno-osmophoresis of the bean strain of southern bean mosaic virus. The wells on the cathode side A, B, C, and D contained 560 ng, 280 ng, 140 ng, and 70 ng, respectively, of purified virus applied in 3 μl of sample Wells D, E, F, and G on the anode side contained antiserum to the bean strain, 3 μl at a 1/10 dilution. Electrophoresis was done at 75 volts (20 ma) for 1.5 hr at room temperature in a 0.7% agarose gel in 0.02 M Tris-phosphate pH 7.0.

2. After the agarose bed is covered with buffer in the submarine electrophoresis cell, the antigen is underlayed in the well of the pair closest to the cathode and the antiserum in the well of the pair closest to the anode. Twelve tests were done on the 65 x 100 mm bed of the Bio-Rad Mini-Sub cell; four are shown in Fig. 1.

3. Conditions of electrophoresis, voltage, amperage, and duration will vary depending upon the ionic strength of the buffer, the size of the electrophoresis bed and the properties of the antigen. The voltage can be kept as high as possible without denaturation of the virus or antigen. The amperage in the run shown in Fig. 1 was 20 ma and the heat derived did not affect the test. The electrophoresis cell is placed over black paper and development of the precipitates can be observed with a white light through the agar bed. The electrophoresis cell cover connecting the power supply to the electrodes may be removed to wipe off condensed water, or the gel tray may be removed from the electrophoresis cell for observation of precipitate formation. Continued electrophoresis after precipitates are formed will not

affect the test. The method is dependent upon the virus or antigen moving to the anode and this migration can be checked by electrophoresis and staining as outlined for immunoelectrophoresis. The sensitivity of the method for southern bean mosaic virus is 35 ng for precipitate visualization as compared to 218 ng for tests stained as outlined for immunoelectrophoresis (**Section III.G.**, Fig. 1).

Applications

Immuno-osmophoresis has three advantages over the gel diffusion serological method. It is more rapid, more sensitive, and can be used with elongated viruses that diffuse slowly in agar gel (2,3). It has been used successfully with tobacco mosaic, alfalfa mosaic, southern bean mosaic, white clover mosaic, clover yellow mosaic and tobacco ringspot viruses. The method is dependent on the virus having a sufficiently fast anodic movement by electrophoresis in the gel to meet with the antibodies moving toward the cathode by electroendosmosis. Viruses or any antigen with a low anodic mobility in liquids are moved to the cathode by the electroendosmosis in agar and are not detected. This can be circumvented by two methods. A simple nitrophenylsulfonation reaction with the antiserum can produce highly negatively-charged, reactive antibodies which migrate to the anode, allowing detection of potato virus X which migrates to the cathode (3). Alternatively the electroendosmotic flow in the agar can be reversed by introducing basic DEAE groups into the agar (4).

Literature Cited

1. Kohn, J. 1971. Immunoelectroosmophoresis. Pages 277-279 in: Methods in Immunology and Immunochemistry. Vol. III. Reactions of Antibodies with Soluble Antigens. C. A. Williams and M. W. Chase, eds. Academic Press, New York. 515 pp.
2. Ragetli, H. W. J., and Weintraub, M. 1964. Immuno-osmophoresis, a rapid and sensitive method for evaluating viruses. Science 144:1023-1024.
3. Ragetli, H. W. J., and Weintraub, M. 1965. Immuno-osmophoresis adapted for weakly charged antigens. Biochim. Biophys. Acta 111:522-528.
4. Ragetli, H. W. J., and Weintraub, M. 1966. Agar derivatives for gel electrophoresis requiring reduced endosmotic liquid flow. Biochim. Biophys. Acta 112:160-167.

Section IV.A.1

ELISA METHODS FOR PLANT VIRUSES

R. H. Converse
United States Department of Agriculture, Agricultural Research Service,
Horticultural Crops Research Unit, Corvallis, OR 97330

R. R. Martin
Agriculture Canada Research Station,
Vancouver, B. C., Canada V6T 1X2.

Principles

Enzyme-linked immunosorbent assays (ELISA or EIA) were first applied to virus detection by Avrameas (2). He showed that covalently linking an immunoglobulin molecule to an enzyme could retain the immunospecificity of the former and the catalytic properties of the latter. Applications with such reagents bound to a solid phase allowed enzyme-mediated amplification of detection sensitivity of antigen-antibody reactions, pushing serological detection to limits previously only achieved by radioimmunoassays (see **Section IV.B.1.** for discussion of radioimmunoassays). Use of this principle in medical applications (17,39) demonstrated its enormous potential, and application of ELISA to the detection of plant pathogens soon followed (10,39). Since the work of Clark and Adams (10) on ELISA protocols for use with plant viruses, ELISA has been used for the detection and quantitative estimation of a large number of antigens from plants, including those of many plant pathogenic viruses, mycoplasmalike organisms (MLO), bacteria and fungi. A treatment of ELISA for MLOs and bacteria appears in **Section IV.B.2**.

Discussion of ELISA as it relates to plant viruses will be presented in this section. Despite its sensitivity, ELISA, typically detects only viral antigen and does not measure infectivity. Bioassays are still necessary to determine whether an ELISA-positive is infective.

For theoretical considerations and a more complete range of applications of ELISA see: Bar-Joseph and Garnsey (5); Clark (9); Clark et al (12); Jones and Torrance (20); Kurstak (24); and Miller and Martin (29). In addition, there are many papers dealing with the detection and quantitation by ELISA of a myriad of specific plant viruses. In this chapter, only non-competitive ELISA assays will be considered, and the reader is referred to Clark and Engvall (8) and Kurstak (24) for a discussion of competitive ELISAs and their uses.

Within the last decade various kinds of ELISA have become the preferred types of serological tests for plant viruses because of their simplicity, adaptability, rapidity, sensitivity and accuracy. If an enzyme is linked covalently to the antiviral immunoglobulin, the test is classified as a type of direct ELISA (Fig. 2, A-D). If the enzyme is linked to a molecule which detects the antiviral immunoglobulin, the test is classified as a type of indirect ELISA (e.g., Fig. 2, E-I). Many other serological tests rely on evaluation of a visible antigen-antibody precipitate. ELISA, on the other hand, registers the occurrence of antigen-antibody complexes by rapid enzymatic development of a distinctly colored product. This is accomplished by successive "layers" of reagents (Fig. 2) culminating in enzymatic hydrolysis of enzyme substrate, converting a colorless substance to a pigment.

Materials

All ELISA methods require a number of components, but the flexibility of ELISA technology has led to the development of many variations in each of these components. The kinds of components often used in ELISA are described under the seven headings that follow.

1. Trapping Antibodies

Antibodies produced in any experimental animal can be used in ELISA. Although crude animal serum can be used and is recommended in some protocols (15), generally the serum globulin (usually = immunoglobulin G, IgG) fraction is preferred. IgGs can be isolated from whole serum by using ammonium or sodium sulfate precipitation followed by dialysis (28,33) and chromatography on DEAE cellulose. IgGs can also be purified from whole antiserum by affinity chromatography on a Protein A-Sepharose CL-4B column (Pharmacia), followed by elution with 0.1 M glycine-HCl buffer, pH 2.5, and dialysis against phosphate-buffered saline (PBS, see **Procedures** for formula). (Also see **Section II.B.1.**). Purified IgGs are usually stored at a concentration of 1 mg/ml [absorbance at 280 nm for rabbit and mouse IgG (1 mg/ml, 1 cm light path): 1.4; and for $F(ab')_2$: 1.06]. Antisera and IgGs usually are stored frozen (-20 C) or at 4 C with sodium azide (a dangerous poison!) at 0.02% (w/v) as a biocide in non-adhering plastic or siliconized glass containers. They also may be stored mixed 1:1 (v/v) with spectral quality glycerol so they remain liquid at -20 C. For use in ELISA the antisera or IgGs must be diluted in suitable buffer (see **Procedures**) to a predetermined optimal concentration [usually 1-10 μg/ml (5)] for the intended test. This optimum is determined by evaluating a series of concentrations of the antibody stock against known antigens.

Use of antiviral IgGs contaminated with IgGs reacting with normal plant antigens ("antiplant antibodies") can lead to false positive ELISA readings. Thorough purification of the viral antigens used to immunize animals for antiserum production usually avoids this problem. Immunization with excessive amounts of immunogen should be avoided (see **Section II.B.1.**). Difficulties in purification of some viruses from some plant hosts may exacerbate this problem. Strong antibody-eliciting plant antigens may be present without being readily detectable by ultraviolet spectrophotometry (30). Various cross-absorption procedures to reduce antiplant antibodies are discussed in **Section II.B.1.** and by Clark (9).

2. Test Antigens

For the majority of viruses, host tissue grinding is required, and optimal buffers and additives must be determined empirically or from literature. Often, however, PBS, pH 7.4, serves as a standard extraction buffer, or at least as a starting point for developing an appropriate extraction buffer. The buffer chosen to use with viral antigen is often called "ELISA virus buffer" (EVB) (also see **Section IV.A.2.**). Nine additives to PBS are listed below. Items a-c generally are included, while the chemicals listed in item d. have been found to be valuable additions in special instances (3,28). (See also **Comments**).

a. Tween-20 (polyoxyethylene sorbitan monolaurate) at 0.5 ml/l is added to PBS to retard non-serological attachment to the ELISA solid phase surface after the trapping antibody or antigen has been incubated.
b. Polyvinylpyrrolidone, M_r 10,000-40,000 (PVP-40), at 20 g/l is added to bind polyphenols that might otherwise modify viral antigenic structures.
c. Various blocking agents at 2-20 g/l: bovine serum albumin (BSA), ovalbumin (OA), gelatin, or nonfat dried milk powder, are added to compete with plant and viral antigens for attachment to unoccupied solid phase support surfaces.
d. 2-mercaptoethanol, sodium diethyldithiocarbamate and sodium ethylenediaminetetraacetate may be added as antioxidants or as chelating agents for undesirable divalent cations.

In the case of some plant hosts, the pH of sap from triturated, infected plant material may be low enough despite the presence of commonly used buffers, to impair the reactivity of immunoglobulins in ELISA or to denature viral antigens. This was apparently the case with sap from highbush blueberry (*Vaccinium corymbosum* L.) infected with various viruses which gave strongly positive ELISA readings when 0.5% nicotine alkaloid (v/v) was added to PBS, Tween 20, OA and PVP-40, pH 7.5 and used as the grinding buffer, but not when nicotine alkaloid was omitted from the buffer system (27).

Viruses may be distributed irregularly throughout the plant, especially in perennial plants. Therefore, multiple sampling or pooled samples must be used to determine virus presence in such cases. Some viruses, such as tobacco mosaic virus, occur in sufficiently high concentrations in infected tissue that virus diffusion from leaf disks simply immersed in buffer for 1-2 hr can produce satisfactory ELISA-test results (32).

Sample preparation is the most labor-intensive step in the entire ELISA process. Various types of equipment are used to triturate virus-infected plant tissues (usually leaves) in EVB. These include a) mortar and pestle, often regarded as the optimum method for quality of preparation, even though it is slow; b) various homogenizing machines like the Polytron PT 1-/35 (Brinkmann Instruments; T-18 Ultra-Turrax, Janke and Kunkel KG); c) manual rollers (hand-held plant tissue disrupter, Agdia) or electrically driven rollers (Firma MEKU), that can express leaf sap with or without the presence of EVB, as desired.

3. Immunoprobes

Antibodies may be linked covalently with any one of several enzymes. The resulting immunoprobes react serologically with target antigens and function enzymatically to amplify the signal of a serological reaction to a detectable level. A discussion of enzymes that are used frequently for this purpose in ELISA, together with substrates and relative merits and demerits has been prepared by Kurstak (24). Two enzymes widely used for conjugation with antibodies are alkaline phosphatase and horseradish peroxidase. Methods for chemically linking these enzymes to immunoglobulins are provided in **Procedures**. Some notes on these two enzymes follow:

a. Alkaline phosphatase - Alkaline phosphatase (AlP) is rather expensive, currently costing about US$80 (Scripps) for enough enzyme to conjugate with 1 mg IgG. It is relatively stable unless frozen, tolerates sodium azide as a preservative, and usually does not react with plant extracts. It has a very high activity: 1 mg of enzyme protein hydrolyzes 550 μM/min of the colorless reactant p-nitrophenyl phosphate (PNPP). AlP quickly hydrolyzes and dephosphorylates PNPP, giving rise to a bright yellow product, p-nitrophenol (PNP). AlP has the added advantage of essentially linear reaction kinetics.

To be effective in dephosphorylating colorless substrate, AlP must be in a medium with an alkaline pH, optimum 9.8. Therefore the enzymatic activity of AlP on substrate usually is stabilized by use of a 1.0 M solution of diethanolamine at pH 9.8. The type of AlP used commonly in ELISA is Type VII from bovine intestinal mucosa, stabilized with 100 μM $ZnCl_2$ and 1 mM $MgCl_2$. However, Type III from the bacterium Escherichia coli, stabilized with 0.5 mM of the same salts, can also be used.

Glutaraldehyde (final concentration 0.06% v/v) is used for linking AlP to the probing IgG. See **Procedures** for protocols and buffer solutions.

For $F(ab')_2$ ELISA (See Procedure 4), usual probes are Staphylococcus aureus A protein (Protein A) conjugated with AlP or with horesradish peroxidase (HRP) which are available commercially (e.g., Sigma). Protein A binds strongly and specifically to the Fc portion of the immunoglobulin.

b. Horseradish peroxidase - HRP is relatively inexpensive. Ten mg of relatively pure (Type VI, RZ=3) HRP (Sigma), enough to conjugate 5 mg of immunoglobulin, currently costs US$14.40. It is a rather stable enzyme mixture that can be stored frozen, but it is inactivated by sodium azide. It usually does not react with plant extracts, but plant extracts may contain polyphenol oxidases or peroxidases which can hydrolyze substrates used with HRP and lead to false positive results in ELISA. HRP has high substrate activity and functions at pH values near neutrality. Two substrates used commonly with HRP are 3,3',5,5'-tetramethyl benzidine dihydrochloride (TMB), which is non-carcinogenic and non-mutagenic, and o-phenylenediamine (OPD) which may be mutagenic. Both require the addition of freshly diluted hydrogen peroxide.

Glutaraldehyde or periodate oxidation may be used to link HRP covalently with desired antibody molecules by procedures paralleling those used for conjugating immunoglobulin and alkaline phosphatase molecules (see **Procedures**).

c. Biotin/avidin - Another useful system is biotin/avidin linking, wherein the immunoprobe is biotinylated and the enzyme is conjugated to the glycoprotein avidin that has a very high binding affinity with biotin (14) (**Sections II.D.** and **IV.A.2.**). Standard reagents are available for biotinylating IgGs (e.g. Zymed), and conjugated avidin is also available commercially (e.g. Zymed).

4. Substrates

With AlP, p-nitrophenyl phosphate (PNPP) is the preferred substrate to produce an ELISA color reaction. The dry chemical should be stored at -20 C. It can be obtained in tablet form for easy preparation of solutions containing 1 mg/ml PNPP and should be made fresh just before use in 10% v/v (=1M) diethanolamine (Caution: toxic, harmful to eyes!) adjusted to pH 9.8 with high normality HCl. A 10% diethanolamine stock solution may be stored, but the pH should be adjusted before each use. Care should be taken to assure that all equipment used to prepare PNPP is free of alkaline phosphatase. Hand and skin surfaces are rich in human alkaline phosphatase, and thus contact with skin or fingerprints must be avoided. Extra unused PNPP substrate solution can be used as a quality control and should remain water-clear throughout the ELISA test. The development of yellow p-nitrophenol (PNP) can be allowed to proceed at room temperature as long as desired (generally 30-60 min, but overnight, if need be). The enzymatic production of PNP can be stopped by the addition and mixing of 50 μl 3 M NaOH/well, or the absorbance at 405 nm can be determined immediately for each well in a rapid ELISA plate reader, without adding NaOH.

For ELISA using HRP, the substrate tetramethyl benzidine (TMB) is first dissolved in dimethylsulfoxide at the rate of 10 mg/ml, and 10 ml of the mixture are added dropwise to 1 liter of 0.1 M sodium acetate/citric acid buffer (pH 6.0). The resulting solution cannot be stored more than 48 hr, and 0.2 ml of 30% H_2O_2 must be added within 2 hr of use. A blue color reaction, indicating a positive result, develops at room temperature within 60 min. The reaction is stopped by addition of 50 μl 2 M sulfuric acid/well, and the absorbance at 450 nm can be evaluated in a plate reader (7). The enzymatic activity of HRP is blocked by sodium azide which should therefore not be used in HRP ELISA.

A number of other substrates can be used in ELISA and are discussed by Kurstak (24). These include fluorescing substrates, discussed in **Section IV.D.**, and ELISA systems that measure enzymatically generated chemical luminescence (40).

5. Microtiter Plates

Although ELISA can be conducted on polystyrene beads as the solid phase (Precision Plastic Ball) and on other polystyrene shapes, rectangular polystyrene plates containing 96 flat- or round-bottomed wells are most generally used. These plates are available with various degrees of antigen or antibody binding capacity. Polyvinyl plates are also used, but tend to give higher non-specific background readings than the somewhat less sensitive polystyrene plates (24). A number of manufacturers make polystyrene ELISA plates. The ability of a given brand of polystyrene plates to bind trapping antibody should be evaluated in the user's laboratory before committing to that brand, and subsequent shipments of ELISA plates should be monitored for uniformity. In a uniformity test the coefficient of variation in an ELISA for a plate should not exceed 5%, and care should be taken to establish that marginal and corner wells produce normal readings.

At all stages of ELISA, temperature uniformity within an ELISA plate is important to avoid variation. For critical work, after each washing step, equilibrating plates to their next incubation temperature before adding the next liquid phase may help reduce within-treatment variation in the final ELISA absorbance values. Uniformity can also be improved if plates are not stacked during the incubation steps.

Complete removal by thorough washing of macromolecules that are not attached specifically on the ELISA plate is of critical importance in reducing background and non-specific ELISA absorbance. Washing is done with PBS-Tween, PBS-Tween-blocking agent, and even tap water has been used successfully. At the end of the incubation period for each step in the ELISA process, the contents of the wells are deftly emptied out in such a way as not to cross-contaminate from well to well. This is of particular importance when emptying antigens from the plate. When washing ELISA plates by hand with a wash bottle, the stream directed into each well should be forceful and directed from a new angle at each of three successive washes. However, care must be taken not to splash from well to well in the process. Excess liquid can be removed from wells after washing by forcefully pounding the plates onto a pad of paper towels. Several manufacturers offer automatic ELISA plate washers (e.g. Dynatech, and Flow Laboratories) that can help reduce within-sample and between-plate ELISA variation. However, careful hand washing can yield similar results.

Once used, plates can be regenerated, if required, to restore part of their original binding activity by special washing procedures, designed to remove previous reactants (6).

6. Plate Readers

Semi-quantitative results of color development in ELISA plates using AlP and PNPP, or HRP and TMB can be obtained by eye with codes assigned to predetermined color intensities (22). For PNP, the unaided eye can determine color at approximately 0.15 O.D. at A_{405nm}. The use of a uniform light source and background will help reduce the variability of such determinations. A small light box (e.g. Glow Box Model GB-11-8, Instruments for Research & Industry) is useful for such readings and also can be used to help decide when the color in a plate is developed sufficiently to determine well absorbances on an ELISA plate reader.

If absorbance values for individual wells are desired in the absence of an ELISA plate reader, the contents of each well may be read individually in a standard colorimeter at the appropriate wavelength in standard or flow-cell cuvettes. Because of the time required for such determinations, the enzymatic reaction should be stopped as noted in 4) above before beginning the readings.

There are now many excellent through-the-plate ELISA readers available commercially. A few examples include manual readers from Bio-tek and Lab Systems, and automatic readers such as Titertek Multiskan from Flow Laboratories, and Veemax Kinetic Microplate Reader, Molecular Devices.

Most ELISA plate readers are able to determine absorbance at several wavelengths that can be sélected by the operator. Some are able to read simultaneously at two wavelengths in the same well and thus correct for optical irregularities in the ELISA plate surface. Many plate readers can be interfaced to a computer to store, analyze, and present absorbance data.

7. Data Recorders and Processors

Machine recording and printing of ELISA absorbance data usually is made available as an optional capability of the equipment supplied by many ELISA plate reader manufacturers. They also generally offer computer software for transmitting absorbance data from the ELISA plate reader to a personal computer and for recording, analyzing, storing, and printing the resulting data. If one is selecting an ELISA reader for the laboratory, it is advisable to have a prior demonstration of its software capabilities as well as its performance in reading plates directly.

Procedures for Plant Virus ELISA

1. Enzyme-immunoglobulin Conjugate Preparation (2) (Also see **Section II.C.**)

a. One-step glutaraldehyde method (alkaline phosphatase)

1) PBS is used throughout ELISA and has the following recipe per liter of distilled water: NaCl, 8.0 g; KH_2PO_4, 0.2 g; $Na_2HPO_4.12H_2O$, 2.9 g; KCl, 0.2 g; NaN_3, 0.2 g; Tween-20 0.5 ml (when called for in PBS-Tween). The pH of PBS is adjusted to 7.4 with HCl.
2) Add 1.0 mg IgG in 1 ml PBS, pH 7.4, to 5 mg Type VII AlP (Sigma, Scripps, Boehringer-Mannheim Biochemicals).
3) Dialyze overnight at 4 C in three changes of PBS, total 1500 ml.
4) Add fresh glutaraldehyde (e.g 25% aqueous, electron microscopy grade, Sigma) to a final glutaraldehyde concentration (v/v) of 0.06%, mixing well. This can be added to the immunoglobulin which is then held at room temperature for 1-2 hr, or the glutaraldehyde can be added to the dialysis buffer to a final concentration of 0.06% and dialyzed overnight at 4 C.
5) Dialyze overnight at 4 C in three changes of PBS, total 1500 ml.
6) Move dialysis sack to 0.05 M Tris-HCl buffer, pH 8.0, with 1 mM $MgCl_2$ and dialyze overnight at 4 C, total 1000 ml.
7) Add bovine serum albumin to 1% (w/v) and sodium azide to 0.02% (w/v) and store at 4 C in the dark.
8) Determine optimum concentration for ELISA by a checkerboard test evaluating a series of conjugate dilutions against a series of trapping globulin dilutions using known virus (homologous antigen) and healthy plant preparations of the same species used to increase the virus (Fig. 1).

b. Periodate oxidation method (horseradish peroxidase) (24,31)

1) Dissolve 4 mg of HRP in 1.0 ml of distilled water.
2) Add 0.2 ml of 0.1 M sodium metaperiodate which must be freshly prepared. If the HRP does not turn a greenish color, start over with fresh periodate.
3) Shake for 20-30 min at room temperature.
4) Dialyze at 4 C for several hr or overnight against distilled water adjusted to pH 4.4 with acetic acid.
5) Transfer to a glass tube and add 50 μl of 0.01 M carbonate buffer, pH 9.6.
6) Add 8 mg of IgG in 1.0 ml of 0.01 M carbonate buffer, pH 9.6.
7) Incubate at room temperature with occasional shaking.
8) Add 0.8 ml of freshly prepared sodium borohydride solution (5 mg/ml freshly prepared in 0.1 mM NaOH) and incubate for 2 hr at 4 C.
9) Add an equal volume of saturated ammonium sulfate and leave at room temperature for 30 min.
10) Collect the precipitate by centrifugation, dissolve in 4 ml of PBS and dialyze against three changes of 500 ml PBS.
11) Store freeze-dried under vacuum or at 4 C.

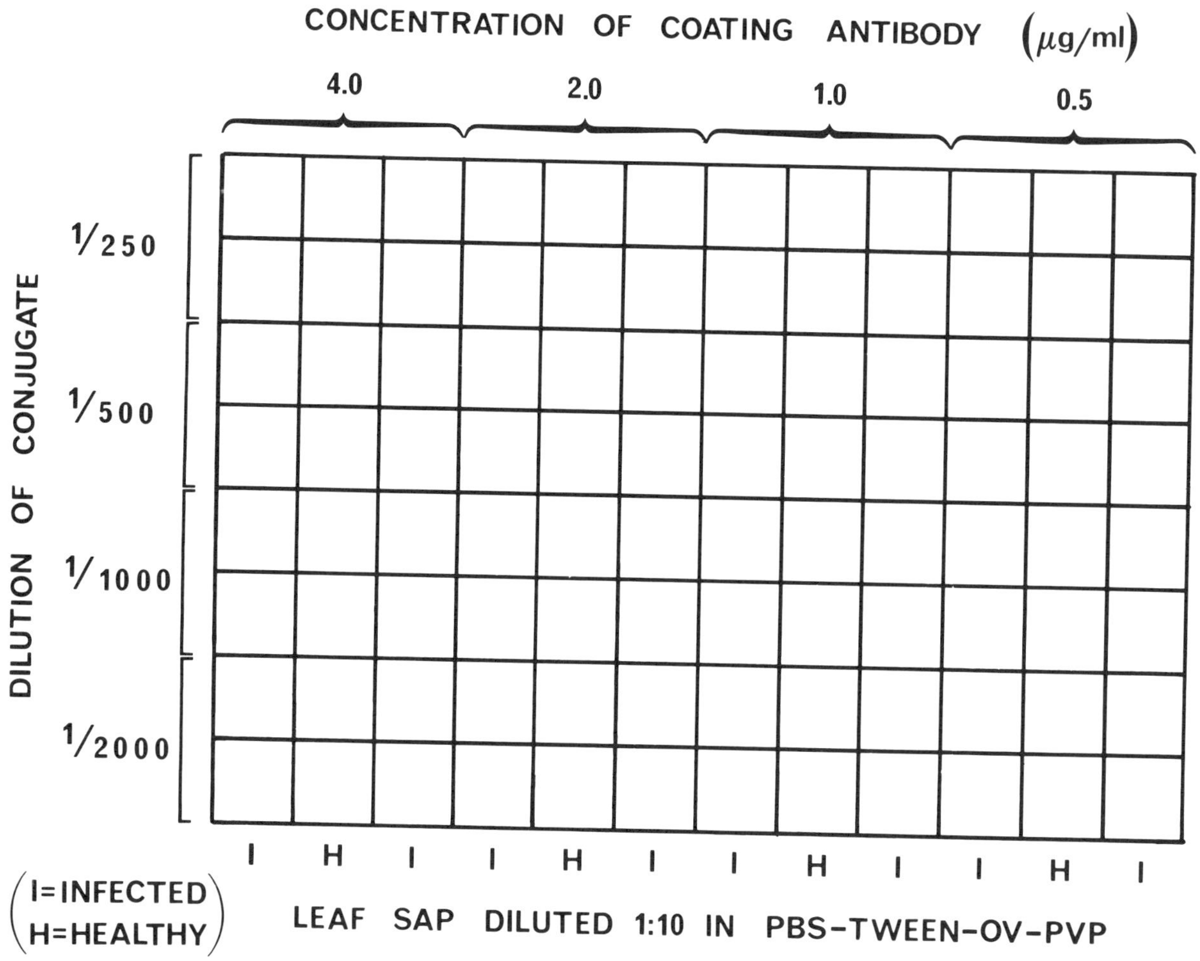

Fig. 1. Schematic diagram of checkerboard arrangement of coating and conjugate concentration on an ELISA test plate to optimize concentrations for ELISA systems.

2. <u>F(ab')$_2$ Preparation (4)</u> (Also see **Section II.B.1.**)

   a. Dialyze 2 mg IgG overnight at 4 C in three changes of 0.07 M sodium acetate plus 0.05 M NaCl, pH 4.0, total 1500 ml.
   b. In distilled water prepare pepsin stock solution (1 mg/ml) directly from high quality frozen dry enzyme powder (e.g. Sigma) and add 90 μl stock pepsin solution to the IgG solution.
   c. Incubate overnight at 37 C.
   d. Dialyze overnight at 4 C in three changes of PBS, total 1500 ml.
   e. Adjust resulting F(ab')$_2$ solution to 1 mg/ml (=1.06 O.D. at A_{260nm} in a 1 cm light path).
   f. Make 0.02% (w/v) with respect to sodium azide.
   g. Determine optimum concentration for use as trapping globulin by a checkerboard test (Fig. 1).
   h. Store at -20 C.

3. Double Antibody Sandwich (DAS) ELISA (Fig. 2 A-D and Fig. 3 A)

a. General notes (10; see Fig. 3A)

1) Add trapping antibodies at appropriate dilution in ELISA coating buffer in ELISA plate and incubate (see DAS ELISA protocol). ELISA coating buffer has the following recipe per liter of distilled water: Na_2CO_3 1.59 g, $NaHCO_3$ 2.92 g, NaN_3 0.2 g. The solution is adjusted to pH 9.6 with 0.1 N HCl and can be stored for use but must be pH-checked just before use. Coated plates can be stored at 4 C, if desired, sealed to prevent water loss, at -20 C, or can be freeze-dried for use at other times or places.
2) By machine or by hand, empty trapping antibody solution from the wells and wash them thoroughly with PBS Tween, at least three times, three min soaking each time. If necessary, add a substance like non-fat dry milk powder (at 1% w/v) to the second wash and incubate for 1-2 hr to help block remaining polystyrene well reactive surfaces. At this and each subsequent washing step be sure the ELISA plate is dry after last wash (by machine aspiration or by pounding the plate on a stack of paper towels by hand). Aspirate out any remaining bubbles but take care not to scratch the well surfaces or the underside of the plate's optical surfaces.
3) Add test antigen samples appropriately diluted in ELISA virus buffer (see **Materials** - 2. Test Antigens). Some test antigen preparations made from infected plant parts can be used directly. In other cases it may be necessary to filter the plant sap in order to pipet it into wells, or the tips of the pipets can be cut off so they are not blocked by plant debris. On each ELISA plate include known virus, healthy plant sap, and buffer checks. Use at least duplicate wells for each treatment and repeat the healthy plant sap treatment in several places on the ELISA plate.
4) Wash plate as in 2) above, taking special care not to cross-contaminate wells with antigens from adjoining wells when emptying or washing the plate.
5) Add appropriately diluted conjugate in ELISA virus buffer and incubate for stable viruses 3-6 hr at 37 C; for unstable viruses overnight at 4 C.
6) Wash plate as in 2) above.
7) Add freshly prepared PNPP 1 mg/ml in substrate buffer (see **Materials** - 4. Substrates). Incubate ELISA plate at room temperature until yellow color becomes visible, generally 10-60 min.
8) Stop or retard the enzymatic action when desired by adding 50 μl of 3 M NaOH/well, stirring each well to mix thoroughly or shaking the plates on a suitable shaker. (This is not necessary for fast-reading automatic plate readers.)
9) Evaluate color visually and score on a predetermined color scale using a uniform light source, or determine the absorbance of the wells at A_{405nm} in a colorimeter designed for the purpose (ELISA reader) or manually well by well in a general-purpose colorimeter or spectrophotometer. The most accurate absorbance readings are usually made near O.D. = 1.0.

10) Assess absorbance data by determining a cutoff value (positive-negative threshold value; see **Comments**, below). The mean absorbance of the healthy plant control wells on that plate plus 3-4 standard deviations of that mean is often used as a cutoff value. Wells that have absorbance values greater than this cutoff value are considered to be positive for viral antigen.

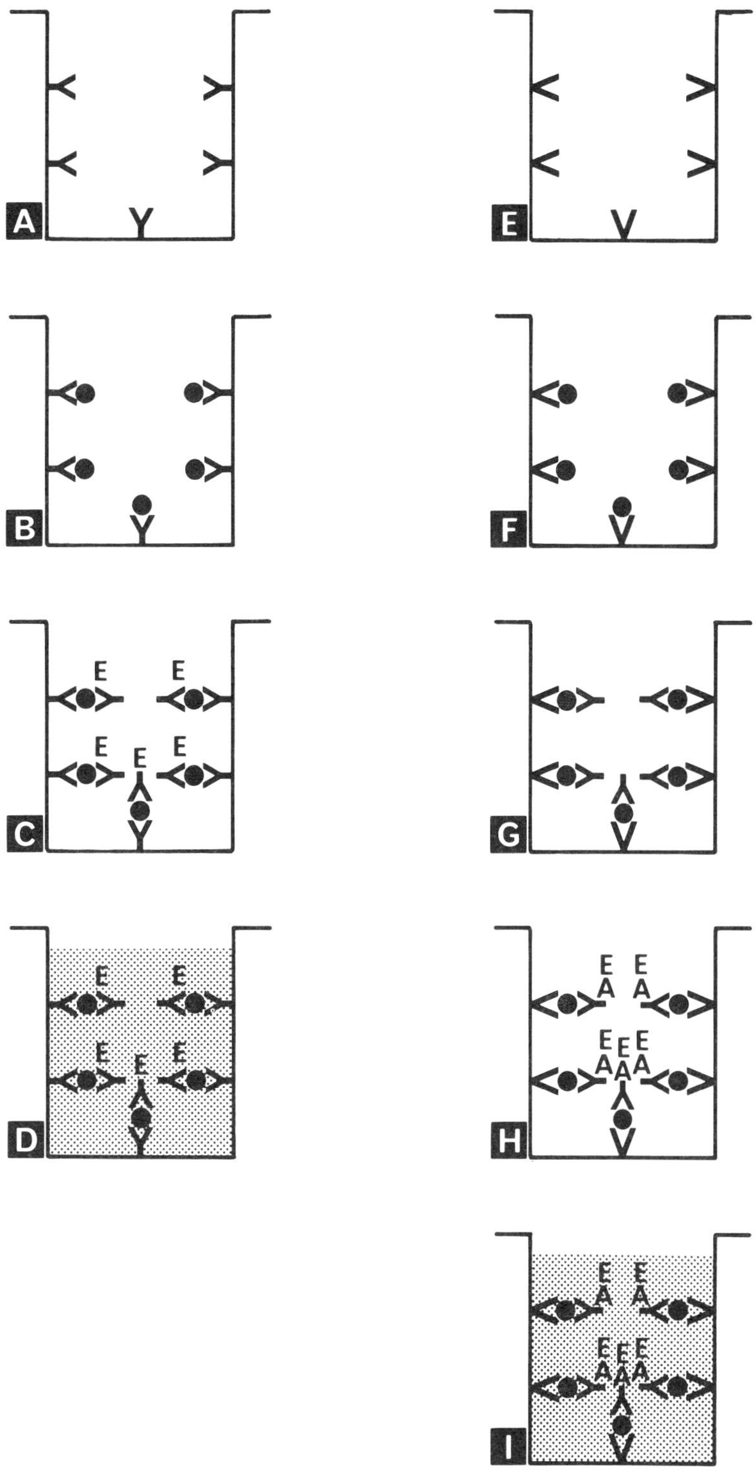

Fig. 2. A-D. Diagrammatic representation of steps for DAS ELISA with alkaline phosphatase and p-nitrophenyl phosphate in a polystyrene ELISA well.
E-I. Schematic representation of the steps involved in $F(ab')_2$ indirect ELISA with alkaline phosphatase and p-nitrophenyl phosphate in a polystyrene ELISA well. Shaded wells D and I represent production of yellow p-nitrophenol. See Fig. 3 for explanation of symbols.

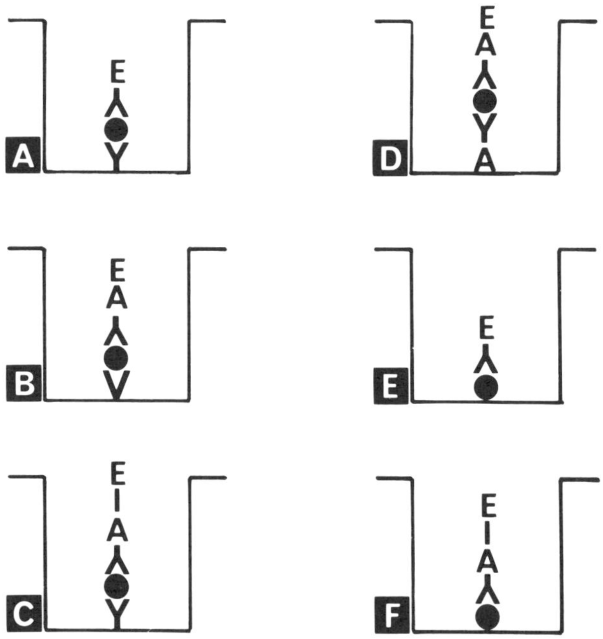

Fig. 3. Representations of several commonly used ELISA formats, with arrangement of reactants shown after the addition of alkaline phosphatase for each format. [A]. Double antibody sandwich direct ELISA. [B]. $F(ab')_2$ indirect ELISA. [C]. Two-animal ELISA used frequently with monoclonal antibodies. [D]. ELISA with Protein A coated plates with Protein A conjugated immunoglobulin. [E]. Direct ELISA with viral antigen attached to the plate. [F]. Indirect ELISA with antigen directly attached to the plate. Within the ELISA well representations: Y = antiviral IgG, V = $F(ab')_2$, YE = antiviral IgG-enzyme conjugate, o = virus, AE = Protein A-enzyme conjugate, A = Protein A, A-E = anti (first animal)-immunoglobulin conjugate.

b. DAS ELISA protocol (Note: All ELISA protocols presented in this chapter are based on use of 200 μl volume/ELISA well of each reactant, but 50 or 100 μl/well also may be satisfactory.)

1) Coat ELISA plates with trapping antibody 200 μl/well diluted in ELISA coating buffer 2 to 4 hr at 37 C or overnight at 5 C.
2) Wash wells as in step 2) of DAS ELISA procedure.
3) Add antigens 200 μl/well diluted in ELISA virus buffer, and incubate 3-6 hr at 37 C or overnight at 4 C.
4) Wash wells as in step 2) of DAS ELISA procedure.
5) Add conjugated antibody 200 μl/well diluted in ELISA virus buffer and incubate 3-6 hr at 37 C or overnight at 4 C.
6) Wash wells as in step 2) of DAS ELISA procedure.
7) Add 200 μl/well of p-nitrophenyl phosphate (1 mg/ml) in 10% diethanolamine, pH 9.8, and incubate at room temperature until yellow color develops (10-60 min or longer).
8) Stop color development if desired by stirring 50 μl 3M NaOH into each well.
9) Determine color intensity by eye or by machine (ELISA plate reader).

4. F(ab')$_2$ Indirect ELISA (Fig. 2 E-I and Fig. 3 B)

a. General notes (4)

Commercially available Staphylococcus aureus Protein A conjugated with alkaline phosphatase can be used to detect the Fc portions of intact IgGs used as second antibody coating in an ELISA plate system in which F(ab')$_2$ of the same IgG prepared by pepsinization is used to trap viral antigen. The method detects a broader spectrum of viral antigenic determinants than DAS ELISA and does not require conjugation of homologous antibodies, but F(ab')$_2$ ELISA does require two more steps than DAS ELISA (see Fig. 3 B)

The method of Edwards and Cooper (15) depicted in Fig. 2 D is discussed under Alternative Systems and Shortcuts.

b. F(ab')$_2$ ELISA protocol

1) In an ELISA plate add 200 μl/well of F(ab')$_2$ antiviral immunoglobulin fragment (see **Procedures** - 2.) that has been appropriately diluted in ELISA coating buffer, and incubate 4 hr at 37 C, or overnight at 4 C. Plates can be stored, if desired, as for plates coated in DAS ELISA (see **Procedures** - 3.).
2) Wash as in 2) of DAS ELISA procedure.
3) Add test antigen as in 3) of DAS ELISA procedure.
4) Wash as in 4) of DAS ELISA procedure.
5) Add 200 μl/well of antiviral IgG diluted in ELISA virus buffer and incubate 3-6 hr at 37 C if the antigens are heat-stable, overnight at 4 C if they are not.
6) Wash plate as in 2) of DAS ELISA procedure.
7) Add 200 μl/well of commercial or laboratory-prepared S. aureus A protein conjugated with AlP appropriately diluted in ELISA virus buffer. Incubate at 37 C 3-6 hr or overnight at 4 C.
8) Wash plate as in 2) of DAS ELISA procedure.
9) Add PNPP as in 7) of DAS ELISA procedure.
10) Stop reaction if desired as in step 8) of DAS ELISA procedure.
11) Evaluate color as in 9) of DAS ELISA procedure.

5. Indirect ELISA, antigen bound to plate (Fig. 3, E, F)

a. General notes (39).

Viral antigens can be bound to a polystyrene plate as a first step in an ELISA test. After appropriate washing and incubation steps, the viral antigen can be detected after adding conjugated homologous antibody and finally substrate (Fig. 3 E). Alternatively, after the plate-bound viral antigen is incubated, antiviral antibody of one animal species (e.g. rabbit) is added, incubated, and washed, and then commercial anti-first animal antibody conjugate (e.g. goat anti-rabbit antibody conjugated to alkaline phosphatase) is added, incubated, washed, and finally substrate is added (Fig. 3 F).

When available, antiviral antibodies derived from two different animal species can also be used for more complex versions of indirect ELISA. The first animal antibody traps viral antigen, the second antibody from a different animal species detects viral antigen. Anti-second-animal-antibody conjugate can then be used to identify the second antibody (Fig. 3 C). This method is used frequently with monoclonal antibodies from mouse or rat as the probing antibodies and rabbit-anti-mouse or rabbit-anti-rat AlP conjugates are then used to detect the probing antibodies. These conjugates are available commercially, (e.g. Sigma, Zymed). Despite the increased number of steps in such 2-animal ELISA systems, they have an advantage over the simple indirect ELISA system in not exposing viral antigen to the distortion that may occur when it is directly bound to a polystyrene surface (1).

b. Indirect ELISA protocol with plate-bound antigen (Fig. 2 F). (Also see **Section II.B.1.**)

1) Add test antigens to ELISA plate 200 μl/well in ELISA coating buffer and incubate 3-6 hr at 37 C or overnight at 4 C.
2) Wash wells with PBS Tween as in step 2) of DAS-ELISA procedure.
3) Add anti-viral antibody 200 μl/well in PBS and incubate 3 hr at 37 C or overnight at 4 C.
4) Wash wells as in step 2) of DAS ELISA procedure.
5) Add conjugated second animal species (anti-first-animal-immunoglobulin) antibody in PBS 200 μl/well and incubate for 1-2 hr at 37 C or overnight at 4 C.
6) Wash wells as in step 2) of DAS ELISA procedure
7) Add p-nitrophenyl phosphate (1 mg/ml) in 10% diethanolamine, pH 9.8, 200 μl/well and incubate at room temperature until color develops.

Comments

1. Choice of ELISA System

Indirect ELISAs, wherein the immunoprobe is not conjugated but is itself identified by another enzyme-conjugated probe, retain the native avidity of the antiviral immunoglobulin molecules. Therefore indirect ELISA systems are sometimes more sensitive and less specific than DAS ELISA, wherein the immunoprobe itself is conjugated to an enzyme (23). Indirect ELISAs, like systems directly attaching antigen to ELISA plates in coating buffer (Fig. 3 E and F), or using the $F(ab')_2$ system, are useful in broad-spectrum studies to determine even remote serological relationships among plant viruses. DAS ELISAs are best used for deciding if the test antigens have many of the same antigenic determinants as those against which the test antibodies were made. In cases where DAS ELISA and $F(ab')_2$ ELISA are both suitable, DAS has the advantage of requiring one less serological step than $F(ab')_2$, while the $F(ab')_2$ system does not require the preparation of a specific conjugate for each antibody. A further discussion of the relative merits of the two systems is given by Lommel et al. (26) and Koenig and Paul (23).

2. Alternative ELISA Systems and Shortcuts

Many standard ELISAs like those already described entail numerous steps and labor-intensive virus sample preparation. A number of investigators have attempted to reduce the number of required steps and work involved in ELISA. A few of these suggested shortcuts are listed below, and many more can be found in the rapidly growing literature. Not all shortcuts are likely to be useful with all plant viruses because of differences in virus concentration, stability, and plant extract composition.

a. Addition of trapping antibodies and conjugated antibodies to ELISA wells before addition of virus samples. Stobbs and Barker (35) outline a procedure whereby trapping antibodies are coated onto ELISA wells and then conjugated antibodies mixed with 5% gelatin (Sigma, Type B, 225 Bloom) are added to the wells in small volumes (e.g., 12 μl of 1:25 conjugate dilution in ELISA virus buffer/well). Such plates, after being sealed, can be stored at -20 C for at least 2 months without activity loss and can subsequently be used to test antigen samples after the gelatin is liquified at 38 C. The number of washing steps is reduced considerably by this method. Flegg and Clark (18) found that incubating antigen and conjugate together in ELISA wells could greatly improve the detection of apple chlorotic leafspot virus (ACLSV). They attributed this success to the instability of the ACLSV particles and their disruption in the standard DAS ELISA steps. Kaniewski and Thomas (21) took this Flegg-Clark procedure a step further by comparing it with DAS ELISA to test for six potato viruses in plant sap. They found that the modification of mixing and incubating plant sap and conjugate for a few hours at 37 C in ELISA plate wells before proceeding was more sensitive than standard DAS ELISA for all six viruses and gave results in a shorter time with the elimination of an intercurrent washing step.

b. Edwards and Cooper (15) proposed a modified $F(ab')_2$ ELISA method in which Staphylococcus Protein A is used two times, first to coat the ELISA plate in order to bind and orient the antiviral IgG, and second as an AlP conjugate to be used to probe for virus-bound second antiviral Ig (Fig. 3 D).

c. For one type of indirect ELISA, Ehlers and Paul (16) proposed covalently binding virus test samples to ELISA plates coated with 3-(triethoxysilyl)-propylamine by use of glutaraldehyde. They found that virus binding was improved by this method as compared to binding of viral antigen to ELISA plates in standard coating buffer.

d. Romaine et al. (32) found that a stack of disks cut from infected leaves and placed in an ELISA well was a good substitute for comminuting or homogenizing leaf tissue for ELISA detection of some viruses.

3. Increasing the Sensitivity of ELISA

The ELISA absorbance values obtained for virus samples can only be evaluated in terms of the mean comparable background absorbance value on that plate. Several factors contribute to the background level: a) antiplant antibodies in the antiserum used, b) non-specific binding of antigens and immunoprobes to ELISA plates, c) presence of contaminating alkaline phosphatase in the system (see discussion this chapter, Section 4. Substrates), and d) random error. The following are some ways to try to reduce background due to these three sources:

a. Antibodies having undesirably high antiplant reactions can be diluted (usually the immunoprobe) in ELISA virus buffer containing filtered crude sap (ca. 1:100 final dilution, w/v) from healthy plants of the test species being used. The resulting mixture may be incubated at room temperature 15 min - 2 hr if necessary before being placed in ELISA plate wells (25).

b. Non-specific binding can be reduced by empirical addition of a number of chemicals to ELISA virus buffer. Some of these additives are mentioned in **Materials** - 2. See the review by Kurstak (24) for a further discussion.

c. McLaughlin et al. (28) suggested that random error in ELISA plates could be reduced by appropriate statistical design of treatment replicate placement within ELISA plates. They described replicated lattice squares and Youden squares appropriate to ELISA.

d. The enzymes and substrates used for ELISA have their own inherent limits to colored product detection. Torrance (37) used the alkaline phosphatase DAS-ELISA system to produce nicotinamide adenine dinucleotide (NAD) from its phosphorylated precursor (NADP). NAD was then converted to its reduced form (NADH) in the presence of alcohol dehydrogenase. NADH in turn caused a violet to red color shift of p-iodonitrotetrazolium in the presence of the enzyme diaphorase (34). This secondary amplification system of DAS ELISA for the monoclonal antibody detection of barley yellow dwarf virus in crude sap diluted 1:256 resulted in a 17-25-fold increase in absorbance compared to that in healthy sap versus a 1.5-2-fold increase in parallel DAS ELISA. The NADP system permitted detection of this virus from extracts of single vector aphids (37).

4. Interpretation of ELISA Data

The question of how to establish the cutoff value or positive-negative threshold in ELISA data analysis has no absolute answer. The matter was discussed by Sutula et al. (36).

The use of ELISA data in a quantitative way to compare serological relationships and establish serological differentiation indices (SDIs) (38) among virus isolates has been considered by Jaegle and Van Regenmortel (19) and by Clark and Barbara (11). The latter authors presented a system for linearizing portions of dosage-response ELISA data for reciprocal tests of several viruses and their homologous antisera, and then deriving SDI values from the linearized data. They also developed computer software required to make these calculations.

Applications

At present, there are 35 plant virus groups that infect angiosperms. Almost all of these groups (13) contain at least one virus against which a specific antiserum has been prepared and for which an ELISA test has been developed. In most cases, if an antiserum has been prepared against a plant virus, an effective ELISA test has been developed to use it for detecting that plant virus coat protein. The greatest use of ELISA in plant virology to date has been in disease diagnosis to determine if the coat protein of a given virus is present in a suspect plant, vegetatively propagated clone, or seed lot. There is currently increasing interest in ELISA as a diagnostic tool for epidemiological studies of plant viruses and as a tool in studies of the infection process within the plant.

Literature Cited

1. Al Moudallal, Z, Altschuh, D., Briand, J. P. and Van Regenmortel, M. H. V. 1984. Comparative sensitivity of different ELISA procedures for detecting monoclonal antibodies. J. Immunol. Methods 68:35-43.
2. Avrameas, S. 1969. Coupling of enzymes to proteins with glutaraldehyde. Use of the conjugates for the detection of antigens and antibodies. Immunochemistry 6:43-52.

3. Banttari, E. E., and Franc, G. D. 1981. Single or combined antisera for detection of potato viruses S and X in plants or tubers using enzyme-immunosorbent assay. Phytopathology 71:201.
4. Barbara, D. J., and Clark, M. F. 1982. A simple indirect ELISA using $F(ab')_2$ fragments of immunoglobulin. J. Gen. Virol. 58:315-322.
5. Bar-Joseph, M., and Garnsey, S. M. 1981. Enzyme-linked immunosorbent assay (ELISA): principles and applications for diagnosis of plant viruses. Pages 35-59 in: Plant Diseases and Vectors: Ecology and Epidemiology. K. Maramorosch, and K. F. Harris, eds. Academic Press. New York. 368 pp.
6. Bar-Joseph, M., Moscovitz, M., and Sharafi, R. 1979. Re-use of coated enzyme-linked immunoassay plates. Phytopathology 69:424-426.
7. Bos, E. S., van der Doelen, A. A., van Rooy, N., and Schuurs, A. H. W. M. 1981. 3,3',5,5'-tetramethylbenzidine as an Ames test negative chromogen for horseradish peroxidase in enzyme-immunoassay. J. of Immunoassay 2:187-204.
8. Clark, B. R., and Engvall, E. 1980. Enzyme-linked immunosorbent assay (ELISA): theoretical and practical aspects. Pages 167-179 in: Enzyme-immunoassay. E. T. Maggio, ed. CRC Press, Boca Raton, FL. 295 pp.
9. Clark, M. F. 1981. Immunosorbent assays in plant pathology. Annu. Rev. Phytopathol. 19:83-106.
10. Clark, M. F., and Adams, A. N. 1977. Characteristics of the microplate method of enzyme-linked immunosorbent assay for the detection of plant viruses. J. Gen. Virol. 34:475-483.
11. Clark, M. F., and Barbara, D. J. 1987. A method for the quantitative analysis of ELISA data. J. Virol. Methods 15:213-222.
12. Clark, M. F., Lister, R. M., and Bar-Joseph, M. 1986. ELISA techniques. Pages 742-766 in: Methods in Enzymology, Vol. 118. H. Weissbach and A. Weissbach, eds. Academic Press, New York, NY. 829 pp.
13. Commonwealth Mycological Institute/Association of Applied Biologists. 1970-present (series). Descriptions of Plant Viruses. Assn. Applied Biologists, Wellesbourne, Warwick, U.K.
14. Diaco, R., Hill, J. H., Tachibana, E. K., and Durant, D. P. 1985. Monoclonal antibody-based biotin-avidin ELISA for the detection of soybean mosaic virus in soybean seeds. J. Gen. Virol. 66:2089-2094.
15. Edwards, M. L., and Cooper, J. I. 1985. Plant virus detection using a new form of indirect ELISA. J. Virol. Methods 11:309-319.
16. Ehlers, U., and Paul, H. L. 1984. Binding of viruses from crude plant extracts to glutaraldehyde-treated plates for indirect ELISA. J. Virol. Methods 8:217-224.
17. Engvall, E., and Perlmann, P. 1971. Enzyme-linked immunosorbent assay (ELISA). Quantitative assay of immunoglobulin G. Immunochemistry 8:871-874.
18. Flegg, C. L., and Clark, M. F. 1979. The detection of apple chlorotic leafspot virus by a modified procedure of enzyme-linked immunosorbent assay (ELISA). Ann. Appl. Biol. 91:61-65.
19. Jaegle, M., and Van Regenmortel, M. H. V. 1985. Use of ELISA for measuring the extent of serological cross-reactivity between plant viruses. J. Virol. Methods 11:189-198.
20. Jones, R. A. C., and Torrance, L. 1986. Developments and applications in virus testing. Association of Applied Biologists. Wellesbourne, Warwick. CV3 59EF. U.K. 273 pp.
21. Kaniewski, W. K., and Thomas, P. E. 1988. A two-step ELISA for rapid, reliable detection of potato viruses. Am. Potato J. 65:561-571.
22. Koenig, R. 1978. ELISA in the study of homologous and heterologous reactions of plant viruses. J. Gen. Virol. 40:309-318.

23. Koenig, R. and Paul, H. L. 1982. Variants of ELISA in plant virus diagnosis. J. Virol. Methods 5:113-125.

24. Kurstak, E. 1986. Enzyme Immunodiagnosis. Academic Press, New York, 256 pp.

25. Lister, R. M. 1978. Application of the enzyme-linked immunosorbent assay for detecting viruses in soybean seed and plants. Phytopathology 68:1393-1400.

26. Lommel, S. A., McCain, A. H., and Morris, T. J. 1982. Evaluation of indirect enzyme-linked immunosorbent assay for the detection of plant viruses. Phytopathology 72:1018-1022.

27. MacDonald, S. G., Martin, R. R., Gillett, J. M., and Ramsdell, D. C. 1988. Controlling pH of blueberry sap for efficient detection of viruses in blueberry leaves by ELISA. Acta Hort. 236:37-44.

28. McLaughlin, M. R., Barnett, O. W., Burrows, P. M., and Baum, R. H. 1981. Improved ELISA conditions for detection of plant viruses. J. Virol. Methods 3:13-25.

29. Miller, S. A., and Martin, R. R. 1988. Molecular diagnosis of plant disease. Annu. Rev. Phytopathol. 26:409-432.

30. Mink, G. I., Howell, W. E., and Fridlund, P. R. 1985. Apple tip leaf antigens that cause spurious reactions with tomato ringspot virus antiserum in enzyme-linked immunosorbent assay. Phytopathology 75:325-329.

31. Newell, D. G. 1984. Ultrastructural localization of surface and intracellular immunoglobulin in human lymphoid cell suspensions. Pages 393-404 in: Methods in Enzymology, Vol. 108, Immunochemical Techniques, Part G. Separation and characterization of lymphoid cells. G. di Sabato, J. J. Langone, and H. van Vunakis, eds. Academic Press, New York. 750 pp.

32. Romaine, C. P., Newhart, S. R., and Anzola, D. 1981. Enzyme-linked immunosorbent assay for plant viruses in intact leaf tissue disks. Phytopathology 71:308-312.

33. Saunders, G. C. 1979. The art of solid-phase enzyme immunoassay including selected protocols. Pages 100-118 in: Laboratory and Research Methods in Biology and Medicine, Vol. 3. Immunoassays in the Clinical Laboratory. R. M. Nakamura, W. R. Dito, and E. S. Tucker III, eds. A. R. Liss, Inc., New York. 366 pp.

34. Stanley, C. J., Johannsson, A., and Self, C. H. 1985. Enzyme amplification can enhance both the speed and the sensitivity of immunoassays. J. Immunol. Methods 83:89-91.

35. Stobbs, L. W., and Barker, D. 1985. Rapid sample analysis with a simplified ELISA. Phytopathology 75:492-495.

36. Sutula, C. L., Gillett, J. M., Morrissey, S. M., and Ramsdell, D. C. 1986. Interpreting ELISA data and establishing the positive-negative threshold. Plant Dis. 70:722-726.

37. Torrance, L. 1987. Use of enzyme amplification in an ELISA to increase sensitivity of detection of barley yellow dwarf virus in oats and in individual aphids. J. Virol. Methods 15:131-138..

38. Van Regenmortel, M. H. V. 1982. Serology and Immunochemistry of Plant Viruses. Academic Press, New York. 302 pp.

39. Voller, A., Bartlett, A., Bidwell, D. E., Clark, M. F. and Adams, A. N. 1976. The detection of viruses by enzyme-linked immunosorbent assay (ELISA). J. Gen. Virol. 33:165-167.

40. Wang, H. X., George, J., Thorpe, G. H. G., Stott, R. A., Kricka, L. J., and Whitehead, T. P. 1985. Enhanced luminescence enzyme immunoassay for factor VIII related antigen. J. Clinical Pathol. 38:317-319.

Section IV.A.2.

ELISA METHODS FOR PLANT PATHOGENIC PROKARYOTES

R. J. McLaughlin
USDA ARS Appalachian Fruit Research Station
Kearneysville, West Virginia 25430

T. A. Chen
Rutgers University, Cook College
New Brunswick, New Jersey 88903

Principles

Particularly after the initial report in 1971 by Engvall and Perlmann (7), the enzyme-linked immunosorbent assay (ELISA) became recognized for its potential in the detection of plant pathogenic bacteria, mycoplasma-like organisms and viruses. Several variations of this procedure have been developed. ELISA methods available for the detection of plant pathogenic prokaryotes may be subdivided according to (i) the immunoreactant initially bound to the microtiter plate wells and (ii) whether the enzyme is conjugated to the primary (direct immunoassay) or a secondary antibody (indirect immunoassay) (Fig. 1). In double antibody sandwich assays that use the direct method (Fig. 1b), antibody is initially bound to the plate wells. After incubation and washing, the antigen is added which attaches to affixed antibody to form an immune complex. After incubation and washing, enzyme-labeled antibody is then added which attaches to antigenic sites that remain exposed in the antigen-antibody complex. After further incubation and washing, enzyme substrate is added. Hydrolysis of the substrate results in a color change, the intensity of which is proportional to the amount of enzyme conjugate bound. Antigen may be affixed to plate wells in some applications, eliminating the need for the initial step as described above. Indirect ELISA methods (Fig. 1c-e) use a secondary antibody, usually an antiglobulin conjugated with enzyme, that is added after addition of the primary antibody.

This section describes two ELISA methods: (i) a direct assay using the double antibody sandwich method and (ii) an indirect assay where antigen is initially bound to the plate wells. Both of these techniques are frequently used for the detection of prokaryotic plant pathogens. General procedures that are important to the successful use of these methods are outlined; however, certain applications may require modification of these procedures. Reports using modifications of these procedures are discussed. An important modification of ELISA technology, discussed in this chapter, includes signal amplification with the biotin-avidin reagents (Fig. 1 D-E). This method originally described by Hsu et al (8) can increase the sensitivity of detection several fold. Additional modifications and precautions in the use of ELISA technology are discussed in **Section IV.A.1.** and in recent publications (2,21).

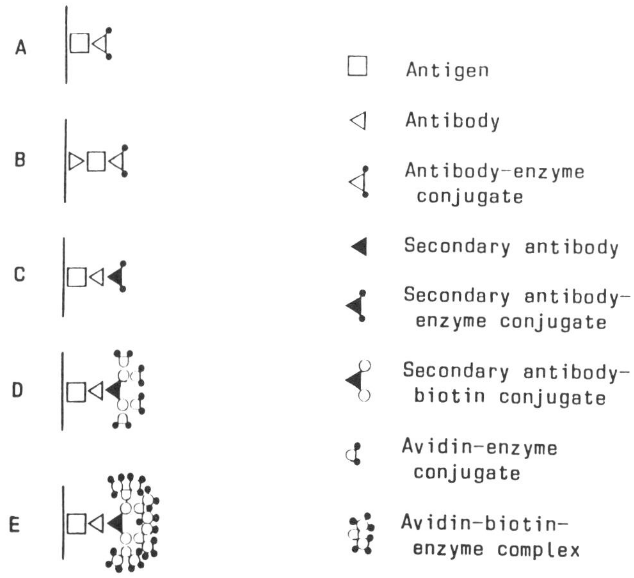

Fig. 1. Non-competitive ELISA methods. In the first, antigen (A, C, D, E) or antibody (B) is immobilized on the solid phase. In the second step, different immunoreactants are added: antigen in the sandwich assay (B), enzyme-labeled primary antibody (A), or unlabeled primary antibody (C, D, and E). In the third step, direct detection with enzyme-labeled primary antibody is commonly done with the sandwich assay (B). Indirect assays require additional steps and may use a variety of detection methods: (C) enzyme-labeled secondary antibody, (D) biotin-labeled secondary antibody plus avidin-enzyme conjugate, or (E) biotin-labeled secondary antibody plus avidin-biotin-enzyme complex. Significant signal amplification is achieved with methods (D) and (E).

Materials (Also see **Section IV.A.1.**)

Reagents

Indirect ELISA

Primary antibody (homologous with test antigen)
Secondary antibody: Anti-Ig:biotin conjugate
Avidin:horseradish peroxidase (HRP) conjugate

Direct ELISA

Homologous antibody:HRP conjugate

Common to Direct and Indirect ELISA

Bovine serum albumin, crystalline grade
Substrate, per 96-well plate:

1. Hydrogen peroxide (33%), 10 μl in 10 ml of distilled water
2. o-phenylenediamine (OPD), 0.8 mg/ml in 10 ml of 0.1 M sodium citrate buffer (0.1 M citrate, pH 5; solution A: 19.21 g citric acid in 1 liter of distilled water (DW); solution B: 29.4 g sodium citrate in

1 liter of DW.. Add 400 ml of A to 600 ml of B, adjust pH to 5.0 if necessary. OPD is light sensitive.

3. Make and mix solutions 1 and 2 immediately before use
4. 3 M H_2SO_4, for stopping the reaction

Buffers

Phosphate-buffered saline (PBS) and carbonate coating buffer may be made as 10X stock solutions. Methods for making final strength buffers are given here. These buffers should be autoclaved if they are to be stored. Sodium azide may be used as a preservative at 0.02% (w:v) when the alkaline phosphatase system is used. Sodium azide is not recommended with the HRP enzyme assay as it will inactivate the enzyme.

1. PBS:
   8.0 g NaCl, 2.9 g Na_2HPO_4. 12 H_2O, 0.2 g KH_2PO_4, 0.2 g KCl. Adjust pH to 7.4 with HCl if necessary. Adjust final volume to 1 liter with DW.
2. Carbonate coating buffer, pH 9.6:
   2.93 g $NaHCO_3$, 1.59 g Na_2CO_3. Dissolve components in 800 ml of DW. Adjust pH if necessary with 0.1 N HCl. Adjust final volume to 1 liter with DW. Store at 4 C.
3. Blocking buffer:
   0.5 to 1% (w:v) bovine serum albumin in coating buffer
4. ELISA plate washing buffer:
   PBS plus 0.05% Tween 20

Equipment and Supplies

1. Polystyrene microtiter plates with 96 U-shaped wells
2. 12-well, 50-200 μl micropipetor
3. Micropipetor tips, 1-200 μl
4. Incubator, 37 C
5. ELISA plate reader (e.g. Titertek Multiscan, Flow)
6. Multi-well ELISA plate washer (e.g. Nunc-Immuno Wash 12, Nunc)

Procedures

1. Double antibody sandwich (DAS) ELISA
   a. Suspend trapping antibody in coating buffer 1-20 μg/ml) and add 50 μl per well.
   b. Incubate for 4 hr at 37 C or overnight (12-16 hr) at 4 C.
   c. Wash the plate three times. (See **Section IV.A.1.**, Procedure 3, DAS ELISA.)
   d. Add 50 μl of antigen sample to each of the wells.
   e. Incubate 2-4 hr at 37 C.
   f. Wash nine times (soak wells for 2 min between each of three sets of three washes).
   g. Add 50 μl antibody:enzyme conjugate per well (1-20 μg/ml).
   h. Incubate 1 hr at 37 C.
   i. Wash 15 times (soak wells for 2 min between three washes of five times each).
   j. Prepare substrate solution. Add 150 μl per well and incubate for several min. Color development may take 15 to 100 min. If long incubation times are necessary, incubate the plates in the dark to

avoid background color development in control wells.

k. When a yellow color reaction is observed in positive control wells, preferably after a predetermined amount of time, stop the reaction by addition of 50 μl of 3M H_2SO_4 per well.

l. Determine the absorbance in plate wells at 492 nm with an ELISA plate reader, if available, or visually score the strength of color development in the wells.

2. Indirect ELISA with biotin-avidin reagents
   a. Prepare bacterial samples for coating. Sonicate in PBS for 2 min in an ice bath. Determine the protein concentration.
   b. Suspend the cells in coating buffer $\leq 10^8$ cells/ml (for pure cultures) or at $\leq$ 60 μg/well (total protein from plant extracts or washings) and add 50 μl to each well.
   c. Incubate for 4 hr at 37 C or overnight (12-16 hr) at 4 C.
   d. Wash the plate three times with washing buffer.
   e. Add 250 μl of blocking buffer to each well.
   f. Incubate for 1 hr at 37 C.
   g. Wash six times (soak wells for 2 min between each of three sets of two washes).
   h. Add antibody to the wells, 50-100 μl at 1-20 μg/ml (9).
   i. Incubate for 1-2 hr at 37 C.
   j. Wash nine times, as in (g).
   k. Add 50 μl of diluted anti-Ig:biotin conjugate to each well. Follow dilution recommendations of the manufacturer, usually 1:1000 in PBS.
   l. Incubate for 45 min at 37 C.
   m. Wash 15 times (soak wells for 2 min between each of five sets of three washes).
   n. Add 50 μl of diluted streptavidin-HRP conjugate. Follow dilution recommendations of the manufacturer, usually 1:1000 in PBS.
   o. Incubate for 15 min at 37 C.
   p. Wash 15 times, as in (m).
   q. Add 150 μl of substrate (OPD) per well. Because of multiple HRP/secondary antibody, incubation time is shorter than the previous assay - usually a sufficient color change is observed within 5 min of adding the substrate.
   r. Follow steps j-l in Procedure 1 to terminate the reaction and determine extent of color development in the microtiter plate wells.

Comments

1. Antigen preparation from plant tissue. Several preparatory methods are useful in assuring successful results in ELISA. Plant bacteria or mycoplasma-like organisms may be present in low levels in infected tissues, requiring homogenization of tissues to release the antigen. Extraction with standard buffers (e.g. PBS) may be inadequate, and it may be necessary to include certain reagents to enhance recovery. Addition of an adsorbent (e.g. polyvinylpyrrolidone 0.6% w:v) to reduce denaturing due to sap phenols, antioxidants (e.g. 2-mercaptoethanol, 0.1-0.5% w:v; cysteine, 4 mM), and/or chelating agents (sodium ethylenediamine-tetraacetate, 10 mM) may be used to enhance recovery of the antigen during maceration of plant tissue (3). These reagents may also be added to crude sap preparations prior to coating (Procedure 1) or prior to adding to antibody-coated wells

(Procedure 2). Isolation of bacteria from macerated, infected tissue can be enhanced by low-speed centrifugation (1000 g, 5 min) to pellet large plant debris.

2. Antigen preparation for adsorption to plate wells. In sandwich assays, adsorbed antibodies facilitate adhesion of intact bacteria to microtiter plate wells. Indirect ELISA methods, in our experience, require cell breakage by sonication to facilitate and improve adhesion of bacterial antigens to microtiter plate wells.

3. Choice of microtiter plates. ELISA plates with U-shaped wells are used in these assays. This type of well configuration reduces the volume of reagents needed. We have concluded that thorough washing of the plates is better facilitated with U-shaped wells than with flat-bottom wells (McLaughlin and Chen, unpublished). This is evidenced by lower background reactions in control wells.

4. Interruption of ELISA procedure. After incubation in blocking buffer, plates may be stored overnight in washing buffer at 4 C.

5. Washing of microtiter plates. Plates should be washed a consistent number of times in order to reduce variability between assays. Thorough washing, as advised in the procedures, is also recommended to reduce background color development in negative control wells. This is critical when you are using signal amplification reagents, particularly with assays as shown in Fig. 1e.

6. Incubation conditions. Incubation times and temperatures for different steps in the procedure can be varied, depending on the specific application and on the affinity of the antibody to the antigen in each system. The following incubation times and temperatures have been suggested for conjugate and substrate steps (2):
   Conjugate: 2-4 hr, 37 C, or overnight at room temperature.
   Substrate (in systems without signal amplification): for AP, 30-100 min, 37 C; for HRP 30-100 min, room temperature.
   Incubation times with signal amplification systems may take as little as 1-5 min.

7. Choice of signal enzyme system. Other substrates or enzyme systems can be substituted for signal detection. If the HRP system is to be used, 5-aminosalicylic acid can be substituted for OPD. This substrate is preferred over OPD as it is not photosensitive and it is not a suspected carcinogen (1,6). Alkaline phosphate (AP) is frequently used in DAS-ELISA. Background signal intensity from endogenous plant peroxidase may be reduced by the use of the AP enzyme system in either direct or indirect ELISA systems. Consult **Section IV.A.1.** for the substrate and buffer system used with the AP system.

8. Specificity of antisera. Background reactions can occur if antiserum has low specificity. Specific antisera may be obtained by use of cross-absorbed serum (**Section II.B.1.**) or monoclonal antibodies (**Section II.B.2.**).

9. Optimization of antigen and antibody concentrations. High concentrations of bacterial (> 10^8 cfu/ml or > 15 μg/well protein) antigen may also result in unacceptable levels of background reaction. The optimal concentrations of antigen and antibody should be empirically determined with each assay

system. For general information on the optimization of antibody and antigen concentrations, consult **Sections II.B.1.** and **IV.A.1.**.

10. Modification of DAS-ELISA for signal amplification. This assay uses primary antibody that is conjugated with enzyme. Methods of enzyme conjugation are discussed in **Section II.C.** The sandwich assay can be done using indirect methods with the biotin-avidin reagents, as described in the indirect ELISA procedure, in order to take advantage of the increased sensitivity of detection that can be achieved by this method. Primary antibody may also be conjugated with biotin (**Section II.C.**) for signal amplification with an avidin-peroxidase or -alkaline phosphatase conjugate.

Applications

The procedure most commonly used for *in planta* detection of plant pathogenic prokaryotes is the DAS-ELISA (1,4,5,19,20). Indirect ELISA methods have been employed for *in planta* detection and serological differentiation of mycoplasma-like organisms (9,10,14), for in vitro screening for antibody specificity toward plant pathogenic bacteria (15,17), and for detection of bacteria in the rhizosphere (13). Indirect ELISA is chosen over DAS-ELISA in some cases because the investigator can commercially obtain secondary antibody-enzyme conjugate and save the time required to produce the secondary antibody and its enzyme conjugate. Commercially available secondary antibody-biotin reagents are also available for amplification. Indirect ELISA may also be preferred in situations where antibodies do not react well as trapping antibodies.

The limits of DAS-ELISA for *in planta* detection of bacteria have been established in several studies (1,5,11,12). A background noise level of 0.1 absorbance units is commonly agreed to be the detection limits of the assay (3). *Rhizobium* spp. have been detected at populations of 3-5 log cfu/ml with polyclonal antibodies (11,16). A detection limit of 5.3 log cfu per well (0.1 absorbance units) has been reported with polyclonal antibodies against *Xanthomonas campestris* pv. *campestris* (1). The potato ring rot bacterium, *Clavibacter michiganense* subsp. *sepedonicum*, has been detected in infected tubers at an effective dilution of 10^6 with a monoclonal antibody (5). Detection of *Azospirillum basilense* at 10^4 cfu/g root has been reported in indirect ELISA with polyclonal antibodies (13).

The detection limits of ELISA can be improved in some instances. Signal amplification techniques, better antibodies, and/or prior treatment of antigen can result in a stronger ELISA reaction. Enzyme amplification methods can be adapted to the DAS-ELISA procedure as discussed earlier. Antibodies can be obtained in some instances that have better specificity and stronger reaction in ELISA. Monoclonal antibodies may react weakly in ELISA since they are directed against a single epitope. DeBoer et al (5) developed a monoclonal antibody that reacted strongly in ELISA toward an extracellular polysaccharide antigen produced by the ring rot bacterium. McLaughlin et al (17) found that a mixture of two monoclonal antibodies with different epitope specificities could improve in vitro detection of *Erwinia amylovora* in indirect ELISA. Prior treatments of antigen can also enhance reactivity in ELISA. Several studies have shown improved reactivity if the antigen is heat treated and/or sonicated prior to absorption to microtiter plate wells (1,11,12).

Improved detection limits may not be essential in some instances, particularly when the assay is used to identify bacteria that are present in large numbers in plant lesions (1). It should also be noted that antisera may detect non-viable cells (1,4). In response to this concern, Alvarez and Lou (1) noted that the number of false positives produced by ELISA was not significantly greater than that produced by direct isolation tests on agar media or by pathogenicity tests. Another study used ELISA to follow loss of viability in Rhizobium populations (18).

The ELISA procedure has been recognized to have some advantages over other serological procedures and traditional methods of disease agent detection. The assay is most effective when large numbers of samples need to be assayed in a timely manner. Under these circumstances, this method is more efficient than immunofluorescence assays. In addition, the test may be performed without the need of an expensive fluorescent microscope. The ELISA procedure may also be of use in circumstances where other serological procedures seem to have inadequate specificity, such as latex agglutination (5). Alvarez and Lou compared the relative effectiveness of ELISA to that of plating and pathogenicity assays (1). They reported that viable plate counts on two media, pathogenicity tests, and DAS-ELISA provided comparable sensitivity, specificity, ability to determine apparent disease incidence, and predictive value. The ELISA method is also more efficient than traditional methods of plant pathogen detection.

Literature Cited

1. Alvarez, A. M., and Lou, K. 1985. Rapid identification of Xanthomonas campestris pv. campestris by ELISA. Plant Dis. 69:1082-1086.
2. Cassell, G. H., and Brown, M. B. 1983. Enzyme-linked immunosorbent assay (ELISA) for detection of anti-mycoplasmal antibody. Pages 457-469 in: Methods in Mycoplasmology, Vol. II. Mycoplasma Characterization. S. Razin and J. G. Tully, eds. Academic Press, New York. 440 pp.
3. Clark, M. F. 1981. Immunosorbent assays in plant pathology. Annu. Rev. Phytopathol. 19:83-106.
4. DeBoer, S. H., and McNaughton, M. E. 1987. Monoclonal antibodies to the lipopolysaccharide of Erwinia carotovora subsp. atroseptica serogroup I. Phytopathology 77:828-832.
5. DeBoer, S. H., Wieczorek, A., and Kummer, A. 1988. An ELISA test for bacterial ring rot with a new monoclonal antibody. Plant Dis. 72:874-878.
6. Ellens, D. J., and Gielkens, A. L. J. 1980. A simple method for the preparation of 5-aminosalicylic acid. Application of the product as a substrate in enzyme-linked immunosorbent assay (ELISA). J. Immunol. Methods 37:325-332.
7. Engvall, E., and Perlmann, P. 1971. Enzyme-linked immunosorbent assay (ELISA). Quantitative assay of immunoglobulin G. Immunochemistry 8:871-874.
8. Hsu, S., Raine, L., and Fanger, H. 1981. Use of avidin-biotin-peroxidase complex in immunoperoxidase techniques: A comparison between ABC and unlabeled antibody (PAP) procedures. J. Histochem. Cytochem. 29:577-580.
9. Hung, S., Chen, T. A., Whitcomb, T. A., Tully, J. G., and Chen, Y. X. 1987. Spiroplasma culicicola sp. nov. from the salt marsh mosquito Aedes sollicitans. J. Syst. Bacteriol. 37:365-370.
10. Jiang, Y. P., Lei, J. D., and Chen, T. A. 1988. Purification of aster yellows agent from diseased lettuce using affinity chromatography. Phytopathology 78:828-831.
11. Kishinevsky, B., and Bar-Joseph, M. 1978. Rhizobium strain identification in Arachis hypogaea nodules by enzyme-linked immunosorbent assay. Can. J. Microbiol. 24:1537-1543.
12. Kishinevsky, B., and Gurfel, D. 1980. Evaluation of enzyme-linked immunosorbent assay (ELISA) for serological identification of different

Rhizobium strains. J. Appl. Bacteriol. 49:517-526.

13. Levanony, H., Bashan, Y., and Kahana, Z. E. 1987. Enzyme-linked immunosorbent assay for specific identification and enumeration of Azospirillum basilense Cd. in cereal roots. Appl. Environ. Microbiol. 53:358-364.
14. Lin, C. P., and Chen, T. A. 1986. Comparison of monoclonal antibodies and polyclonal antibodies in detection of aster yellows mycoplasmalike organism. Phytopathology 76:45-50.
15. Lin, C. P., Chen, T. A., Wells, J. M., and van der Zwet, T. 1987. Identification and detection of Erwinia amylovora with monoclonal antibodies. Phytopathology 77:376-380.
16. Martensson, A. M., Gustafson, J., and Ljunggren, H. D. 1984. A modified, highly sensitive enzyme-linked immunosorbent assay (ELISA) for Rhizobium meliloti strain identification. J. Gen. Microbiol. 130:247-253.
17. McLaughlin, R. J., Chen, T. A., and Wells, J. M. 1989. Monoclonal antibodies against Erwinia amylovora: Characterization and evaluation of a mixture for detection by ELISA. Phytopathology 79:610-613.
18. Nambiar, P. T. C., and Anjaiah, V. 1985. Enumeration of rhizobia by enzyme-linked immunosorbent assay (ELISA). J. Appl. Bacteriol. 58:187-193.
19. Saillard, C., and Bove', J. M. 1983. Application of ELISA to spiroplasma detection and classification. Pages 471-476 in: Methods in Mycoplasmology, Vol. II. Mycoplasma Characterization. S. Razin and J. G. Tully, eds. Academic Press, New York. 504 pp.
20. Sinha, R. C., and Benhamou, N. 1983. Detection of mycoplasmalike organism antigens from aster yellows-diseased plants by two serological procedures. Phytopathology 73:1199-1202.
21. Tijssen, P. 1985. Practice and theory of enzyme immunological assays. Vol. 15 of: Laboratory Techniques in Biochemistry and Molecular Biology. R. H. Burdon and P. H. Knippenberg, eds. Elsevier, New York. 549 pp.

Section IV.B.1.

SOLID PHASE RADIOIMMUNOASSAY

J. H. Hill
Iowa State University
Ames, Iowa

Principle

Radioimmunoassay (RIA) has received relatively little attention for plant virus identification and detection. RIA has been considered slightly more sensitive than conventional double-sandwich enzyme-linked immunosorbent assay (ELISA). However, with the recent development of numerous amplification systems for ELISA, the sensitivity advantage of RIA appears to be questionable. One significant advantage, relative to various versions of ELISA, is that variability of assays performed with RIA is usually less than that which occurs with ELISA because the necessity of dealing with enzyme kinetics is obviated. This may be of considerable significance for certain kinds of assays. For other applications, it is unimportant. Another advantage of RIA is that its potential for quantitative assay of a potentially large range of naturally occurring viral strains is greater than ELISA (2). Conjugation of an antibody molecule with a large enzyme may alter the molecular conformation of the antibody and induce altered specificity for heterologous antigens (2).

Success of the RIA is based upon the reaction between immunoreactive antigen and antibodies highly specific to the antigen. Most assays used in plant virology have been designed in a solid-phase format in which the antigen or the capture antibody is bound to the solid phase. After addition of the sample, a detection antibody, either directly labeled with radioactive isotope or with biotin, is added. Biotin-conjugated specific antibody involves the use of isotopically-labeled -avidin or -streptavidin (see **Section IV.A.2.**). Alternatively, if the detection antibody is not directly conjugated, specific isotopically-labeled immunoglobulins or -Protein A can be added.

Radioactive iodine (^{131}I or ^{125}I) is most frequently used for RIA. However, it can alter biological and physicochemical properties of proteins and it gives products with relatively short half-lives. Alternatively, tritiation is a gentle reaction which yields labeled reagents that can be stored for longer periods (ca. 6-9 months) without significant losses in specific activity (2). In addition, ^{3}H is a beta rather than a gamma emitter and fewer safety precautions are needed during its use. Unfortunately, the specific activity that can be achieved with ^{3}H is low compared with that possible with ^{125}I or ^{131}I. A second disadvantage involves the greater instability of ^{3}H with respect to stable hydrogen atoms. During storage, tritiated compounds tend to lose their label through exchange with environmental hydrogen.

Materials

1. Coating buffer. The coating buffer is used to bind either the antigen (indirect assay) or capture antibody (double-sandwich assay) to the solid phase. The buffer of choice depends, in part, upon the molecular conformation desired for the antibody/antigen and the solid phase to be used in the assay. In most applications, 0.05 M carbonate, pH 9.6, or 0.1 M sodium phosphate, pH 7.2, can be used.
2. Wash buffer. Generally 0.01-0.05 M phosphate buffer, pH 7.2, containing 0.85% NaCl and 0.05% Tween-20 (PBS-Tween).
3. Blocking reagent. Several blocking reagents can be used. The choice and concentration depend upon characteristics of the antibody, antigen, and solid phase and is usually determined empirically by experimentation. The reagent is typically prepared in PBS-Tween. Common blocking agents are BLOTTO (9) [5% Carnation non-fat dried milk and 0.03% antifoam A, (Sigma)], 1-3% gelatin (Bio-Rad), 0.2-2.0% ovalbumin (Sigma), 1% bovine serum albumin (Sigma), or 10-50% horse or fetal calf serum (e.g. HyClone).
4. Chemicals. HEPES (Sigma), pyridoxal-5-phosphate (Calbiochem), NaB^3H_4 (Research Products International), Sephadex G-25 (Pharmacia), chloramine T (Amersham), L-tyrosine (Sigma), $Na^{125}I$, ^{125}I-labeled immunoglobulins, ^{125}I-Protein A, ^{125}I-avidin or -streptavidin for biotin labeled antibody, (e.g. Amersham).
5. Solid phase. "Frosted" polystyrene beads (Precision Plastic Ball), Removawell strips (Dynatech).
6. Contained washing system. "Penta Wash" (Abbot).

NOTE: Suppliers are not recommended to the exclusion of others that may be satisfactory. The ones listed have proven satisfactory.

Procedure

Labeling Methods

Several methods can be used for direct labeling of antibody. A procedure for labeling antibody with ^3H or ^{125}I, respectively, is presented. Labeling with ^3H has been used for both polyclonal (2) and monoclonal IgG (7) but not immunoglobulin M (IgM). All labeling procedures should be performed in an exhaust hood while wearing two pairs of gloves.

A. Labeling of antibody with ^3H (2)

1. Incubate antibody [4 to 5 mg/ml in 0.05 M HEPES (N-2-Hydroxyethyl-piperazine-N'-2-ethanesulfonic acid) buffer, pH 7.5] at 37 C for 30 min. Remove insoluble material by centrifugation.
2. Add 0.01 M pyridoxal-5-phosphate in HEPES to antibody at a ratio of 6 μ moles of pyridoxal phosphate per μ mole of antibody [for immunoglobulin G (IgG) equivalent to 4 μl of 0.01 M pyridoxal phosphate/mg of IgG]. Incubate at 37 C for 30 min.
3. Cool reaction mixture for 15 min in an ice-salt water bath.
4. Add 100-125 mCi of NaB^3H_4 sequentially in 4-5 steps with a 5 min incubation at 0-4 C following the first addition and a 15 min incubation terminating the sequence.
5. Inactivate residual NaB^3H_4 by adding excess 0.1 M pyridoxal phosphate.

6. Separate unbound NaB^3H_4 from ^3H-anti-viral (e.g., SMV) -IgG by gel filtration on Sephadex G-25 which has been equilibrated with 0.05 M sodium borate, pH 7.2, containing 0.85% NaCl and presaturated with 1% bovine serum albumin in the same buffer. Elute with 0.05 M sodium borate, pH 7.2, containing 0.85% NaCl and collect and pool ^3H-anti-viral-IgG fractions. Store at 4 C or at -20 C until used.

B. Labeling of antibody with ^{125}I. This procedure, courtesy of Dr. R. A. Consigli, Kansas State University, Manhattan, KS 66506, may be used for labeling polyclonal and monoclonal IgG.

1. Add 50 μl of antibody (ca. 0.5 mg/ml), 10 μl of 1.0 M phosphate buffer, pH 7.0-7.5, and 2.5 μl $Na^{125}I$ (0.25 mCi) to a disposable plastic test tube.
2. Add 10 μl of chloramine T (the initiator of the reaction).
3. After 3 min terminate the reaction by adding either a) 10 μl sodium bisulfite (12.5 mg/10 ml); or b) 75 μl L-tyrosine 200 μg/ml (time of the reaction is variable and depends on sample concentration; 3 min is appropriate for samples of 25 μg in 50 μl). L-tyrosine may be preferable as a more gentle procedure for terminating the reaction.
4. Separate "bound" from "unbound" ^{125}I by "dry" column centrifugation using Sephadex G-25 (centrifuge water-swelled Sephadex G-25 to remove water before loading the ^{125}I-labeled sample). Dilute the labeled sample to 0.4 ml with buffer and place on a 2.0 ml dry column of Sephadex (it is appropriate to make columns in 5 ml disposable syringes for this purpose). Centrifuge the column at 1000 rpm for 5 min. The protein sample comes through and the free ^{125}I is retained.

NOTE: Labeling IgM while maintaining antibody activity using this or other labeling methods such as "Iodogen" or "Iodobeads" has yielded little success. A previous report (11) suggests that the Bolton-Hunter reagent may be useful for labeling IgM. All labeling procedures should be performed in an exhaust hood while wearing two pairs of gloves.

Alternatively, several labeled immunological reagents are available from commercial sources which can be used in RIA procedures. For example, specific ^{125}I-labeled immunoglobulins, ^{125}I-protein A, and ^{125}I-avidin or -streptavidin.

Assay Methods

The protocol for a successful solid phase RIA frequently follows the same principles used in ELISA (6). As such, several different protocols (as discussed elsewhere in this manual) for direct and indirect immunosorbent assays can be utilized effectively in RIA procedures. Examples of several protocols can be found in references 2,3,4,5,7,8, and 10.

Alternative radioimmunoassay formats have also been used for radioimmunoassay of plant viruses. These include radioimmunoprecipitation (2), competition (1,7), and instances where the antigen rather than the antibody is labeled (1). They have, however, been utilized to a very limited extent. Details concerning their use can be found in the references cited.

Comments

Because of potential hazards in isotopic assays, the manipulation of the solid phase imposes restrictions not present in non-isotopic assays. This imposes practical restrictions on the conformation of the solid phase that is utilized as well as procedures that are used for washing the solid phase during the various steps of the assay. Many users of RIA procedures find that polystyrene "frosted" balls [3.4 mm (better for RIA in our experience) or 6.5 mm diameter, Precision Plastic Ball] or IMMULON 1 Removawell Strips (Catalog No. 011-010-6301, Dynatech) are highly satisfactory for use as the solid phase. Because of the necessity for containment of wash and certain liquids associated with the RIA procedure, an automated contained washing system is highly desirable. Our laboratory has found the "Penta Wash" System (Abbot) to be most useful.

Effective use of RIA requires that all immunological reagents be used at optimal concentrations. One effective way of doing this is to maximize the P/N ratio (6) for all immunological reagents under conditions of constant antigen concentration. For example, to optimize the concentration of capture antibody, antigen and detecting antibody are used at constant concentrations and various concentrations of capture antibody are reacted in the presence and absence of antigen. The ratio of positive to negative response (P/N), or binding ratio, is calculated and the concentration at which the maximum P/N ratio is obtained is used in the assay. This should be done with all commercially obtained reagents as well as those produced in your own laboratory. Concentrations recommended by manufacturers are often inaccurate.

Applications

RIA can be used for most assays for which ELISA has been used. These include detection of viral antigens in seed (2,4,6), leaf tissue (2,3), and insect vectors (2,3,5,10). RIA is most appropriate for development of antigenic signatures of viral strains (11, and unpublished) because of less tendency toward repetitive variability with RIA than with ELISA.

Literature Cited

1. Ball, E. M. 1973. Solid phase radioimmunoassay for plant viruses. Virology 55:516-520.
2. Bryant, G. R., Durand, D. P., and Hill, J. H. 1983. Development of a solid-phase radioimmunoassay for detection of soybean mosaic virus. Phytopathology 73:623-629.
3. Childress, A. M., and Ramsdell, D. C. 1987. Bee-mediated transmission of blueberry leaf mottle virus via infected pollen in highbush blueberry. Phytopathology 77:167-172.
4. Ghabrial, S. A., and Shepherd, R. J. 1980. A sensitive radioimmunosorbent assay for the detection of plant viruses. J. Gen. Virol. 48:311-317.
5. Gillet, J. M., Morimoto, K. M., Ramsdell, D. C., Baker, K. K., Chaney, W. G., and Esselman, W. J. 1982. A comparison between the relative abilities of ELISA, RIA, and ISEM to detect blueberry shoestring virus in its aphid vector. Acta Hort. 129:25-29.
6. Hill, J. H., Bryant, G. R., and Durand, D. P. 1980. Detection of plant virus by using purified IgG in ELISA. J. Virol. Methods 3:27-35.

7. Hill, E. K., Hill, J. H., and Durand, D. P. 1984. Production of monoclonal antibodies to viruses in the potyvirus group: Use in radioimmunoassay. J. Gen. Virol. 65:525-532.
8. Hill, J. H., Bailey, T. B., Benner, H. I., Tachibana, H., and Durand, D. P. 1987. Soybean mosaic virus: Effects of primary disease incidence on yield and seed quality. Plant Dis. 71:237-239.
9. Johnson, D. A., Gaetsch, J. A., Sportsman, J. R., and Elder, J. H. 1984. Improved technique utilizing nonfat dry milk for analysis of proteins and nucleic acids transferred to nitrocellulose. Gene Anal. Techn. 1:3-8.
10. Morimoto, K. M., Ramsdell, D. C., Gillett, J. M., and Chaney, W. G. 1985. Acquisition and transmission of blueberry shoestring virus by its aphid vector *Illinoia pepperi*. Phytopathology 75:709-712.
11. Wands, J. R., Wong, M. A., Shorley, J., Brown, R. D., Marciniak, R. A., and Isselbacher, K. J. 1984. Hepatitis B and viral antigenic structure: Signature analysis by monoclonal radioimmunoassays. Proc. Natl. Acad. Sci. U.S.A. 2237-2241.

Section IV.B.2.

RADIOIMMUNOASSAY FOR PLANT PATHOGENIC BACTERIA

A. A. Benedict and A. M. Alvarez
University of Hawaii
Honolulu, HI 96822

Principle

Radioimmunoassay (RIA) was first introduced as a useful clinical technique by Yalow and Berson (12). Their method was an extension of basic procedures described earlier by Farr (3). Since that time a variety of RIA procedures have been developed. Solid-phase RIA is a common method in which either antibodies or ligands are adsorbed to plastic surfaces, e.g. polystyrene tubes or microtiter plates. The free ligand is easily separated from the immobilized antibody-bound ligand by simply decanting the fluid phase. Until recently, the procedure required the use of either labeled antibody or antigen; however, the procedure used to radiolabel may result in products that are unstable or have lost their immunoreactivity. To avoid these problems, an alternative method is based on the ability of Protein A of Staphylococcus aureus to bind to the Fc region of IgG antibodies of many mammals (6). Protein A binds poorly or not at all to the IgM of all species and poorly to some mouse IgG subclasses, but binds strongly to rabbit IgG (8). Therefore, as a universal reagent rabbit anti-immunoglobulin serum is used in an intermediate incubation for detection of any immunoglobulin (Ig) class, followed by incubation with ^{125}I-labeled Protein A.

The method described here is in principle the same as the enzyme-linked immunosorbent assay (ELISA) (2). ELISA is preferred, where feasible, to avoid handling and disposing of radioactive materials. However, RIA is preferred in instances where plant enzymes (e.g., peroxidases) interfere with ELISA. With bacterial antigens, RIA has proven two- to four-fold more sensitive than ELISA. The following method is recommended for detection of phytopathogenic bacteria with monoclonal antibodies (McAbs) (1).

Materials (Also see **Section IV.B.1.**)

1. Solid phase: Flexible microtiter plates are needed for the gamma counter. Bacteria do not bind equally well to all types of plates, so the user must first test plates for reproducibility of results prior to running assays. Consistent results may be obtained with polyvinyl chloride 96-well flexible plates (Costar, Dynatech).
2. Coating buffer: A 0.05 M carbonate-biocarbonate buffer at pH 9.6 is used for optimum binding of bacteria to the solid surface. Lower binding occurs with 0.1 M sodium phosphate, pH 7.2, a reagent that is commonly used to extract bacteria from plant tissues.
3. Stock solutions: a) 0.01 M phosphate buffer (pH 6.8) containing 0.85% NaCl (PBS) is kept refrigerated until use; b) 0.16 M borate buffered-saline (pH 8.3) is stored at room temperature.
4. Washing buffer: Borate buffered-saline is diluted 1:4 prior to use.

5. Blocking reagent: BLOTTO, 5% Carnation non-fat dried milk in PBS (5) used without antifoam A. Alternative blocking reagents are described by Hill (see **Section IV.B.1**).
6. Antisera: Stored undiluted or diluted 1:1 with glycerol at -20 C.
7. Antiserum diluent: 1:3 dilution of BLOTTO in PBS.
8. Rabbit-anti-mouse globulin must react with all mouse Ig classes and subclasses; stored in the refrigerator and diluted just prior to use.
9. ^{125}I-labeled-Protein A; may be purchased (Amersham) or labeled as described below:

   Labeling Reagents:
   a) Protein A (Sigma); 50 μg in 50 μl in PBS
   b) Enzymobead Radioiodination Reagent (Bio-Rad): 50 μl
   c) ^{125}I: 0.5 mC in approximately 5 μl (Amersham)
   d) 1% B-D-glucose: 50 μl

10. Equipment: Microcentrifuge; gamma spectrometer; pipets capable of accurate delivery within the ranges of 1-20 μl, 50-200 μl, and 100-1000 μl (a multichannel pipet is a great convenience).
11. Supplies: Microcentrifuge tubes (1.5 ml); disposable pipet tips.

Procedures

A. Labeling methods

The chloramine-T technique (4,9) may be used (see **Section IV.B.1.**); however, to avoid the damaging effect of the oxidant and to obtain high iodination efficiency, a solid phase radioiodination employing the Enzymobead Radioiodination reagent (Bio-Rad) is preferred. The method is based on enzymatic iodination rather than oxidative substitution (10,11) in which the enzymes (lactoperoxidase and glucose oxidase) are immobilized.

Protein A, Enzymobeads, ^{125}I, and β-D-glucose are incubated in a 37 C waterbath for 30 min with gentle mixing every 10 min. The solution is transferred to a G-75 Sephadex column (40 x 1.5 cm) that has been equilibrated in borate buffer and saturated with 1 ml of 1% bovine serum albumin (BSA) in PBS. Two ml fractions are collected to recover the bound ^{125}I. Each fraction is counted in the gamma-counter; the first peak contains bound ^{125}I. The two tubes with highest activity are pooled and tested for protein bound ^{125}I by adding 10 μl to 100 μl of PBS to which an equal volume of 20% trichloroacetic acid is added. Following centrifugation in a microfuge, 100 μl of the supernatant fluid is removed, and the counts per minute (cpm) in the supernatant fluid and in the precipitate (ppt) are determined. The percentage of bound ^{125}I is determined by the following calculation: (cpm ppt/cpm supernatant fluid + ppt) x 100. Using this procedure, approximately 94% or greater of the protein is labeled.

B. Growth of bacteria and preparation of antigens

1. Test bacterial strains for purity using a general growth medium containing 0.001% triphenyl tetrazolium chloride (TZC), which helps to differentiate colonies and detect contaminants but does not retard growth of bacteria, even the xanthomonads.
2. After examining cultures for purity, streak colony heavily onto yeast-glycerol (YGA) growth medium (yeast extract, 5 g; K_2HPO_4, 1 g; $MgSO_4$,

0.5 g; glycerol, 16.7 ml; agar, 20 g) that contains no sugar and thus reduces the amount of extracellular polysaccharide produced by the culture. Other media can be used, but alternatives should be checked to see that a given antigen is consistently detected on the medium of choice.

3. Incubate bacteria for 3-4 da on YGA then harvest in saline containing 0.5% formalin. The cell suspension should be concentrated (e.g., A_{600} >.3 OD). Formalin-killed preparations may be refrigerated for several months with no change in serological reactivity.

C. Preparation of antigen

1. Centrifuge 1.0 ml of the bacterial suspension in a microcentrifuge tube for one min to form a pellet. Pour off supernatant fraction, add saline, and repeat process 3 times to wash the cells.
2. After the third wash, resuspend the pellet in carbonate bicarbonate (CBC) coating buffer and adjust to 0.1 OD on spectrophotometer.
3. Adjust absorbance values within a few minutes of suspending cells in CBC, to preclude cell lysis before standardization. After cell lysis, absorbance values underestimate the concentration of antigen.

D. Coating of microtiter plates

1. Coat antigens onto microtiter plates in desired formats.
2. Include appropriate controls so that values for maximum binding as well as background readings can be calculated.
3. Dry coated plates at room temperature or 37 C in a circulating-air incubator to dry antigen onto the plate. Coated plates can be used immediately or may be covered with Parafilm and stored in the refrigerator for several months prior to use.

E. Radioimmunoassay

1. Prepare BLOTTO solution for use as blocking solution and antibody diluent. Calculate amount needed for blocking (200 μl/well), and dilutions of coating antibody, rabbit anti-mouse antibody, and ^{125}I-labeled-Protein A (100 μl/well).
2. Wash plates 3 times by injecting borate buffer into the wells, then emptying them forcefully into sink. Firmly tap plates several times, wells downward, on paper towels to remove remaining liquids.
3. Pipet 200 μl of blocking solution into each well and incubate at room temperature for 30 min. The blocking reagent occupies binding sites not occupied by bacterial antigen.
4. Wash plates once after the blocking step.
5. Prepare final dilutions of antisera or McAbs in 1:3 BLOTTO/PBS, 100 μl/well, and pipet reactants into each well. Incubate plates 1 hr at room temperature.
6. Wash plates 3 times.
7. Dilute rabbit anti-mouse serum, previously titrated for activity,in 1:3 BLOTTO/PBS, dispense at 100 μl/well and incubate plate 1 hr.
8. Wash plates 3 times.
9. Dilute ^{125}I-labeled Protein A in 1:3 BLOTTO/PBS to provide approximately 25,000 cpm/100 μl. Dispense into plates and incubate for 1 hr.
10. Wash plates 3 times.

11. From microtiter plates excise individual wells, and place each well into a separate tube for counting with a Gamma spectrometer.

Comments

For each antibody and bacterium assayed, the following controls are included: a) buffer control (no antigen or antibody); b) a negative antiserum or ascites; and c) an unrelated bacterium. All controls should give values of approximately $\leq$ 1000 cpm. The highest control value is subtracted from the test value.

Although RIA is more sensitive (2- to 4-fold) and less subject to variation than ELISA, ELISA is preferable in most instances to avoid handling, storing, and disposing of radioactive materials. RIA is particularly useful when pigmentation and/or plant enzymes interfere with ELISA data.

Literature Cited

1. Alvarez, A. M., Benedict, A. A., and Mizumoto, C. Y. 1985. Identification of xanthomonads and grouping of _Xanthomonas campestris_ pv. _campestris_ strains with monoclonal antibodies. Phytopathology 75:722-728.
2. Alvarez, A. M., and Lou, K. 1985. Rapid identification of _Xanthomonas campestris_ pv. _campestris_ by enzyme-linked immunosorbent assay (ELISA). Plant Dis. 69:1022-1026.
3. Farr, R. S. 1957. Antibody response of rats, mice and guinea pigs to bovine serum albumin as measured by antigenic binding capacity test. Fed. Proc. 16:412.
4. Hunter, W. M. 1978. Radioimmunoassay. Pages 14.1-14.40 in: Handbook of Experimental Immunology, 3rd Ed. D. M. Weir, ed. Blackwell Scientific Publications, Oxford. 1232 pp.
5. Johnson, D. A., Gautsch, J. A., Sportsman, J. R., and Elder, J. H. 1984. Improved technique utilizing nonfat dry milk for analysis of proteins and nucleic acids transferred to nitrocellulose. Gene Anal. Techn. 1:3-8.
6. Kronvall, G., Seal, U. S., Finstad, J., and Williams, Jr., R. C. 1970. Phylogenetic insight into evolution of mammalian Fc fragment of globulin using Staphylococcal protein A. J. Immunol. 104:140-147.
7. Langone, J. J. 1978. ^{125}I protein A: A tracer for general use in immunoassay. J. Immunol. Methods 24:269-285.
8. Lindmark, R., Thoren-Tolling, K., and Sjöquist, J. 1983. Binding of immunoglobulins to protein A and immunoglobulin levels in mammalian sera. J. Immunol. Methods 62:1-13.
9. McConahey, P. J., and Dixon, F. J. 1966. A method of trace iodination of proteins for immunologic studies. Int. Arch. Allergy Appl. Immunol. 29:185-189.
10. Marchalonis, J. J. 1969. An enzymatic method for the trace iodination of immunoglobulins and other proteins. Biochem. J. 113:299-305.
11. Morrison, M., Bayse, G. S., and Webster, R. G. 1971. Use of lactoperoxidase catalyzed iodination in immunochemical studies. Immunochemistry 8:289-297.
12. Yalow, R. S., and Berson, S. A. 1960. Immunoassay of endogenous plasma insulin in man. J. Clin. Invest. 39:1157-1175.

Section IV.C.1.

COMPETITIVE BINDING BETWEEN MONOCLONAL ANTIBODIES TO DEFINE EPITOPE RELATIONSHIPS OR CHARACTERIZE IMMUNOGLOBULIN IDIOTYPES

J. H. Hill
Iowa State University
Ames, Iowa

Principle

Competition among different antibodies for the same epitope on a viral antigen has been used to define relationships among epitopes on antigens and to characterize specific immunoreactivity of different antibodies to the antigen. In a competition experiment, one monoclonal antibody carries a marker and unlabeled antibodies are examined for their ability to inhibit binding of labeled antibody with antigen. If competition is seen, the antibodies can be placed in a related binding group. The data also provide evidence for the topological arrangement of specific epitopes. Because monoclonal antibodies are specific to a single defined epitope, the usefulness of competition experiments is primarily confined to monoclonal antibody applications. Therefore, the discussion of this chapter addresses only monoclonal antibodies.

Consideration of the mechanism of binding inhibition may enhance interpretation of competition experiments. Inhibition may occur by one of four mechanisms: (1) inhibition may be due to competition for the same epitope; (2) inhibition of binding of a second antibody specific for another epitope may occur if the epitopes are in close proximity or overlapping; (3) inhibition may be due to steric hindrance caused by binding of the first antibody; and (4) the first antibody may prevent access of a second antibody by altering the conformation of the second epitope or even the entire antigen.

In the competition experiment, varying amounts of unlabeled antibody, relative to a fixed concentration of labeled antibody, are added to the antigen either consecutively or simultaneously with the labeled antibody. The experiments reveal that monoclonal antibodies can be divided into two basic groups; i.e., those that do not compete and those that do compete. The characteristic dilution profile of non-competing antibodies is shown in Figure 1A.

The second basic group is defined by monoclonal antibodies that do compete. Competition curves from these experiments demonstrate whether affinity or concentration of the unlabeled antibody, relative to the labeled antibody, impacts most significantly on its ability to compete with the labeled antibody. The data can be interpreted on the basis of the equilibrium equation for an antigen-antibody reaction designated as $SL \rightleftharpoons S+L$, where S = antibody binding site, L = antigen binding site, and SL = the antigen-antibody complex (1,2,11). Greater competition of the unlabeled antibody in the consecutive assay relative to the simultaneous assay reveals the significance of affinity in the competition (1,5). In the consecutive assay, the unlabeled antibody can initially bind to all available sites on the antigen. The dissociation of the unlabeled antibody is affected by its affinity which, in turn, affects the equilibrium equation for the antigen-antibody reaction. If the affinity of the unlabeled antibody is lower than that of the labeled antibody, the equilibrium equation favors dissociation and the labeled antibody replaces the unlabeled antibody. In this situation, the labeled antibody may replace the unlabeled antibody and little competition by the unlabeled antibody occurs (Fig. 1B).

The significant dilution effect on competition of the lower affinity unlabeled antibody with the labeled antibody in the simultaneous assay is demonstrated in Figure 1B. At high concentrations of unlabeled antibody, more unlabeled antibody binds and competition becomes significant (1,8). As concentration of the unlabeled antibody decreases, fewer of these antibodies bind and competition is diminished. This demonstrates that an unlabeled antibody of relatively low affinity will compete only at high concentrations in the simultaneous assay, yielding a dilution curve with a steep slope. Alternatively, an antibody of higher affinity is characterized by a dilution curve with a shallow slope. Thus, the consecutive and simultaneous curves demonstrated in Figure 1B suggest that the unlabeled antibody is of lower affinity relative to the labeled antibody. Little competition occurs in the consecutive assay. Competition does occur in the simultaneous assay but the curve is characterized by a moderately steep increase in slope.

In contrast, high affinity unlabeled antibodies will replace the labeled antibody in the simultaneous assay and significant competition occurs, even at high dilutions of the unlabeled antibody (Fig. 1C). In this illustration, the equilibrium equation favors binding of the unlabeled antibody. For this antibody, both the simultaneous and consecutive assays suggest this is an antibody of high affinity relative to the labeled antibody. Affinity has a greater impact on the ability of the antibody to compete than does concentration and the slope of the competition curve is very gradual.

The shape of the curve in Figure 1D, in which both dilution and affinity are important, is diphasic in contrast to the monophasic curves of Figures 1B and 1C. This suggests that the antibody competes for two sites rather than a single site on the antigen. Several potential explanations exist for this phenomenon. The antibody has a higher affinity for one of the two epitopes, and initial binding is to the high affinity site followed by binding to a site of lower affinity. Alternatively, the binding sites may have slightly different conformations, and initial binding occurs at the site of best fit followed by binding at a related site. Additionally, when antibodies bind to a site, they may induce a conformational change in the antigen which results in exposure of a second site recognized by the remaining antibodies. At low antibody concentrations, antibodies bind to epitopes on the antigen for which they have high affinity and at high concentrations they may bind to epitopes for which they have lower affinity.

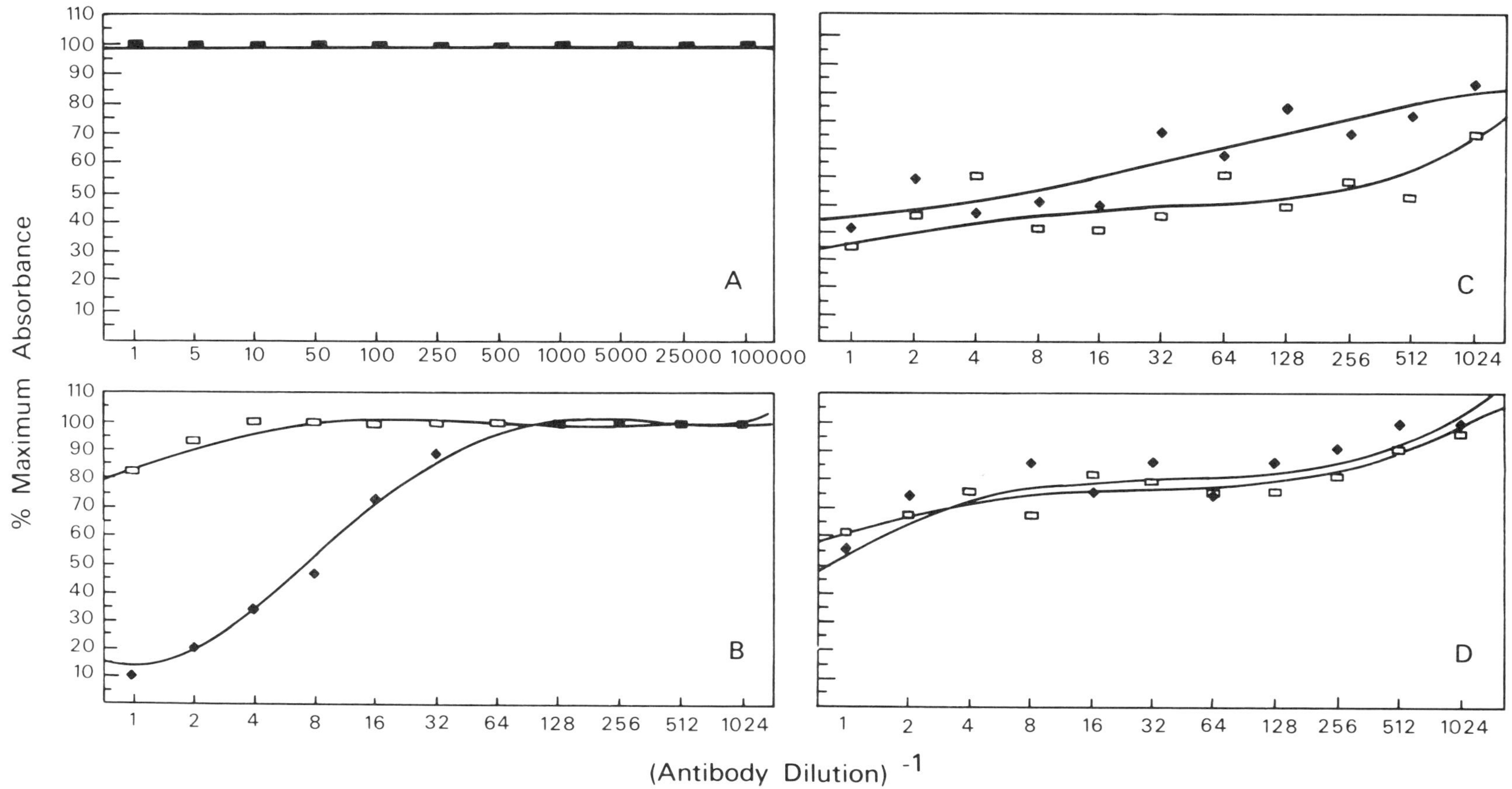

Figure 1 A-D. Competition simultaneous (■—■) or consecutive (□—□) ELISA assays of four different unlabeled monoclonal antibodies (A, B, C & D) against a constant concentration (0.125 μg) of labeled (biotinylated) monoclonal antibody. Note that measured inhibition ranged from zero (A) to substantial (B, simultaneous) at low dilutions of unlabeled antibodies. All antibodies were made to the same viral antigen. Assays were performed in Immulon I 96-well plates (Dynatech). Wells were coated with viral antigen at 0.200 μg/well. Data from Ref. 9.

Materials

1. Coating buffer. The coating buffer is used to bind either the antigen (indirect assay) or capture antibody (double-sandwich assay) to the solid phase. The buffer of choice depends, in part, upon the molecular conformation desired for the antibody/antigen and the solid phase to be used in the assay. In most applications, 0.05 M carbonate, pH 9.6, or 0.1 M sodium phosphate, pH 7.2, can be used.
2. Wash buffer. Generally 0.01 - 0.05 M phosphate buffer, pH 7.2, containing 0.85% NaCl (PBS) and 0.05% Tween-20 (PBS-Tween) is used.
3. Blocking reagent. Several blocking reagents can be used. The choice and concentration depend upon characteristics of the antibody, antigen, and solid phase and is usually determined empirically by experimentation. Usually, the reagent is prepared in PBS-Tween. Common blocking agents are BLOTTO [(6) 5% Carnation non-fat dried milk and 0.03% antifoam A (Sigma), 1-3% gelatin (Bio-Rad), 0.2 - 2.0% ovalbumin (Sigma), 1% bovine serum albumin (Sigma), or 10-50% horse or fetal calf serum (e.g. HyClone).
4. Solid phase. We find Immulon I or Immulon II microtiter plates (Dynatech) or numerous other plastics which are commercially available to be useful in competition assays.
5. Chemicals. CM Affi-Gel Blue (Bio-Rad), Protein A Sepharose CL-4B (Pharmacia), anti-mouse IgM (μ-chain specific) (Sigma), N-hydroxy-succinimidobiotin (Sigma), N,N-dimethylformamide (Fisher), p-nitrophenyl phosphate (Sigma), diethanolamine (Sigma).

Note: Suppliers are not recommended to the exclusion of others that may be satisfactory. We have found the ones listed to be satisfactory in our laboratories.

Procedure

Competition experiments can be performed either by using an antibody to capture the antigen or by binding the antigen directly to the solid phase. Results obtained by direct antigen binding are easier to interpret. Therefore, the procedure described employs antigen bound directly to the solid phase.

1. Bind antigen to the solid phase by using an appropriate buffer. Antigen can generally be bound to the solid phase at 37 C for one hr or at 4 C overnight.
2. Wash solid phase three times with wash buffer.
3. Block solid phase with appropriate blocking reagent (generally 1 hr at 20 C is sufficient).

<u>Simultaneous assay</u>

4s. Select antibody that is to be used as the labeled antibody. Purify selected antibody by affinity chromatography [CM Affi-Gel Blue (use according to manufacturers instructions, Bio-Rad) followed by Protein A Sepharose CL-4B (Pharmacia) for immunoglobulin G (IgG) (4) or anti-mouse IgM (μ-chain specific, Sigma) agarose for immunoglobulin M (IgM)]. Alternatively, highly purified IgM can be obtained as described by Jones et al. (7).

5s. Label purified antibody with biotin. Add N-hydroxysuccinimidobiotin (BNHS) (Sigma) dissolved in N,N-dimethyl formamide (Fisher) to purified antibody in a 1:50 (v/v), 10:1 (mol/mol) ratio (for IgG, this would be accomplished by adding 10 μl BNHS solution at 5 mg/ml, to 1.0 ml of IgG solution at 2 mg/ml). Rotate mixture for 4 hr at 20 C, using an end-over-end mixer, followed by overnight incubation at 4 C. Remove unconjugated biotin by dialysis against 0.01 M sodium phosphate containing 0.15 M NaCl, pH 8.0. Remove any precipitate that occurs by centrifugation (5,000 g for 10 min). Store preparation with NaN_3, 0.2 g/l, at 4 C.

6s. Prepare serial dilutions of unlabeled monoclonal antibody in culture medium (if ascitic fluid is used, use an initial dilution of ca. 1:600) using PBS as the diluent. Add each dilution simultaneously with a fixed concentration [determined by calculation of maximum P/N (positive to negative response) (3)] of labeled (biotinylated) antibody to the antigen-coated solid phase.
7s. Incubate solid phase at 37 C for 90 min.
8s. Wash solid phase three times with wash buffer.
9s. Add avidin conjugated with alkaline phosphatase [concentration determined by calculation of maximum P/N (3)] to solid phase.
10s. Incubate solid phase at 37 C for 90 min.
11s. Add substrate (p-nitrophenyl phosphate, 1.0 mg/ml in 10% diethanolamine, pH 9.6) to solid phase.
12s. Incubate 60-120 min at 20 or 37 C.
13s. Terminate reaction by addition of NaOH to final concentration of 0.27 M.
14s. Read reaction products at 405 nm.

Consecutive Assay

4c. Repeat step 4s.
5c. Repeat step 5s.
6c. Prepare serial dilutions of unlabeled monoclonal antibody in culture medium (if ascitic fluid is used, use an initial dilution of ca. 1:600) with PBS as the diluent. Add each dilution to the antigen-coated solid phase.
7c. Incubate solid phase at 37 C for 60 min.
8c. Wash solid phase three times with wash buffer.
9c. Add fixed concentration [determined by calculation of P/N (3)] of labeled (biotinylated) antibody to the antigen-coated solid phase.
10c. Repeat steps 7s - 14s.

Calculation of Results

Labeled (biotinylated) antibody is added to the solid phase containing no competing antibody as a positive control in which no competition occurs. These wells are designated as having an absorbance of 100% and are used as a basis for comparison to quantitate the competition of the unlabeled antibody. Results are graphed by using a third order polynomial. Antibody dilution is expressed on the "X" axis and competition expressed as percentage of maximum absorbance is expressed on the "Y" axis.

Negative controls to assess non-specific binding of the labeled (biotinylated) antibody should also be incorporated into the experiment. Concentrations of the labeled antibody should be sufficient to compete with the unlabeled form of the same antibody at maximum concentrations of the unlabeled antibody used in the experiment.

Comments

Results of competition assays are subject to interpretation based on the equilibrium equation for an antigen-antibody reaction. Competition experiments performed at only one antibody concentration or in only a simultaneous or consecutive manner can lead to erroneous conclusions. Interpretation of data must account for both truly competitive binding for identical sites and prevention of binding by steric blocking of distinct but close sites. These two possibilities are not readily distinguished, but do imply proximity of sites. It is possible that a monoclonal antibody bound to a single epitope may have considerable "swivel" movement making steric effects unimportant. For example, Stone and Nowinski (10) reported that two IgG molecules (M_r = 150,000) could bind non-competitively to a retrovirus protein (M_r = 15,000). Failure to differentiate steric hindrance resulting from antigen-antibody binding (e.g., two epitopes in close proximity) and true antibody competition (i.e., unrestricted competitive advantage in binding to a specific determinant) can be resolved by testing varying concentrations of unlabeled antibody. Data may show progressive inhibition of labeled antibody by increasing unlabeled antibody if inhibition is due to identical or overlapping epitopes. Additionally, individual monoclonal antibodies may share identical specificity but differ in affinity with the high affinity antibody favored in an experiment. This potential artifact is identified by performing the competition experiments in a sequential and simultaneous manner to distinguish between affinity and dilution effects upon competition.

Applications

Although there is considerable literature in animal virology dealing with this subject, it is still an emerging field for plant virology. It may be possible to delineate biologically active sites on viral proteins in terms of epitopes differentiated by a panel of monoclonal antibodies. The construction of an operational epitope map (12) of certain plant viruses may have particular relevance to those viruses for which the basis of virus-vector specificity is thought to reside, at least in part, in the coat protein.

Competition assays are also relevant to the selection of panels of monoclonal antibodies for differentiation of virus strains or isolates. Different antibodies demonstrating little or no competition are of particular interest.

Literature Cited

1. Berzofsky, J. A., and Berkower, I. J. 1984. Antigen-antibody interactions. Pages 595-644 in: Fundamental Immunology. W. E. Paul, ed. Raven Press, New York. 830 pp.
2. Campfield, L. A. 1983. Mathematical analysis of competitive protein binding assays. Pages 125-148 in: Principles of Competitive Protein-Binding Assays. W. D. Odell and P. Franchimont, eds. John Wiley and Sons, Inc., New York. 311 pp.
3. Hill, J. H., Bryant, G. R., and Durand, D. P. 1980. Detection of plant virus by using purified IgG in ELISA. J. Virol. Methethods 3:27-35.
4. Hill, E. K., Hill, J. H., and Durand, D. P. 1984. Production of monoclonal antibodies to viruses in the potyvirus group: Use in radioimmunoassay. J. Gen. Virol. 65:525-532.

5. Hosang, M. 1985. Surinam binds to platelet-derived growth factor and inhibits its biological activity. J. Cell. Biochem. 29:265-273.
6. Johnson, D. A., Gaetsch, J. A., Sportsman, J. R., and Elder, J. H. 1984. Improved technique utilizing nonfat dry milk for analysis of proteins and nucleic acids transferred to nitrocellulose. Gene Anal. Techn. 1:3-8.
7. Jones, F. E., Hill, J. H., and Durand, D. P. 1988. Detection and differentiation of maize dwarf mosaic virus, strains A and B, by use of different class immunoglobulins in a double-antibody sandwich enzyme-linked immunosorbent assay. Phytopathology 78:1118-1124.
8. Kalmakoff, J. L., Parkinson, A. J., Crawford, A. M., and Williams, B. R. G. 1977. Solid phase radioimmunoassays using labeled antibodies: A conceptual framework for designing assays. J. Immunol. Methods 14:73-84.
9. Kubanek, C-.A. 1987. Antigenic analysis of cowpea severe mosaic virus and other selected viruses using monoclonal antibodies. M.S. Thesis. Iowa State University. 161 pp.
10. Stone, M. R., and Nowinski, R. C. 1980. Topological mapping of Murine Leukemia virus proteins by competition binding assays with monoclonal antibodies. Virology 100:370-381.
11. Wide, L. 1983. Noncompetitive versus competitive binding assays. Pages 243-254 in: Principles of Competitive Protein-Binding Assays. W. D. Odell and P. Franchimont, eds. John Wiley and Sons, Inc., New York. 311 pp.
12. Yewdell, J. W., and Gerhard, W. 1981. Antigenic characterization of viruses by monoclonal antibodies. Annu. Rev. Microbiol. 35:185-206.

Acknowledgments

Journal paper No. J-13503 of the Iowa Agriculture and Home Economics experiment Station, Ames, Iowa. Project No. 2700.

Section IV.C.2.

ANTIBODY BINDING INHIBITION ASSAY FOR DETECTION OF BACTERIA IN PLANT TISSUES

A. M. Alvarez
University of Hawaii
Honolulu, Hawaii 96822

Principle

Detection and quantitation of a substance either by competitive or noncompetitive radioimmunoassays (RIA) have been employed since the first clinical application of RIA (6). Similar assays may be performed using ELISA. In solid phase immunoassays for antibody inhibition, the antigen is fixed to the solid surface, such as microtiter plates. The inhibitor may be mixed with antibody before dispensing into the plates or added directly to plates before addition of antibody. In either case this assay depends on the inhibition of antibody binding which is detected by the usual RIA or ELISA techniques described elsewhere (see **Sections IV.A.2., IV.B.2.**). Isotope assays probably have a competitive edge in terms of sensitivity and have more predictable assay parameters; nevertheless, ELISA is preferred for diagnostic work because it is easily performed at a field station with minimal equipment.

The inhibition assay is particularly useful for detecting bacteria in mixed microbial populations or in plant extracts when the bacteria in question are difficult to isolate into pure culture. Maximum sensitivity of the inhibition assay is achieved when both the bound antigen and the antibody are used at the appropriate dilution. If the antibody concentration is too high, it will not be neutralized by the competing antigen in the test solution. If it is too low, maximum binding will be indistinguishable from background readings; thus, antibodies must be titrated with the solid phase antigen prior to setting up the inhibition assay.

Materials

1. Solid phase: Immulon-II, polystyrene flat-bottom microtiter plates (Dynatech) or plates that achieve similar binding of bacterial antigens (Costar; Nunc).
2. Carbonate-bicarbonate (CBC) coating buffer, stock solutions and washing buffer are those described for RIA (see **Section IV.B.2.**).
3. Blocking reagent: 0.01 M phosphate-buffered saline, pH 6.8, (PBS) and 0.05% Tween-20 (PBS-Tween) or BLOTTO (5).
4. Antisera (stored undiluted or diluted 1:1 with glycerol at -20 C).
5. Antibody diluent: 1% Bovine serum albumin (BSA) (Sigma) in PBS or 1:3 dilution of BLOTTO in PBS.
6. Rabbit-anti-mouse globulin, used as described previously (see **Section IV.B.2.**) for assays with monoclonal antibodies (McAbs).
7. Protein A conjugated to horseradish peroxidase (Protein A-HRP) (Bio-Rad). Each lot must be titrated prior to use.

8. Substrate: 5-amino salicylic acid (5-AS), recrystallized in ethanol (4); 3% hydrogen peroxide.
9. Substrate buffer: 0.01 M phosphate buffer, pH 6.8, containing EDTA (ethylenediamine tetraacetic acid disodium salt, 35 mg/l).

Equipment and supplies - As described for RIA, except that ELISA readings are made with a microtiter plate reader equipped with a 450 nm filter. Since a dark purple-brown color develops with the chromophore of 5-AS, visual evaluations are easily made using a two-fold dilution series of a known antigen as standard.

Procedure

Growth of bacteria and preparation of antigens - Bacterial strains are tested for purity using a general growth medium containing 0.001% triphenyl tetrazolium chloride (TZC). Bacteria are transferred and incubated for 3-4 da on yeast glycerol agar (YGA) then harvested and washed three times in saline (0.85% NaCl in distilled water). A suspension of the strain to be used as coating antigen is adjusted to A_{600}=0.1 OD in CBC and coated immediately onto plates as described previously (see **Section IV.B.2.**).

Antigen-antibody titrations - Binding curves are determined for the antibody or ascites fluid in a checkerboard titration with homologous antigen. The antigen (bacterial cell suspension) is heated for 1 hr in a boiling waterbath if McAbs directed to lipopolysaccharide epitopes are to be used for the inhibition assay (provides greater assay sensitivity). After heating, the suspension is centrifuged and two-fold dilutions of the supernatant liquid are coated onto microtiter plates (100 µl/well). With polyclonal antisera the heating step is eliminated. Plates are dried at room temperature or at 37 C in a circulating-air incubator. A direct binding assay is performed as described below (eliminating Steps 5 and 6) and binding curves are determined for the antibody or ascites. The final dilution of antibody for inhibition assays is 50% of maximum binding. For antibody used in this example, Xanthomonas-specific McAb, (X11) (1), the final antibody dilution was 1:70,000. For the inhibition assay, the solid phase antigen is dried onto plates at the lowest concentration that gives 100% binding with relatively high dilutions of antibody (the solid phase antigen in this example was applied at 5 x 10^7 cells/ml). Plates can be used immediately or may be covered with Parafilm and stored at 4 C for several months prior to use.

ELISA inhibition assay

1. Coated plates are washed 3 times in borate buffer, pH 7.2.
2. Free binding sites are blocked by adding 200 µl of blocking solution to each well and incubating plates for 30 min at room temperature.
3. Plates are washed once after the blocking step. All subsequent washes should be performed 3 times.
4. Final dilutions of antisera or McAbs are made in BSA-PBS or 1:3 BLOTTO/PBS.
5. A standard curve of the homologous competitive antigen is made in separate tubes, starting with a known suspension of cells (A_{600}-0.1 OD) and making two-fold dilutions in PBS for 10 to 11 steps (final volume 200 µl/tube). Unknown samples are prepared similarly or tested at a

single dilution for a qualitative evaluation. Antigens are incubated for 1 hr at room temp with an equal volume of antiserum or ascites.

6. 100 μl of the antigen-antibody mixture is dispensed into each well of the microtiter plate. Alternatively, the competitive binding step may be performed in the microtiter well by dispensing the antibody-antigen mixture immediately after mixing and incubating in the microtiter wells. Incubation of inhibitor with antibody is preferable to competitive inhibition on the solid phase because binding of the immobilized antigen is favored.
7. Plates are washed 3 times.
8. When McAbs are used in the inhibition assay, rabbit anti-mouse serum previously titrated for activity is appropriately diluted, dispensed (100 μl/well) and incubated 1 hr. Otherwise this step is eliminated.
9. Plates are washed 3 times
10. Protein A-HRP is appropriately diluted in BSA-PBS or 1:3 BLOTTO/PBS. This is dispensed, 200 μl per well, and plates are incubated for 1 hr at room temperature.
11. The substrate is prepared just before use by adding 0.04% 5-AS to phosphate-EDTA buffer. The mixture is warmed to 50 C to dissolve the 5-AS and cooled to room temperature.
12. Plates are washed 3 times.
13. Add 0.003% or 0.006% H_2O_2 to the chromogen to accelerate substrate conversion. The mixture is dispensed (100 μl/well) and plates are incubated 45 min to 1 hr. ELISA values are determined with a microtiter plate reader equipped with a 450 nm filter or evaluated visually, comparing test values with known controls. Note: A stopping reagent is not needed because the reaction with 5-AS normally goes to completion within 1 hr, and further changes are insignificant. Furthermore, stopping reagents, such as strong acids, denature the chromogen, and readings are not reproducible.

<u>Standardization of inhibition curves</u> - A standard inhibition curve is drawn by plotting OD_{450} against the log number of inhibitor cells in the reaction mixture (Fig. 1). At low cell numbers, OD readings are near maximum for the binding reaction between antibody and the antigen bound to the solid phase. As the amount of competitive antigen increases, OD readings are reduced, reaching a minimum when inhibitor cell numbers are greater than 5×10^8 cells/ml. The dilution of inhibitor that inhibits binding of antibody to antigen by 50% is I_{50}. Using the procedures described above, the binding reaction with rabbit-anti-*Xanthomonas campestris* pv. *campestris* (strain A249)-antiserum was inhibited by 50% at 0.498 absorbance units, and the bacterial cell concentration at that point was approximately 4.9×10^7 cells/ml (Fig. 1). Closely related bacteria gave binding curves with similar slopes and I_{50} values, whereas unrelated bacteria showed little or no inhibition (Fig. 2).

<u>Analysis and interpretation of results</u> - The number of cells/ml of competitive antigen is plotted against percentage inhibition calculated as follows:

$$\%I = 1 - \frac{Bt - Bc}{B\,max - Bc} \times 100$$

where Bt and Bc represent the ELISA values for test and control samples, respectively, and B max is the ELISA value for antibody bound in the absence of inhibitor. The control values were obtained by adding an unrelated antibody to wells coated with bacterial antigen. The slopes and 50% inhibition points are calculated. Reproducibility may be determined in repeated ELISA tests on separate plates. In the example provided (Fig. 3), the slope of the binding curve for the homologous antigen (C-0), cells of *X. campestris* pv. *campestris* strain A249, was 24.7 ± 2.1 (standard deviation). Binding curves of other strains of the same pathovar (C-1, C-2, C-3) had slopes of 22.7 ± 1.6, 20.6 ± 1.8, and 16.6 ± 1.3, respectively, whereas an unrelated bacterium, *Erwinia herbicola*, had a slope of 0.8 ± 0.06. The latter could also be used as a negative control in calculating inhibition values. Inhibition curves using McAb X11 with these strains are shown in Figure 4.

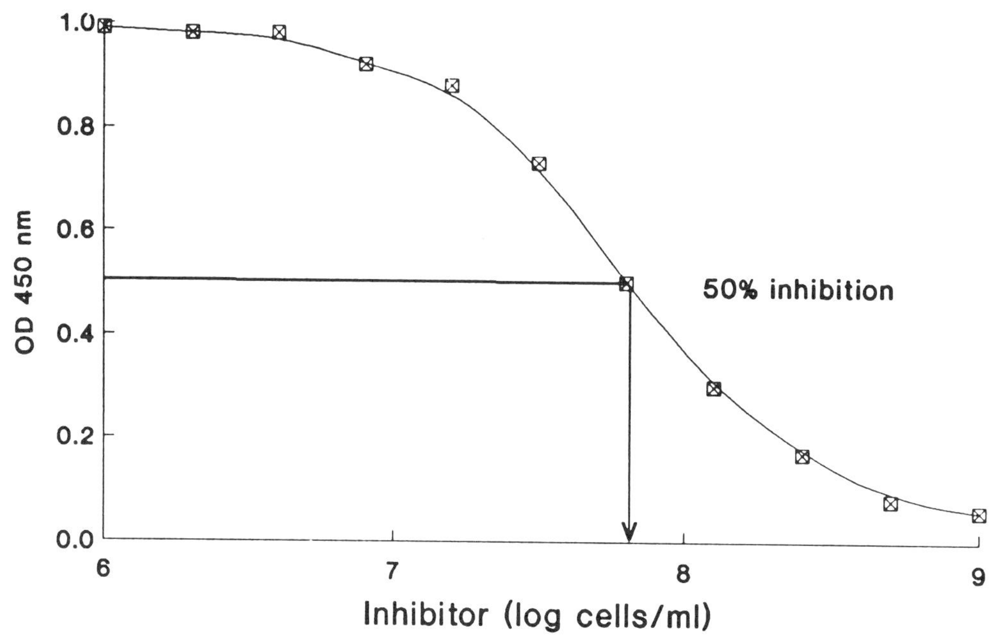

Fig. 1. Standard inhibition curve. Two-fold serial dilutions of bacterial cells, *Xanthomonas campestris* pv. *campestris* strain A249 were mixed with equal volumes of antiserum (rabbit-anti-A249) prior to addition of antibody to microtiter plates coated with strain A249.

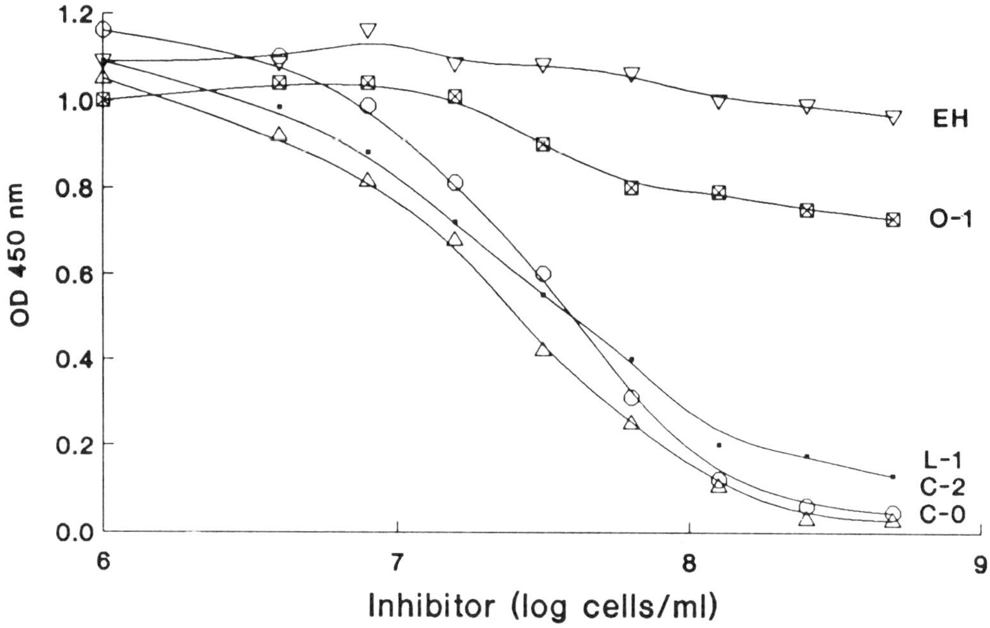

Fig. 2. Inhibition curves using antigen preparations of Xanthomonas campestris pathovars isolated from cabbage, lettuce, and onion. C-0, strain A249 from cabbage; C-2, strain KC-3-1-2 from cabbage; L-1, strain 10TB10 from lettuce; O-1, strain A255-4 from onion; EH, Erwinia herbicola.

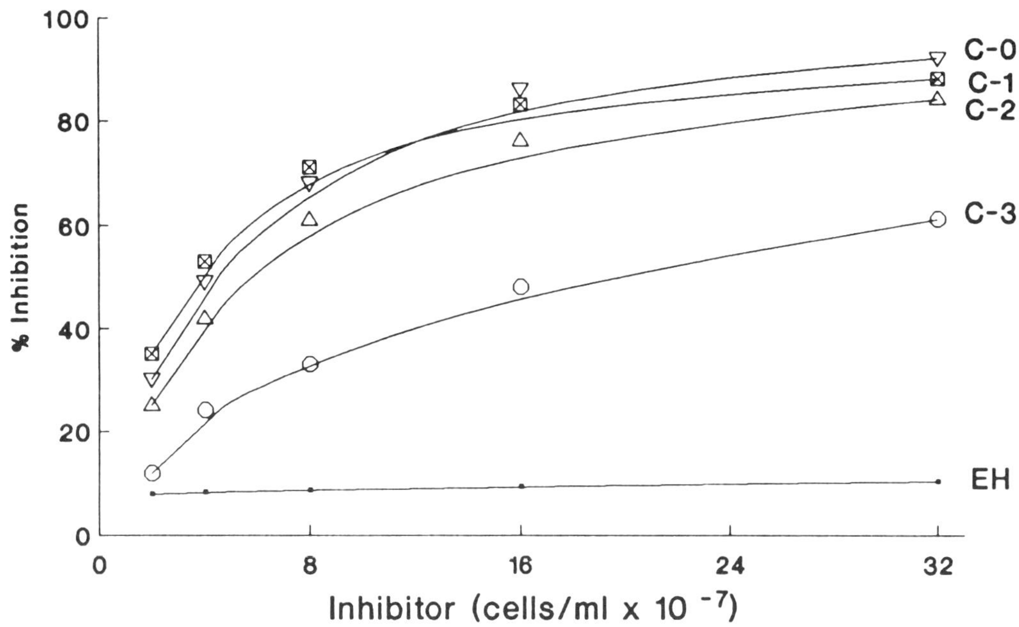

Fig. 3. Percentage inhibition of antibody binding (rabbit anti-Xanthomonas campestris pv. campestris strain A249) with the solid phase homologous antigen at various concentrations of inhibitor. C-0, strain A249 from cabbage; C-1, strain OK2 from cabbage; C-2, strain KC-3-1-2 from cabbage; C-3, strain EEXC 118 from cabbage; EH, Erwinia herbicola.

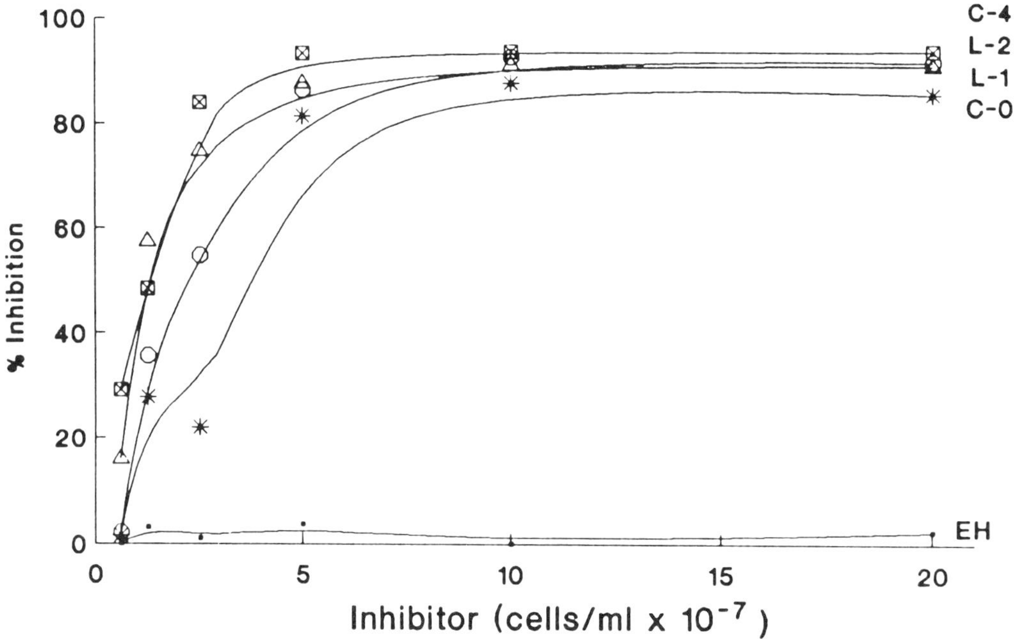

Fig. 4. Percentage inhibition of antibody binding (Xanthomonas-specific monoclonal antibody X11 with solid phase antigen Xanthomonas campestris pv. vitians strain A674) at various concentrations of inhibitor. C-0, X. campestris strain A249; L-1, X. campestris pv. vitians strain 10TB10; L-2, X. campestris pv. vitians strain A674; C-4, X. campestris pv. campestris strain EEXC 114; EH, Erwinia herbicola.

Detection of bacteria in plant samples - Leaf or stem disks (ca. 5 mm diam) are suspended in 0.5 ml sterile water for 2 to 4 hr to release bacteria from plant tissues. One hundred microliter samples of the fluid are incubated in separate tubes with an equal volume of antibody. The inhibition assay is performed as described previously, except that in Step 5, duplicate samples of the unknown are tested at a single dilution. It is useful to prepare a standard curve for the homologous competitive antigen for each assay to aid in the interpretation of results. The percentage inhibition of antibody binding is calculated for the plant sample. Based on the examples provided, inhibition values greater than 10% indicate that at least 2 x 10^7 cells/ml of the inhibitor are present in the plant sample. In practice, inhibition values ranging from 70-100% were commonly obtained for plant samples in which presence of the pathogen also was demonstrated by other methods.

Applications

Practical applications of the inhibition assay are; 1) comparing serological relationships of various bacteria; 2) identification of pathogenic bacteria in plant extracts; and 3) detection of a specific antigen (eg., lipopolysaccharide) in soluble extracts. The assay can be performed using either polyclonal or monoclonal antibodies. Sensitivity of the inhibition assay is similar to that of the double antibody sandwich-ELISA when used with polyclonal antibodies (2,3). However, with McAbs greater sensitivity is achievable in detecting bacteria in plant extracts with the competitive inhibition assay than with a sandwich assay.

Literature Cited

1. Alvarez, A. M., Benedict, A. A., and Mizumoto, C. Y. 1985. Identification of xanthomonads and grouping Xanthomonas campestris pv. campestris strains with monoclonal antibodies. Phytopathology 75:722-728.
2. Alvarez, A. M., Kuniyuki, S. R., and Lou, K. 1981. Antigenic relationships between Xanthomonas strains causing diseases of cabbage, lettuce, and onion in Hawaii. Pages 122-128 in: Proc. 5th Int. Conf. Plant Pathol. Bact. Cali, Colombia. 640 pp.
3. Alvarez, A. M., and Lou, K. 1985. Rapid identification of Xanthomonas campestris pv. campestris enzyme-linked immunosorbent assay (ELISA). Plant Dis. 69:1022-1026.
4. Ellens, D. J., and Gielkens, A. L. J. 1980. A simple method for the preparation of 5-aminosalicylic acid. Application of the product as substrate in enzyme-linked immunosorbent assay (ELISA). J. Immunol. Methods 37:325-332.
5. Johnson, D. A., Gautsch, J. A., Sportsman, J. R., and Elder, J. H. 1984. Improved technique utilizing nonfat dry milk for analysis of proteins and nucleic acids transferred to nitrocellulose. Gene Anal. Techn. 1:3-8.
6. Yalow, R. S., and Berson, S. A. 1960. Immunoassay of endogenous insulin in man. J. Clin. Invest. 39:1157-1175.

Section IV.D.

ENZYME-LINKED FLUORESCENT ASSAY (ELFA)

J. H. Hill
Iowa State University
Ames, IA 50011

Principle

The desire to enhance sensitivity of ELISA (enzyme-linked immunosorbent assay) for detection of antigens led to the development of enzyme-linked fluorescent assay (ELFA). In this assay, the utilization of fluorescent substrates such as 4-methylumbelliferyl phosphate (MUP) or 4-methylumbelliferyl-β-D-galactopyranoside (MUG) has enhanced detection of several diverse antigens (1,3,7,8,9,10,11). Success of the assay is based upon the reaction between immunoreactive antigen and antibodies highly specific to the antigen. The assay is designed in a solid-phase format in which either the antigen or the capture antibody is bound to the solid phase. After addition of the sample, a detection antibody, either directly conjugated with an appropriate enzyme or with biotin, is added. When biotin conjugated specific antibody is used, enzyme labeled avidin or streptavidin is added before addition of the final substrate. Alternatively, if the detection antibody is not directly conjugated, specific enzyme-labeled immunoglobulins or -Protein A can be added before addition of the substrate.

The use of MUP as a substrate in assays has the advantage that it can be substituted for p-nitrophenylphosphate in ELISA-based assays that use alkaline phosphatase as the enzyme label. MUP has the significant disadvantage, however, that it may undergo spontaneous degradation, which can result in high background levels (8). This can be resolved by using MUG which is a substrate for β-galactosidase. However, procedures for direct labeling of antibody with β-galactosidase are somewhat more laborious than those used for conjugation of antibody with alkaline phosphatase.

Materials

1. Coating buffer - The coating buffer is used to bind either the antigen (indirect assay) or capture antibody (double-sandwich assay) to the solid phase. The buffer of choice depends, in part, upon the molecular conformation desired for the antibody/antigen and the solid phase to be used in the assay. In most applications, 0.05 M carbonate, pH 9.6, or 0.1 M sodium phosphate, pH 7.2, can be used.
2. Wash buffer - generally 0.01 - 0.05 M phosphate buffer, pH 7.2, containing 0.85% NaCl and 0.05% Tween-20 (PBS-Tween).
3. Blocking reagent - several blocking reagents can be used. The choice and concentration depend upon characteristics of the antibody, antigen, and solid phase and is commonly determined empirically. Usually, the reagent is prepared in PBS-Tween. Common blocking agents are BLOTTO [(6) 5% non-fat dried milk and 0.03% antifoam A (Sigma)], 1 - 3% gelatin (Bio-Rad),

0.2 - 2.0% ovalbumin (Sigma), 1% bovine serum albumin (Sigma), or 10 - 50% horse or fetal calf serum (e.g. HyClone).

4. Buffer used for dilution of enzyme-conjugated antibody. Polyethylene glycol (M_r 6000, Fisher) incorporated into PBS-Tween is effective for enhancement of fluorescence. It is important to establish the concentration of polyethylene glycol (generally 1-2%) by calculation of P/N ratio (positive-to-negative-response (5) for maximum enhancement.
5. Chemicals - N-succinimidyl 3-(2-pyridyldithio) propionate (SPDP) (Pharmacia); Sephadex G-25 (Pharmacia); Sepharose CL-6B (Pharmacia); dithiothreitol (Sigma); β-galactosidase (>12 free -SH groups/molecule, Boehringer Mannheim); alkaline phosphatase (Sigma); 4-methylumbelliferyl phosphate (MUP) (Sigma); 4-methylumbelliferyl-β-D-galactopyranoside (MUG) (Sigma); specific β-galactosidase-labeled immunoglobulins, -Protein A, -avidin or -streptavidin (Zymed).
6. Solid phase - "Frosted" polystyrene beads, 3.4 mm or 6.5 mm diameter (Precision Plastic Ball), Immulon I or Immulon II plates (Dynatech) or numerous other plastics which are commercially available.
7. Fluorimeter - Turner Model 110 fluorimeter (Turner) equipped with #7-60 and #2A primary and secondary filters, respectively. With this instrument, it is convenient to read reaction products in 12 mm x 75 mm borosilicate glass culture tubes to obtain a quantitative measurement of the reaction product. Alternatively, the reaction products can be effectively visualized with a "black light."

Note: Suppliers are not recommended to the exclusion of others that may be satisfactory. The ones listed have proven satisfactory.

Procedure

Labeling Methods

<u>Labeling of antibody with alkaline phosphatase</u>

ELFA, which employs the substrate MUP, uses standard procedures outlined elsewhere in this manual (see **Section II.C.** and **Section IV.A.1.**) to directly label antibody with the enzyme alkaline phosphatase.

<u>Labeling of antibody with β-Galactosidase</u>

Several methods have been outlined by Neurath and Strick (8) for direct labeling of antibody. The method which is most desirable is presented here.

1. Mix 5 mg antibody (5 - 10 mg/ml in 0.02 M sodium phosphate containing 0.15 M NaCl, pH 7.2) with 25 μl SPDP solution in ethanol (10 mg/ml). Incubate at 20 C for 30 min.
2. Chromatograph mixture on Sephadex G-25. Elute with 0.1 M sodium acetate containing 0.1 M NaCl, pH 4.5. Combine fractions containing the majority of protein.
3. Add solid dithiothreitol (7 mg/ml) to pooled fractions. Incubate at 20 C for 20 min.
4. Chromatograph mixture on Sephadex G-25. Elute with 0.1 M sodium phosphate containing 0.1 M NaCl, pH 7.5. Combine fractions containing the majority of protein. Note that no residual odor of dithiothrietol is present.
5. Add 3 to 5 mg β-galactosidase and incubate at 20 C overnight.

6. Chromatograph mixture on Sepharose CL-6B. Elute with 0.01 M sodium phosphate, 0.14 M NaCl, 0.003 M NaN_3, pH 7.0. Use fractions eluting after the void volume of the column and before the unbound β-galactosidase peak for maximum activity and assay sensitivity. These fractions do not correspond with the major protein peaks of the elution profile.

Note: Difficulty may be encountered when attempting to directly label certain monoclonal antibodies using this technique. Antibody activity may be substantially reduced or lost due to putative disulphide bridges in close proximity to paratopes. Alternatively, specific β-galactosidase-labeled immunoglobulins, -Protein A, or -avidin or -streptavidin (for biotin labeled antibody) may be used.

Preparation of substrate

1. MUP - 0.1 mM in 1.0 mM diethanolamine, pH 9.8, containing 0.01mM $MgCl_2$.
2. MUG - 0.05 mM in 0.01 M sodium phosphate, pH 7.0.

Assay Methods

The procedure for a successful solid phase ELFA follows procedures similar to those used in all immunosorbent assays including ELISA. As such, several different protocols (as discussed in other chapters of this manual) for direct and indirect immunosorbent assays can be utilized effectively in ELFA procedures. In general, the only difference is substitution of a substrate which yields a fluorescent product for one which yields a colored product. Obviously, an enzyme appropriate to the substrate must be used. Examples of several protocols can be found in references 1, 3, and 7-11. Enzymatic reactions are terminated by addition of K_2HPO_4 - KOH, pH 10.4, to a final concentration of 0.2 M. Dilutions for reading of reaction products are made in the appropriate substrate buffer containing K_2HPO_4 - KOH, pH 10.4, at a final concentration of 0.2 M.

Comments

Effective use of ELFA requires that all immunological reagents be used at optimal concentrations. One effective way of doing this is to maximize the P/N ratio (5) for all antibody reagents used in the assay under conditions of constant antigen concentration. For example, to optimize the concentration of capture antibody, antigen and detecting antibody are used at constant concentrations, and various concentrations of capture antibody are reacted in the presence and absence of antigen. The P/N ratio, or binding ratio, is calculated, and the concentration at which the maximum P/N ratio is obtained is used in the assay. This should be done with all commercially obtained reagents as well as those produced in your own laboratory. Concentrations recommended by manufacturers are often inaccurate. In addition, because autofluorescence of extracts from some plant species can minimize effectiveness of ELFA, it is important to include proper negative controls in experiments as well as determine the most effective blocking reagents for the assay.

Polyethylene glycol (M_r 6000) has previously been reported to enhance specific antigen-antibody interaction (7). Benner et al (1) found that 1-2% polyethylene glycol incorporated into PBS-Tween used to dilute the enzyme conjugated antibody provides enhancement of fluorescence (1). Because high concentrations of polyethylene glycol (i.e., 4-10%) can significantly increase background, it is important to determine the concentration of polyethylene glycol at which maximum enhancement of fluorescence occurs for each different source of polyethylene glycol. This is normally done by calculation of the maximum P/N ratio (5) in experiments employing various concentrations of polyethylene glycol.

The choice of the solid phase involves principles shared in common between ELFA and ELISA. These include such issues as the format of the assay, ease of manipulation, and "stickiness" of the plastic.

ELFA has the potential for detecting relatively small amounts of antigen; however, ELFA has not yet obtained the sensitivity levels that theoretical considerations predict (1,3,8). Data suggest that, for detection of plant viruses, ELFA may have sensitivity levels equivalent to biotin-avidin ELISA (2) or radioimmunosorbent assay (RISA) (4) and is approximately 10- to 25-fold more sensitive than double-antibody sandwich ELISA in which the second antibody is conjugated with alkaline phosphatase by using glutaraldehyde (3).

Applications

ELFA can be used for most assays for which ELISA has been used. It is particularly useful for sensitive assays of purified virus antigen (1,3). The limiting factor is the potential for autofluorescence of tissue from certain plant species. ELFA has been used for detection of virus antigen in seed (1,3) and leaf tissue (10).

Literature Cited

1. Benner, H. I., Hill, J. H., and Durand, D. P. 1990. Detection of soybean mosaic virus by enzyme-linked fluorescent assay (ELFA). Seed Science and Techn. (In press)
2. Diaco, R., Hill, J. H., Hill, E. K., Tachibana, H., and Durand, D. P. 1985. Monoclonal antibody based biotin-avidin ELISA for the detection of soybean mosaic virus in soybean seeds. J. Gen. Virol. 66:2089-2094.
3. Dolores-Talens, A. C., Hill, J. H., and Durand, D. P. 1989. Application of enzyme-linked fluorescent assay (ELFA) to detection of lettuce mosaic virus in lettuce seeds. J. Phytopathol. 125:149-154.
4. Ghabrial S. A., and Shepherd, R. J. 1980. A sensitive radioimmunosorbent assay for the detection of plant viruses. J. Gen. Virol. 48:311-317.
5. Hill, J. H., Bryant, G. R., and Durand, D. P. 1980. Detection of plant virus by using purified IgG in ELISA. J. Virol. Methods 3:27-35.
6. Johnson, D. A., Gaetsch, J. A., Sportsman, J. R., and Elder, J. H. 1984. Improved technique utilizing nonfat dry milk for analysis of proteins and nucleic acids transferred to nitrocellulose. Gene Anal. Techn. 1:3-8.
7. Konijn, A. M., Levy, R., Link, G., and Hershko, C. 1982. A rapid and sensitive ELISA for serum ferritin employing a fluorogenic substrate. J. Immunol. Methods 54:297-307.
8. Neurath, A. R., and Strick, N. 1981. Enzyme-linked fluorescence immunoassays using β-galactosidase and antibodies covalently bound to polystyrene plates. J. Virol. Methods 3:155-165.
9. Shalev, A., Greenberg, A. H., and McAlpine, P. J. 1980. Detection of attograms of antigen by a high-sensitivity enzyme-linked immunoabsorbent assay (HS-ELISA) using a fluorogenic substrate. J. Immunol. Methods 38:125-139.
10. Torrance, L., and Jones, R. A. C. 1982. Increased sensitivity of detection of plant viruses by using a fluorogenic substrate in enzyme-linked immunosorbent assay. Ann. Appl. Biol. 101:501-509.

11. Yolken, R. H., and Stopa, P. J. 1979. Enzyme-linked fluorescence assay: Ultrasensitive solid-phase assay for detection of human rotavirus. J. Clinical Microbiol. 10:317-321.

Acknowledgments

Journal paper no. J-13507 of the Iowa Agriculture and Home Economics Experiment Station, Ames Iowa. Project No. 2700.

Section IV.E.1.

DOT BLOTS (VIRUSES) AND COLONY SCREENING

J. Hammond and R. L. Jordan
USDA ARS, Florist and Nursery Crops Laboratory
Beltsville, MD 20705

Introduction

Membranes are suitable for several types of serological assays. Depending upon the application, nylon may be more suitable than nitrocellulose, and dissociated antigen may be preferable to "native" antigen. The type of antibody available (polyclonal or monoclonal; cross-reactive or highly specific) may determine the assay format, and will certainly influence the choice of detection system (e.g. enzyme-, gold-, fluorescent-, or radiolabeling of either the primary antibody, a secondary antibody, or Protein A). Antisera that are adequate for other purposes may require absorption of non-specific antibodies before use in membrane-based methods because of the high sensitivity of these techniques.

As presented herein, membranes are used for: 1) "Dot-blots", in which antigen is applied to the membrane either by hand or by manifold (including "slot-blots") and 2) colony screening, in which bacteria or other organisms are grown on a membrane applied to the surface of nutrient gel prior to lysis of cells, for *in situ* assay of antigens (e.g., proteins expressed by cloned cDNA). "Western blots" (5) consisting of proteins separated by SDS-polyacrylamide gels and transferred to membranes, with subsequent serological detection and signal amplification, are discussed in **Sections IV.C.1.** and **IV.C.2.**

Nylon membranes have a higher protein binding capacity than *nitrocellulose* (NC); Nylon 66 is modified by the incorporation of tertiary amino groups which result in increased binding of polyanions. This formulation binds ca. 480 $\mu g/cm^2$ compared to 80 μg of protein for unmodified NC (6) Nylon has significantly greater mechanical strength than NC, which is relatively fragile. Polyvinylidine difluoride (PVDF) membranes are sold by Millipore under the name "Immobilon". These membranes are hydrophobic, have high mechanical strength, high protein binding capacity, and good solvent resistance (19). PVDF must be wetted in methanol before use because of the hydrophobic nature of the material (19). *Diazobenzyloxymethyl* (DBM) or *diazophenylthioether* (DPT) modified cellulose papers trap negatively charged proteins in a way that may be more efficient than NC. The initial reaction with DBM and DPT papers is probably an electrostatic effect through the positively charged diazonium groups; this is followed by the formation of stable azo derivatives that covalently link proteins to the matrix (7). These papers are less stable than NC or nylon, and are usually supplied in a form that must be stored refrigerated and then chemically activated before use. Glycine may interfere with the blotting of protein to DBM paper, and there may be lower resolution on DBM paper than on NC or nylon because of the coarser matrix structure of DBM paper (7). The binding capacity is also lower (25-30 $\mu g/cm^2$ for DBM) (4). Thus, although the stability of the protein bonding allows elution of antibody probes and successive reprobing of blots made on these matrices (not usually possible on NC or nylon), DBM and DPT papers are less commonly used.

The surface properties of nylon that enhance protein binding also necessitate use of more protein to block the membranes to prevent non-specific adsorption of antibody, and none of the common stains can be used on nylon to reveal total bound protein independent of antibody reactions. However, nylon membranes can be stained with Ferridye (Janssen), which can also be used on NC. A starch/iodine method for visualizing total protein (14) may have broad usefulness, as the stain can be completely decolorized and the same blot subsequently probed with antibody and developed with a chromogenic substrate system, thus allowing "before-and-after" pictures of the same blot. A "Biotin-blot Protein Detection Kit" (BioRad) can also be used to detect total protein on NC, nylon or PVDF. This procedure involves biotinylating the proteins bound to the membrane, using N-hydroxysuccinimide-biotin, and detection via an avidin-horseradish peroxidase system. Aurodye (Janssen), a colloidal gold stain, can be used on NC and PVDF but not on nylon, as it interacts with the positive charge. NC and PVDF membranes can also be stained for total protein with Coomassie brilliant blue, Amido black, India ink, and a number of other stains (7,19); background destaining of PVDF may be more complete than with NC because of the solvent resistance of PVDF.

Polyclonal antisera usually contain some antibodies reactive only with "native" protein configurations, some reactive only with denatured protein, and some that are conformation-independent. Thus polyclonal antisera are almost always suitable for all types of assay on membranes. However, the apparent specificity of the antiserum may not be the same as in other applications; greater cross-reactivity to related proteins may be observed because the binding of antigens to membranes may expose conserved structural sequences not accessible in other types of assay. Many monoclonal antibodies will function well in dot-blot assays but not on Western blots. This is often an indication that the epitope recognized is "structural" or conformation-dependent rather than "sequential" or conformation-independent. Enzyme conjugation of some monoclonal antibodies may influence the reactive site folding and affect antibody-antigen binding. For this reason, and because of the additional amplification of signal, second antibody or Protein A techniques are often superior to labeling of the primary antibody.

Several reviews have provided considerable detail regarding advantages and disadvantages of particular systems. These include Gershoni and Palade (7), Towbin and Gordon (25), Bers and Garfin (4), Beisiegel (3), and Oberfelder (17). This description will therefore be relatively specific and these reviews may be consulted for alternatives.

Dot-blots and Slot-blots

Principle

Solutions containing the antigens are applied to a membrane, usually in a grid pattern; the samples may be applied either by hand or with a manifold. The remaining membrane surface is then "blocked" by incubation in a solution of a non-reactive heterologous protein, to prevent non-specific adsorption of antibody to the membrane. After incubation in the primary antibody solution (specific for the target antigen), unreacted antibody is rinsed away and the location of bound antibody revealed by use of a suitable detection system. Typically this consists of a species-specific second antibody conjugated to an enzyme. The enzyme is reacted with a soluble substrate to form an insoluble colored product which precipitates at the site of the reaction, indicating which samples contained the antigen of interest (Fig. 1).

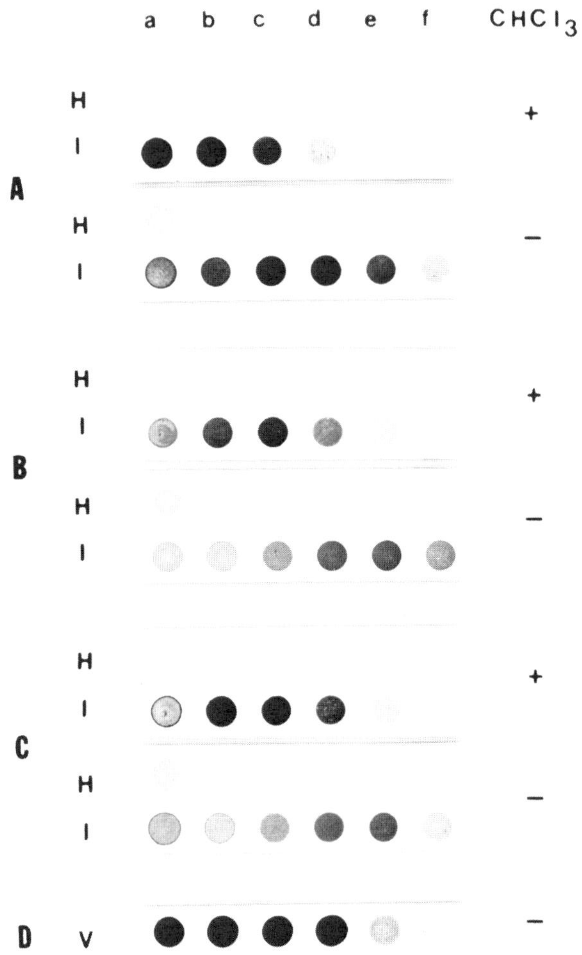

Fig. 1. Example of an evaluation of different extraction buffers (A-C), with (+) and without (-) chloroform clarification, for the detection of bean yellow mosaic potyvirus (BYMV) in sap extracts of Nicotiana benthamiana using a monoclonal antibody in a dot-blot immunoassay. Healthy (H) and BYMV-infected (I) leaves were extracted 1:10 (w/v) in A) TBS + 50 mM DIECA; B) 0.5 M sodium carbonate/bicarbonate, pH 9.6, + 0.5% sodium sulfite; or C) 0.5 M Tris-HCl, pH 8.6, + 0.5% sodium sulfite. Samples were diluted in the appropriate buffer and spotted to a nitrocellulose membrane in 100 μl aliquots at final dilutions of: a) 1:25; b) 1:100; c) 1:400; d) 1:1600; e) 1:6400; and f) 1:25600. D) Purified virus (V) was diluted in TBS and spotted in 100 μl aliquots containing: a) 200 ng; b) 67 ng; c) 22 ng; d) 7.4 ng; e) 2.5 ng; and f) 0.8 ng. After blocking the remaining protein binding sites, the membrane was reacted sequentially with monoclonal antibody, goat anti-mouse antibody conjugated to alkaline phosphatase, and NBT-BCIP substrate as described in the text. Color responses in (H) are due to the green color of the sap extracts (from Jordan and Hammond, unpublished).

Materials

1. Mortar and pestle; or tissue disrupter; or blender.
2. Cheesecloth.
3. Pipetor and pipet tips.
4. Microcentrifuge and tubes.
5. Dot-blot (or slot-blot) apparatus.
6. Membrane cut to fit apparatus; filter paper support of same size as membrane if required in dot-blot apparatus.
7. Buffers: TBS; TBS + 100 mM diethyldithiocarbamate (DIECA); TBS + 50 mM DIECA.
8. Chloroform.
9. Vortex mixer.

TBS = 0.02 M Tris, 0.15 M NaCl, pH 7.5; a 5x TBS stock is useful. To prepare a 5x TBS stock, dissolve 12.1 g Tris base and 43.8 g NaCl in 800 ml distilled water. Adjust the pH to 7.5 with HCl and then bring the volume to 1000 ml. TBS + 100 mM DIECA can be prepared by dissolving 2.25 g DIECA in 20 ml 5x TBS and bringing the volume to 100 ml with distilled water; TBS + 50 mM DIECA by dissolving 1.125 g DIECA in 20 ml 5x TBS and bringing to 100 ml with distilled water. DIECA is not stable for long periods in solution, so these solutions should be prepared as needed.

Procedure

1. Extract tissue 1:10 (w/v) in TBS + 50 mM DIECA. Filter through cheesecloth.
2. Pipet 0.8 ml of filtrate into a 1.5 ml microcentrifuge tube.
3. Add 0.4 ml of chloroform, cap, and vortex to mix.
4. Centrifuge (ca. 12000 *g*) for 2 min.
5. Pipet 200 µl of the aqueous (upper) layer into 800 µl of TBS + 100 mM DIECA, vortex to mix. (Eject pipet tip into the tube and use it for sample application.)
6. Cut an appropriately sized piece of membrane and wet it by floating it on TBS. To apply samples by hand, first allow membrane to air-dry (or use an infrared lamp); for use in a manifold do not allow to dry.

7.a. To hand spot samples, hold pipet with tip in gentle contact with the desired location; apply up to 10 µl. If the membrane is not dry, or the sample is applied too fast, the spot will spread and likely develop into a star instead of a circle.

b. Manifold systems vary in their gasketing arrangements; some may require a wetted sheet of filter paper (Whatman 3MM, or Schleicher and Schuell blot-block paper, or equivalent) to support the membrane. If so, wet the filter paper in TBS and place in the manifold, and put the wet membrane on top; Parafilm may be used to block off any unneeded wells. Assemble the manifold according to the manufacturer's instructions, taking care not to over-tighten. Pipet 10-400 µl samples into the wells as desired (duplication is recommended), and allow drainage to proceed unaided before applying gentle vacuum. Wells can be washed with TBS to remove non-bound proteins. Allow the manifold to stand for a few minutes before disassembly. Use forceps to remove the membrane, and mark orientation, if necessary, using a ballpoint pen.

8. Allow the membrane to air-dry, and store between sheets of filter paper until ready to continue with membrane development. If stored dry in a sealed plastic bag or suitable container, membranes can be stored for several months before developing. Dry storage at room temperature is superior to storage at 4 C in blocking solution.
9. Develop the membrane (see below).

Comments

1. ALWAYS USE FORCEPS OR GLOVES WHEN HANDLING MEMBRANES. FINGER MARKS RESULT IN FALSE REACTIONS THAT MAY ALSO OBSCURE RESULTS.
2. Purified samples that do not require clarification can be diluted as appropriate directly in TBS + 100 mM DIECA.
3. The antigens may be applied to the membrane either by hand or with a manifold. The volume of material that can be applied by hand is limited because of diffusion of the extraction medium from the point of application. Repeated applications to the same spot are possible, (allowing the membrane to air-dry between applications), but introduce greater opportunity for error. Ten μl or less per application is recommended. Use of a manifold in either dot (circular) or slot (rectangular) format allows more uniform application of a larger volume, and hence use of a more dilute antigen solution. The slot-blot format makes it easier to quantitate the results by use of a densitometer.
4. Application of too much vacuum, or too soon, may result in material being drawn through instead of being maintained on the membrane. If the membrane is allowed to dry before addition of the samples, an air bubble may form between the membrane and filter paper, resulting in doughnut shaped deposits instead of solid circles.
5. Clarification of the sample is essential for either method of application; the presence of particulates will partially block the membrane surface, preventing antigen attachment. Moreover, particulates may wash off during subsequent steps, leaving potential sites for non-specific interactions. Clarification of sap extracts to remove chlorophyll and other pigments is also recommended to reduce the background color and increase sensitivity. Chloroform is commonly used for this purpose. Higher dilution of sap extracts may be necessary to determine a positive response in unclarified samples because of high heterologous plant protein binding relative to virus binding, and because of interference from the chlorophyll coloration. As a rule, chloroform-clarified samples may be diluted 1:100 to 1:400, and unclarified samples may need to be diluted 1:500 to 1:1000 for effective discrimination between infected and healthy samples.
6. Equally important are the pH and ionic strength of the extraction buffer and the presence of additives that may significantly affect results. TBS (Tris-buffered saline, pH 7.5) containing DIECA (method suggested by J. W. Moyer) is appropriate for a wide variety of plant species and several virus groups. Figure 1 shows the comparative sensitivity achieved with different sample buffers and treatments. Other buffers and additives have also been used successfully with plant viruses (1,16,20,21) and it is suggested that different buffers should be examined if satisfactory results are not obtained with the system outlined here.
7. The condition of the antigen and the reactivity of the antibody are significant concerns. It has been suggested (23) that heating of hydrated membranes prior to incubation with antibody can significantly improve both sensitivity and signal:noise ratio with certain antibody:antigen

combinations. This is true for many antibodies that recognize conformation-independent epitopes, but decreased sensitivity was observed with antibodies specific for "native" structures (23).

Applications

Dot-blot assays have been used to detect a variety of plant viruses, including members of the carlavirus (1), potexvirus (1), potyvirus (1), tobamovirus (13,18,20), rhabdovirus (16), luteovirus (22), and nepovirus (20) groups. At least some of these viruses occur in very low concentrations in infected plants, and dot-blot tests were usually at least as sensitive as plate ELISA assays where direct comparisons have been made. Monoclonal antibodies have been shown to be useful for detection of the potyviruses peanut mottle virus (21) and bean yellow mosaic virus (9), although not all monoclonal antibodies may be suitable for this application.

The use of dot-blots to detect bacterial infections is described in **Section IV.E.2.**

Membrane Development

Principle

The basic requirements for membrane development are the same for dot-blots, Western blots, and colony screening, so a general protocol is given here. One of the simplest methods is the use of alkaline phosphatase-labeled second antibody, as there are species-specific conjugates available from a number of manufacturers, and indeed kits containing all of the necessary reagents. The NBT-BCIP substrate system (15) is relatively safe, and unlike some of the other substrates available, it is relatively resistant to fading. This is the system that will be described. Other systems are described in **Section IV.E.2., IV.E.3.**

Materials

1. A shaker.
2. Containers slightly larger than the membranes to be developed: these should be of a material that does not adsorb significant amounts of protein, or should be treated to prevent protein adsorption. Suitable materials include glass and polypropylene, and the lids of disposable pipet-tip boxes from a number of manufacturers are an excellent size for this purpose. Polystyrene lids of microtiter plates adsorb proteins and thus may cause reduced sensitivity or high background reactions. Although some people successfully use sealable microwave pouches, some brands have been reported to cause problems due to protein adsorption or because of sustained contact between the bag and the membrane preventing reagent distribution.
3. Blocking solution: 1% non-fat dried milk + 0.5% BSA in TBS is usually quite adequate for nitrocellulose unless the antiserum contains antibodies that react with milk or serum proteins. Blocking requirements for nylon include higher concentrations of protein (5% dried milk, 3-10% BSA, or 10% serum), often at elevated temperatures and for longer times (2 hr to overnight). Gershoni and Palade (6) recommended 10% BSA overnight at 45-50 C.
4. TBS: this is used both as wash solution and diluent for reagents.
5. Antiserum: typical dilutions are equivalent to ELISA dilutions, in the range of 1 μg/ml of immunoglobulin. Dilution of monoclonal antibodies must be determined empirically because of differences between individual

hybridoma cell culture and ascites fluids in concentration of antibody, and differences in affinity and avidity between individual monoclonal antibodies.

6. Second antibody conjugated to alkaline phosphatase: Commercially available preparations can typically be diluted 1:1000-1:5000. TBS is an adequate diluent; addition of $ZnCl_2$ and/or $MgCl_2$ may enhance enzyme activity.
7. Substrate solution: the final preparation of substrate should be made immediately before use. Stock solutions of each component can, however, be prepared in advance and stored for many weeks. Nitro Blue Tetrazolium (NBT) can be dissolved in N, N'-dimethylformamide (DMF) at a concentration of 75 mg/ml, and Bromo-chloro-indolyl phosphate (BCIP) at 50 mg/ml in DMF (**CAUTION**: DMF is toxic and probably carcinogenic, and can be absorbed through the skin. Avoid breathing the fumes, and wear gloves to dispense); store the stock solutions refrigerated, in amber bottles. Add NBT to 0.33 mg/ml and BCIP to 0.175 mg/ml, to the substrate buffer (0.1 M Tris-HCl, 0.1 M NaCl, 5 mM $MgCl_2$)just before use.
8. Stop solution: this is 10 mM Tris-HCl, 1 mM EDTA pH 7.5.

Procedure

1. Prepare a suitable volume of blocking solution; 40-50 ml is sufficient for an 8 x 12-cm NC membrane. Float the membrane on the solution until rehydrated, and then place on the shaker with gentle oscillation for at least 1 hr at room temperature (or overnight at 4 C).
2. Rinse once in TBS for ca. 10 min.
3. Incubate in the primary antibody diluted in TBS (or absorbed with a bacterial lysate for colony screening) for 1-4 hr at room temperature; 1.5 hr is usually sufficient.
4. Rinse three times for ca. 10 min each in TBS.
5. Incubate for 1-4 hr at room temperature in the secondary antibody; typically 1.5 hr (or overnight at 4 C).
6. Rinse four times in TBS as above.
7. Incubate in substrate and watch for development; the time for adequate color development will vary with concentration of reactants, amount of antigen present, the desired sensitivity, and the presence of background reaction in the antiserum. Typically good color will have developed within 5-10 min; if no significant reaction is observed within 20 min, it is unlikely to occur. Inclusion of a test piece of membrane spotted with dilute conjugate as a control for normal reaction speed may be helpful.
8. Stop the reaction by transfer to the stop solution before background coloration becomes significant; this is a question of judgment. Rinse twice in the stop solution for at least 10 min per rinse. Color intensity often appears to increase in the stop solution; the reaction intensity may also appear greater once the membrane is dried. The reaction may be photographed either wet or dry.
9. Store protected from the light.

Comments

1. Species-specific second antibodies conjugated to a variety of ligands are available. These include other enzymes such as horseradish peroxidase, as well as fluorescent dyes and colloidal gold. Primary or secondary antibodies can also be labeled with radioisotopes. Protein A can be substituted for second antibody, and can be labeled by the same

techniques. For alternative methods see Burnette (5) and reviews by Towbin and Gordon (25), Bers and Garfin (4), Beisiegel (3) and **Section IV.E.2.** Some of the alternatives require special handling and disposal of potentially hazardous reagents. The reagents for the alkaline phosphatase system (15) described here are relatively safe, may be the most sensitive (17), and the product is quite resistant to fading.

2. Because assays on membranes have very high sensitivity, the presence of non-specific antibodies may be more apparent than in some other types of assay. These unwanted antibodies can be absorbed from the antiserum as described below.
3. Tween 20 (2,3,24) and other non-ionic detergents have also been used as blocking agents, and can allow staining for total protein on the same blot (2).

Colony Screening: Immunological detection of proteins expressed from cloned DNA

Principle

Proteins can be expressed from cloned DNA in appropriate bacterial phage or plasmid vectors either from complete genes, or as fusion proteins if the cloned sequence is fused in-frame with another gene. Immunological screening can be used either as a primary method of selecting for clones containing a particular gene, or for secondary identification of clones that express the desired gene.

A membrane replica of an existing plate can be prepared, or bacterial colonies can be picked from one plate with sterile toothpicks and streaked onto a membrane. The bacteria are allowed to grow (usually overnight) before the membranes are removed from the plates. The cells are treated in chloroform vapor chamber prior to digestion with lysozyme and DNase, with simultaneous blocking. A rinsing step follows for removal of the bacterial debris, and the membranes are then ready for incubation with the primary antibody. The following protocol is essentially from Helfman et al. (11,12).

Materials

1. Nitrocellulose membranes; circles of Petri plate size are available from several suppliers.
2. Petri plates with appropriate selective media.
3. Chloroform vapor chamber; a suitable chamber is easily improvised. Place a glass rod across a large beaker, into which ca. 50 ml of chloroform is poured, and cover the beaker with foil. Allow to equilibrate before use.
4. Binder clips to hold membranes in vapor chamber.
5. Digestion/blocking solution: 3% bovine serum albumin, 1 μg/ml DNase, 40 μg/ml lysozyme in 50 mM Tris-HCl, 150 mM NaCl, 5 mM $MgCl_2$, pH 7.5. Make fresh before use.
6. Shaker.
7. Rinse solution: 50 mM Tris-HCl, 150 mM NaCl, pH 7.5.
8. Petri dishes for digestion and blocking; one per membrane.
9. TBS.
10. Primary antibody absorbed with a bacterial lysate; prepare the lysate beforehand by pelleting a late log-phase or steady-state culture of cells containing plasmid without insert, grown in rich medium. Resuspend the cells in distilled water or TBS, heat in a boiling water bath 5-10 min, cool and use to dilute the antibody. The lysate will effectively absorb diluted

serum 1-5 times the original volume of the bacterial culture. Incubate the serum and lysate for 2 hr at 4 C before incubating with the membrane. The antibody dilution should be equivalent to ELISA usage, about 1 μg/ml of immunoglobulin.

11. Secondary antibody conjugated to horseradish peroxidase.
12. Substrate solution: 4-chloro-1-naphthol at 3 mg/ml in methanol; the stock solution is stable at 4 C in the dark for 1-2 wk or can be prepared fresh.
13. Hydrogen peroxide (30% aqueous solution).

Procedure

1. a. Prepare a replica of an existing plate by placing a nitrocellulose circle over the colonies, orienting the membrane by means of reference marks, and then carefully transferring the membrane (colony side up) to a fresh plate of selective medium to allow bacterial growth, OR:
   b. Mark a grid on a nitrocellulose circle and lay it over a dish of selective media. Pick candidate colonies from existing plates with sterile toothpicks and gently streak them on the gridded membrane. Allow bacterial growth (usually overnight at 37 C).
2. Remove the membranes from the plates, and suspend them from a binder clip placed over the glass rod in the chloroform vapor chamber. Incubate in the vapor for 15-20 min to lyse the cells.
3. Place each membrane in a separate Petri dish with 10-15 ml of digestion/blocking solution, and shake gently at room temperature several hr or overnight.
4. Rinse gently but thoroughly in several changes of rinse buffer (50 mM Tris-HCl, 150 mM NaCl, pH 7.5).
5. Incubate in lysate-absorbed antiserum, about 1.5 hr.
6. Rinse three times in TBS.
7. Incubate in secondary antibody conjugate (about 1.5 hr).
8. Rinse four times in TBS.
9. Dilute 4-chloro-1-naphthol solution in 5 volumes of TBS and add hydrogen peroxide to 0.01% by volume. Incubate in substrate until color development reaches suitable level. Keep away from bright light.
10. Rinse twice in TBS or distilled water, dry and store in the dark.

Comments

1. Membranes are incubated with primary antibody as for other systems, with a few provisos. Use of lysate-absorbed antisera avoids non-specific reactions, which are otherwise quite common. The primary antibody could be adsorbed onto membranes incubated with bacterial lysates instead of absorbed directly by the lysate as described here.
2. For <u>primary</u> screening of colonies the maximum sensitivity possible may be necessary, and hence a radio-labeled secondary antibody or Protein A may be used. For secondary screening, where bacteria are streaked on a gridded membrane and bacterial growth can be allowed to proceed longer without danger of colonies coalescing, there should be sufficient antigen present in "positive cultures" to allow use of non-radioactive methods.
3. It is possible to use alkaline phosphatase labeled second antibody, and be able to distinguish positive from negative results, but there will likely be significant background from the reaction of <u>bacterial</u> alkaline phosphatase with the substrate. This might be avoided by heating the membranes to denature the bacterial enzyme, if the primary antibody will react with the

desired antigen in the denatured state. Horseradish peroxidase, or some other enzyme that is lacking in bacteria would be more suitable as a conjugate with the secondary antibody. If horseradish peroxidase is used, none of the wash or antibody solutions may contain sodium azide, as it will inactivate the peroxidase; the enzyme is very sensitive to azide. Alternative substrates are available for horseradish peroxidase, but some are suspected carcinogens and offer no advantage for sensitivity (10).

4. Some cloning vectors have cloning sites in genes that are inducible; i.e. the gene is "turned on" or expressed at higher level in the presence of an inducing agent. Many vectors utilize the lac promoter and lacZ gene, which can be induced by isopropyl-β-D-galactoside (IPTG). Vectors bearing the P_L promoter from bacteriophage λ are repressed at low temperature (28 C) by a thermo-labile product of the λ cI857 gene, and expression is initiated by induction at 42 C. If expression from the cloned DNA is inducible, the detection efficiency will be enhanced if the bacteria are induced before lysis.
5. Phage plaques can be screened in essentially the same way as described for bacterial colonies.
6. Clones identified by colony blotting can be further investigated by Western blotting to determine the size of the expressed protein.

Applications

Immunological screening for identification of clones expressing a particular gene product has become a commonplace technique in molecular biology, and many plasmid and phage vectors are designed to express cloned sequences. In some of these vectors, expression is constitutive, and in others expression must be induced chemically or by a temperature shift. However, to date there has been little reported on the use of immunological screening to select cloned plant virus genes. The colony screening method described here has been used to detect clones expressing the bean yellow mosaic virus coat protein gene (Fig. 2) (8).

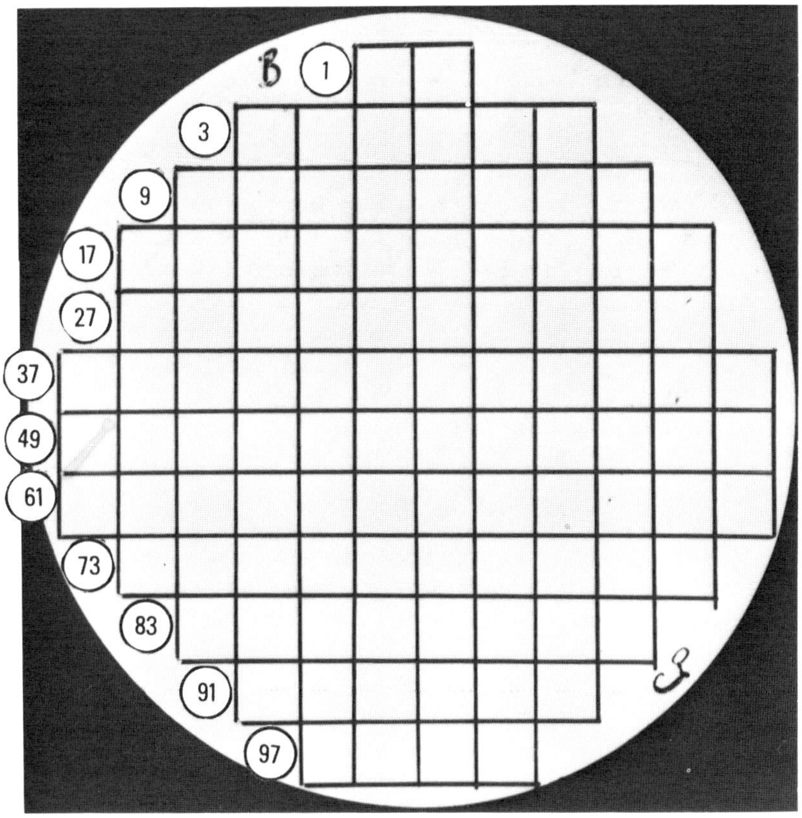

Fig. 2. Detection by colony screening of a recombinant DNA clone expressing the bean yellow mosaic virus coat protein gene. Note diagonal line indicating positive substrate reaction, in Grid No.49. A rabbit polyclonal antiserum raised against whole virus and absorbed with a bacterial lysate as described in the text was used (at ca. 2.5 μg/ml IgG) to screen the filters. The reaction was detected via goat anti-rabbit antibody conjugated to horseradish peroxidase, and 4-chloro-1-naphthol substrate.

Literature Cited

1. Banttarri, E. E., and Goodwin, P. H. 1985. Detection of potato viruses S, X, and Y by enzyme-linked immunosorbent assay on nitrocellulose membranes (Dot-ELISA). Plant Dis. 69:202-205.
2. Batteiger, B., Newhall, W. J., and Jones, R. A. 1982. The use of Tween 20 as a blocking agent in the immunochemical detection of proteins transferred to nitrocellulose membranes. J. Immunol. Methods 55:297-307.
3. Beisiegel, U. 1986. Protein blotting. Electrophoresis 7:1-18.
4. Bers, G., and Garfin, D. 1985. Protein and nucleic acid blotting and immunobiochemical detection. Biotechniques 3:276-288.
5. Burnette, W. M. 1981. "Western blotting": Electrophoretic transfer of proteins from sodium dodecyl sulfate-polyacrylamide gels to unmodified nitrocellulose and radiographic detection with antibody and radioiodinated protein A. Anal. Biochem. 112:195-203.
6. Gershoni, J. M., and Palade, G. E. 1982. Electrophoretic transfer of proteins from sodium dodecyl sulfate-polyacrylamide gels to a positively charged membrane filter. Anal. Biochem. 124:396-405.

7. Gershoni, J. M., and Palade, G. E. 1983. Protein blotting: Principles and application. Anal. Biochem. 131:1-15.
8. Hammond, J. and Hammond, R. W. 1989. Molecular cloning, sequencing and expression in Escherichia coli of the bean yellow mosaic virus coat protein gene. J. Gen. Virol. 70:1961-1974.
9. Hammond, J., and Lawson, R. H. 1988. A strain of bean yellow mosaic virus is aphid-transmitted from orchid. Acta Hort. 234:365-370.
10. Hawkes, R., Niday, E., and Gordon, J. 1982. A dot-immunobinding assay for monoclonal and other antibodies. Anal. Biochem. 119:142-147.
11. Helfman, D. M., Feramisco, J. R., Fiddes, J. C., Thomas, G. P., and Hughes S. H. 1983. Identification of clones that encode chicken tropomyosin by direct immunological screening of a cDNA expression library. Proc. Natl. Acad. Sci. USA 80:31-35.
12. Helfman, D. M., Feramisco, J. R., Fiddes, J. C., Thomas, G. P., and Hughes, S. H. 1984. Immunological screening of cDNA expression libraries. Focus (Bethesda Research Laboratories) 6:1-5.
13. Hibi, T., and Saito, Y. 1985. A dot immunobinding assay for the detection of tobacco mosaic virus in infected tissues. J. Gen. Virol. 66:1191-1194.
14. Kumar, B. V., Lakshmi, M. V., and Atkinson, J. P. 1985. Fast and efficient method for detection and estimation of proteins. Biochem. Biophys. Res. Comm. 131:883-891.
15. Leary, J. J., Brigati, D. J., and Ward, D. C. 1983. Rapid and sensitive colorimetric method for visualizing biotin-labeled DNA probes hybridized to DNA or RNA immobilized on nitrocellulose: Bioblots. Proc. Natl. Acad. Sci. USA 80:4045-4049.
16. Lockhart, B. E. L., Elmaataoui, M., Carroll, T. W., Lennon, A. M., and Zaske, S. K. 1986. Identification of barley yellow striate mosaic virus in Morocco and its field detection by enzyme immune assay. Plant Dis. 70:1113-1117.
17. Oberfelder, R. 1989. Immunoblotting: Comparison of detection methods. Focus 11:1-5.
18. Parent, J. G., Belanger, F., Desjardin, S., and Brisson, J. D. 1985. Dot-immunobinding for detection of tomato mosaic virus and potato virus X infecting greenhouse tomatoes. Phytoprotection 66:53-57.
19. Pluskal, M. G., Przekop, M. B., and Kavonian, M. R. 1986. Immobilon PVDF transfer membrane: A new membrane substrate for Western blotting of proteins. BioTechniques 4:272-282.
20. Powell, C. A. 1987. Detection of three plant viruses by dot-immunobinding assay. Phytopathology 77:306-309.
21. Sherwood, J. L., Sanborn, M. R., and Keyser, G. C. 1987. Production of monoclonal antibodies to peanut mottle virus and their use in enzyme-linked immunosorbent assay and dot-immunobinding assay. Phytopathology 77:1158-1161.
22. Smith, F. D., and Banttari, E. E. 1987. Dot-ELISA on nitrocellulose membranes for detection of potato leafroll virus. Plant Dis. 71:795-799.
23. Swerdlow, P. S., Finley, D., and Varshavsky, A. 1986. Enhancement of immunoblot sensitivity by heating of hydrated filters. Anal. Biochem. 156:147-153.
24. Tovey, E. R., Ford, S. A., and Baldo, B. A. 1987. Protein blotting on nitrocellulose: Some important aspects of the resolution and detection of antigens in complex extracts. J. Biochem. Biophys. Methods 14:1017.
25. Towbin, H., and Gordon, J. 1984. Immunoblotting and dot immunobinding - Current status and outlook. J. Immunol. Methods 72:313-340.

Section IV.E.2.

THE DOT IMMUNOBINDING ASSAY (DIA) - BACTERIA

G. Lazarovits
Agriculture Canada, London Research Centre
London, Ontario, Canada

Principle

Serological procedures for detection of bacteria immobilized on membrane filters have been undertaken for many years. In most studies the antibody was conjugated to a fluorescent material and the reaction with the antigen identified by observation using a microscope equipped with an ultraviolet light source. This procedure, although reliable and sensitive, required costly instrumentation, tedious microscopic viewing and trained personnel to insure accurate results. The dot immunobinding assay (DIA), which is a modification of this assay, circumvented these limitations but preserved the simplicity, sensitivity and reliablilty of the procedure. Detection limits of 1000-2000 cells per 2 μl spot applied can be expected.

DIA differs from conventional ELISA primarily in that the solid phase matrix used to adsorb the test material is a plastic membrane or paper sheet rather than a well in a polystyrene microtiter plate. Most commonly, DIA involves applying microliter quantities of sample to a membrane and then incubating this in a solution containing specific antibody. The membrane is then incubated with an enzyme conjugated anti-globulin antibody that reacts with serological specificity with bound immunoglobulin. Depending upon design, this assay can be adapted to detect both antigens and antibodies, with several possible amplification systems (Fig. 1). Enzyme activity is revealed by addition of an appropriate substrate, which, after conversion, precipitates to form a colored spot (Fig. 2.).

DIA is the most simple, rapid and sensitive assay for immunological characterization of phytopathogenic bacteria and fastidious prokaryotes, obtained either from cultures or from diseased tissues (1,6,8,12,13,15,16,17,18,21,29; also see **Section IV.E.1.**). A large number of samples can be processed within 2 to 3 hr, rather than the day required with conventional ELISA procedures, Both the antibody and the enzyme conjugate solutions can be reused several times.

Materials and Solutions

1. Nitrocellulose: use 0.45 μm pore size available from a number of manufacturers; filter paper may be a suitable alternative adsorbent.
2. Containers which allow for complete coverage of membranes and minimize volume of reactants, i.e. covers of microtiter plates, Petri dishes, plastic baggies.
3. Distilled or deionized water.
4. Tris buffered saline (TBS): [50 mM Tris (hydroxymethyl)-aminomethane, 150 mM sodium chloride, 0.01% Thimersal, pH 7.4].

		SCHEMATIC PRESENTATION	Number Of Steps To Completion									Ref.
			1	2	3	4	5	6	7	8	9	
(A)	Direct Assay For Antigen	I-Ag····^{1}Ab Ⓔ	Add Ag	Block untreated sites	Add enzyme conjugated specific ^{1}Ab Ⓔ	Wash membrane	Add substrate					7
(B)	Indirect Assay For Antigen	I-Ag····^{1}Ab····^{3}Ab Ⓔ I-Ag····^{1}Ab····ProtAⒺ I-Ag····^{1}Ab····^{2}AbBio····StreptAⒺ I-Ag····^{1}Ab····^{2}Ab Ⓔ····^{3}Ab Ⓔ	"	"	Add specific antibody ^{1}Ab	"	Add enzyme conjugated secondary ^{2}Ab Ⓔ or similar reactant with ^{1}Ab	Wash membrane	Add substrate			15 20 27
(C)	Direct Assay For Antigen Using Capture Antibody	I-^{1}Ab····Ag····^{1}Ab Ⓔ	Add capture ^{1}Ab	"	Add Ag	"	Add enzyme conjugated primary ^{1}Ab Ⓔ	"	"			1 22
(D)	Indirect Assay For Antigen Using Capture Antibody	I-^{1}Ab····Ag····^{11}Ab····^{2}Ab Ⓔ and as in (B)	"	"	"	"	Add 2nd specific ^{11}Ab made in a different species of animal than capture ^{1}Ab	"	Add enzyme conjugated ^{2}AbⒺ specific for ^{11}Ab	Wash	Add substrate	

I - nitrocellulose membrane or similar solid phase support
Ag - antigen
^{1}Ab - specific antibody made against the antigen in animal species A
^{11}Ab - " " " " " " " " B
ProtAⒺ - protein A-enzyme conjugate
^{2}Ab - secondary antibody made against specific antibody ^{1}Ab or ^{11}Ab
^{3}Ab - " " " " enzyme used as the conjugate
Ⓔ - enzyme used as conjugate - e.g. peroxidase, alkaline phosphatase
^{2}AbBio - secondary antibody conjugated to biotin
StreptAⒺ - streptavidin conjugated to enzyme

Fig. 1. Dot immunobinding assay (DIA) configurations illustrating typical approaches for detecting antigens and amplifying signals. Note that either antigen or antibody can be bound to the nitrocellulose membrane, followed by any one of numerous choices of secondary antibody and conjugates.

5. Homogenization buffer - TBS plus 0.1 M ascorbic acid and/or 0.2% (w/v) sodium sulfite and/or 0.01 M diethyldithiocarbamate and/or 1% (v/v) mercapto-ethanol and/or 0.01 M ethylaminetetraacetic acid (EDTA).
6. Blocking solution and antibody diluent (TBS-BSA-Tween): TBS containing 0.05% Tween 20 or Nonidet 40 and 1.0% BSA or similar protein.
7. Antisera.
8. Enzyme conjugate.
9. Substrate buffer (see **Section IV.E.1.**)
10. Substrate (see **Section IV.E.1.**).
11. Platform shaker.
12. Densitometer, optional (e.g. Model 620 Video Densitometer, Bio-Rad)

Procedure

1. Preparation of nitrocellulose membranes

Cut membranes to the appropriate size and mark as desired with a pencil and a plastic template (holes placed 0.5-1 cm). Use at least 1 cm spacing if volumes applied exceed 2 μl (Fig. 2). Handle membranes only with gloves or forceps. Avoid trapping air in the membrane during wetting, as antigen will not be adsorbed uniformly. Insert one edge of the membrane into distilled water or TBS buffer and allow it to take up liquid by wick action. Rinse with gentle shaking for 5 to 10 min and then air-dry thoroughly on filter paper.

2. Preparation of antigen samples

a. samples from cultures

Suspend and wash bacterial cells in saline and make appropriate dilutions. Apply 1-2 μl of antigen suspension to nitrocellulose membrane (NCM) with a Hamilton type syringe or a micropipeting device. In some instances ordinary filter paper can be substituted for nitrocellulose. Such paper, however, must be blocked in a protein solution prior to application of the bacterial suspension. Bacteria grown on agar media can be transferred directly to NCM. Cut membranes to size, mark at one place for reference and soak in saline. Place membrane carefully on agar surface and leave for 1-60 min (optimal time for transfer should be predetermined). Up to five copies can be made from the same plate without loss of resolution. Media or bacterial products carried along with the transfers can increase the background and lower the sensitivity. If this occurs, make a copy of the first transfer (master) by placing a fresh piece of NCM on top of it (copy) and compressing these between filter paper with a 1 kg weight for 5 min (8). Use the copy for assay and the master for replica plating on fresh medium.

Inactivate native bacterial peroxidase or alkaline phosphatase by either heating the dried membranes to above 80 C or by incubating in 3% aqueous H_2O_2 for 30 min with gentle agitation. Rinse H_2O_2 from the membrane with several changes of Tris buffered saline. For detection of internal antigens lyse bacteria in situ after this step.

b. samples from diseased tissues

Whenever possible, avoid using homogenization as the preparation procedure for detection of bacteria in diseased plant tissues. Diffusates or ooze from the diseased tissue is preferable as the antigen source. To obtain diffusates, excise the lesion, place it in a drop of buffer or water, cut into small pieces with a razor blade and leave for 1 hr. Prepare dilutions of this liquid for assay (Fig. 2) When grinding or homogenization of diseased tissues is necessary, include anti-oxidants in the buffer. Pigments in the homogenate may stain the membrane and should be removed prior to application. Extraction with an equal amount of dichloromethane is an effective way of achieving this. Remove all residual solvent from the sample prior to application by placing it in a well-ventilated area. Alternatively, pellet the bacteria by centrifugation and resuspend the pellet in fresh buffer. However, bacteria added to plant sap and pelleted in this way tend to be less reactive antigenically. Problems caused by pigmented material can also be overcome by substituting filter paper for nitrocellulose (6). First, block the filter paper in blocking solution for 20-30 min and allow to dry. Apply 2-4 µl of sample, allow to dry and then soak in 70% ethanol to remove green pigments. Dry paper or wash in buffer before further processing.

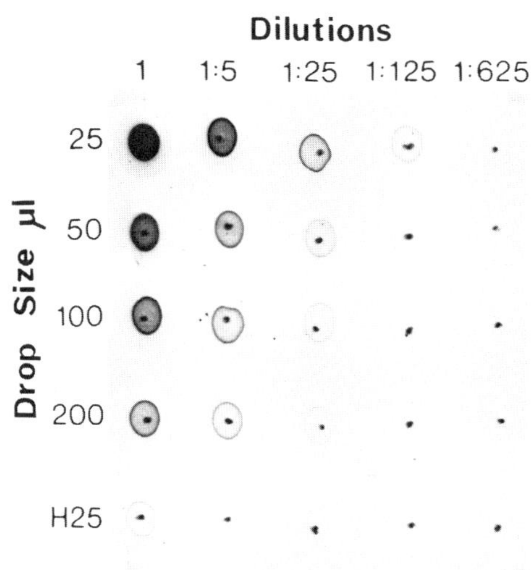

Fig. 2. The effect of drop volume on detection of bacteria released from a single lesion on pepper leaves harvested 7 da after inoculation with Xanthamonas campestris pv. vesicatoria. The membrane was incubated in antiserum prepared against X. campestris pv vesicatoria. Alkaline phosphatase conjugated Protein A was used for detection. The apparent reaction seen with the control sample (H = healthy plant) is due to the small amount of green pigment present in the sample.

3. Blocking of unreacted binding sites

Immerse the membrane in TBS-BSA-Tween for 10 min with gentle shaking. Remove membranes and rinse in generous amount of distilled water and place into blocking solution for a further 5-10 min.

4. Reaction with specific antiserum and enzyme labeled conjugate

Antiserum can be added directly to the blocking solution if the exact volume of the blocking solution is known. Incubation times of 5-15 min with the specific sera at titers of 1:1,000-1:5,000 are often sufficient. For greater sensitivity, however, extend the incubation times to 60-180 min. Both the antiserum and enzyme conjugate can be reused several times. When many different antisera preparations are used, a conjugate which reacts universally with all antibody preparations is recommended, i.e., species specific antibody or Protein A conjugate. If only a few sera are used routinely, the length of the assay can be reduced significantly by preparing an enzyme conjugate of the primary antibody.

5. Washes

Remove antiserum with an aspirator and wash away unbound antibody with 2-3 quick changes of TBS-Tween, followed by two 5-10 min soaks in a liberal amount of wash solution.

6. Incubation in enzyme conjugate

Dilute the enzyme conjugate reagent in blocking solution. Generally, dilutions of 1:1000 and incubation times of 90 min are used with these reagents. Rinse the membranes as previously described except that the last rinse should be in the substrate buffer.

7. Visualization of enzyme activity

Apply appropriate substrate solution and incubate until color develops to desired intensity. After color development rinse membrane thoroughly in distilled water and dry between pieces of filter paper. Protect membranes from strong light as many of the precipitates are light sensitive.

Numerous procedures have been described for visualization of enzymatic activity. Selection of the substrate greatly affects the efficiency of detection and quality of assay results. Some of the protocols that have been reported to provide excellent sensitivity of detection with the two most commonly used enzymes are detailed below.

8. Peroxidase substrates

a. Protocol a

Just before use, dissolve 50 mg of diaminobenzidine tetrahydrochloride (DAB) or O'-dianisidine dihydrochloride (DAN) (Sigma; both are carcinogens; use latter compound if available) in 100 ml of 50 mM phosphate buffer, pH 7.4, and filter. Add dropwise, with agitation, 3 ml of 1% (w/v) cobalt chloride and 1% nickel ammonium sulfate (in

H_2O). Incubate the NCM in this solution for 10 min and then transfer to a fresh preparation of the same solution containing 1 ml of 0.05% H_2O_2. Incubate 30 min at room temperature or until a satisfactory reaction is obtained. Wash in phosphate buffer several times, fix in 0.1% OsO_4 for 5 sec, and then rinse in water. Store dried membranes in the dark (3).

b. Protocol b

Incubate the membrane in TBS (50 mM Tris-HCl, pH 7.4, 200 mM NaCl) containing 0.06% (w/v) 4-chloro-1-naphthol [can be stored in the dark as a 0.3% (w/v) stock solution in methanol] and 0.01% H_2O_2 (9).

c. Protocol c

Dissolve 20 mg of nicotinamide adenine dinucleotide, reduced form, (NADH) in 5 ml of 0.1 M phosphate buffer, pH 7.0. Add to this 4 ml of water containing 1.0 mg/ml phenol and 1 ml with 3.0 mg/ml nitroblue tetrazolium (NBT). Immediately before starting the reaction, add 20 μl of 10% (w/v) H_2O_2. The resulting ten ml of staining solution contains 0.3 mg/ml NBT, 2.0 mg/ml NADH, 0.4 mg/ml phenol and 0.02% (w/v) H_2O_2 in 50 mM phosphate buffer, pH 7.0. Allow the reaction to proceed for 30 min in the dark at 25 C (24).

d. Protocol d

Luminescent visualization: dissolve 4.4 mg of luminol (Sigma) in 1.5 ml of 1 M Tris and add to this 0.5 ml of 1 M Tris-HCL and 18 ml of H_2O giving a solution of pH 8.5 (luminol does not dissolve readily at pH 8.5). Add 25 μl of 10 mg/ml synthetic firefly luciferin (sodium salt, stored as prepared samples at -70 C), followed by 10 mg of 4-methyl-umbelliforme (MUB, sodium salt). Add H_2O_2 (6 μl of 30%) just before use. Photograph luminescent spots with blue-sensitive X-ray film (Fuji RX). This reaction is extremely sensitive as it continues for several hours (14).

9. Alkaline phosphatase

a. Protocol a

Prepare the following solutions: (A) 5 mg/ml 5-bromo-4-chloroindoxyl phosphate (BCIP), (US Biochemicals) in dimethylformamide; (B) 1 mg/ml NBT in veronal acetate; (C) 2 M $MgCl_2$. Then mix 0.1 ml of A, + 1 ml of B, + 20 μl of C and 9 ml of 0.15 M veronal acetate buffer (pH 9.6) (2).

b. Protocol b

Prepare the following solutions: (A) 0.33 mg/ml NBT in Tris-HCl (25 mM, pH 9.55, containing 25 mM diethanolamine, 0.1 M NaCl, 2 mM $MgCl_2$, 10^{-6}M $ZnCl_2$), (B) 2 mg/ml phenazine metho-sulfate in H_2O, (C) 4 mg/ml BCIP in anhydrous dimethylformamide. Immediately before use, to each ml of solution A, add with rapid mixing 6.7 μl of B and 3.4 μl of C.

Solution A and B must be freshly prepared, whereas C is stable for 3-4 wk at 4 C (5).

c. Protocol c

Dissolve 5 mg of BCIP in 20 ml AMPD buffer (10.2 g of 2-amino-2-ethyl-1,3-propanediol; 10 mg $MgCl_2.6H_2O$ in 1 liter distilled water, pH 10.4) and add 0.2 ml $CuSO_4$ solution. Inclusion of thiazolyl blue at 1 mg/ml enhances color formation (26).

d. Protocol d

Prepare the following solutions: (A) 0.4 mg/ml Naphthol AS-MX phosphate in distilled water, (B) 6 mg/ml fast red TR salt in 0.2 M Tris-HCl, pH 8.0 containing 2 mM $MgCl_2$. Mix 1:1 immediately before staining.

e. Protocol e

Working solutions: Prepare these solutions immediately before use: (A) 20 mg Fast blue B or Fast blue RR in 10 ml H_2O, (B) 10 mg β-naphthyl phosphate (sodium salt) in 9.4 ml borate buffer (.05 M boric acid and 0.05 M KCL to pH 9.2 with NaOH) and add 0.6 ml 0.1 M $MgCl_2$. Combine A+B and add immediately to filter (26).

10. Densitometry

Measure the optical density of the spots with a reflectance densitometer. Best results are obtained when the readings are made from photographs of the membrane (15).

Comments

1. DIA is ideal for detection of bacteria and fastidious prokaryotes, or their products, either in purified form or as constituents of plant or animal extracts (1,6,10,15,16,19,20,21,25). It is a rapid, simple assay that consumes very small quantities of the antibody and antigen reagents. Sensitivity and reliability are comparable or superior to that obtained with conventional ELISA. Particulate and water insoluble antigens, difficult candidates in other assays, are easily accommodated with DIA. The results are permanent and can be conveniently inserted into a notebook. On the negative side, since the sample volume must be very small, the assay is restricted to preparations with relatively high antigen concentrations. Difficulties often arise with samples containing pigments, as these stain the membrane and interfere with the interpretation of the histochemical reaction. DIA is not as good as conventional ELISA for quantitative studies.

2. Membrane selection and handling

Membranes of quite diverse chemical and physical composition including nylon (ZetaProbe/Zetabind, Biotrace RP, Nytran), polyvinylidene difluoride (Immobilon), NCM and glass fiber and cellulose sheets can be used in DIA.

The most versatile and widely used of these, by far, is NCM, which acts as both a surface filter and an adsorbing matrix for macromolecules. The binding capacity of NCM varies considerably depending upon the manufacturing process used, the pore size of the membrane and the nature of the material to be bound. As a general rule, the smaller the pore size the greater the binding capacity. Optimal binding capacity for 0.45 μm pore size membrane ranges from 80-250 $\mu g\ cm^{-2}$. Some NCMs, such as those containing a mixed ester matrix, are designed to have low binding properties and should not be used for tests involving adsorption of macromolecules. Nylon membranes offer superior mechanical properties and greater binding capacities than NCM.

3. Diluents

The buffer selected as diluent is critical to the successful outcome of DIA. The two diluents most commonly employed, phosphate and Tris buffered saline (PBS and TBS), have each been identified as the most and least effective diluents (1,7,19,22). The basis for the conflicting claims is not understood. TBS, overall, is used more widely than PBS.

4. Preparation of antigen

Most antigen samples can be applied to the membrane without any special preparative procedures. However, with plant extracts nonspecific reactions between plant sap components and immunoglobulins can be anticipated and eliminated or reduced by incorporating chelating agents such as EDTA (0.01 M) and antioxidants such as ascorbic acid (0.2%, w/v), sodium sulfite (1-10%, w/v), 2-mercaptoethanol (1%, w/v) or diethyldithiocarbamate (0.01 M) into the extraction buffer (1,19,20,22). These agents also enhance detection sensitivity by improving the extraction efficiency for the antigen. Some reagents, such as 0.2 M $MgCl_2$ (25) or 1% KOH (28), achieve similar effects by increasing the binding forces between specific proteins and nitrocellulose. Photosynthetic pigments, which also interfere with the assay can be removed prior to application by extraction with dichloromethane (19) or after application by incubation of the membranes for 1 hr in 2% Triton X-100 (20). The latter procedure, however, may result in loss of other components, including the antigen itself. Partial purification from lipophilic materials can also be achieved by passage of the sample through small Sep Pak cartridges. Organic solvents should not be applied to nylon or nitrocellulose membranes as these solvents will damage membrane integrity. Such materials are best assayed using glass fiber or cellulose sheets as adsorbents (4).

5. Application of antigens or antibodies to the membrane

Hamilton syringes or micropipeting devices offer the most rapid and convenient method of applying antigen solutions. Volumes of 0.1 to 10 μl can be applied, but 1-2 μl drops are optimal. Multiple application to the same spot is permissible providing the membrane is allowed to dry after each drop. It is often useful to prepare a large number of identical membranes and then store the unused ones at or below 4 C for several weeks (22). Filtration manifolds are recommended for tests using the double antibody sandwich procedure or for tests where low antigen concentrations dictate applying milliliter volumes of solution (1,22). Applying sample in this way also provides a more uniform reaction site for densitometric measurements. Commercial manifolds, however, are expensive and using them restricts the number of tests that can be conducted at one time.

6. Controls

Two types of controls are used in DIA; one to monitor the performance of the reagents and the other the performance of the assay. To monitor the performance of the reagents apply a drop of the following solutions to the membranes: (1) the specific antibody - a positive reaction verifies the reaction between the immunoglobulin and the enzyme-labeled conjugate; (2) the enzyme-labeled conjugate - verifies enzyme activity and the proper functioning of the substrate stain; (3) serial dilutions of purified antigen or known infected material - to act as a positive control and for standardization of quantitative procedures; (4) an extract from a healthy plant and/or an unrelated antigen which does not react with the specific antibody - acts as a negative control. The controls for monitoring the proper performance of the immunological reactions require that membranes prepared in an identical manner to the test membrane be subjected to the following conditions: (1) instead of the specific antiserum, incubate the membrane in an antiserum from a pre-immune source - positive reactions indicate nonspecific reactions; (2) incubate in specific antibody but omit incubation in enzyme-labeled conjugate - positive reactions indicate nonspecific precipitation of the substrate; (3) incubate only in the enzyme conjugate - positive reactions indicate nonspecific binding of conjugate or nonspecific precipitation of the substrate.

7. Heat treatment and incubation

Heating the membranes after the antigen has been applied can significantly enhance the sensitivity of the immunoassay for some proteins. Swerdlow et al. (23) obtained the most improvement when rehydrated membranes were boiled or autoclaved for 30 min. Heat treatment also inactivates any enzyme associated with the adsorbed material that may contribute to the activity of the enzyme conjugate (such as peroxidase).

Membrane incubation containers should provide for complete contact between the membranes and the solutions while keeping the reagent volume to a minimum (ca. 0.2 ml/cm^2). Shallow plastic trays such as microtiter plate covers, petri dishes or plastic baggies are satisfactory containers.

8. Blocking of unreacted sites

Free sites on the membrane must be blocked from reacting with reagents used in subsequent steps. Failure to do this will result in unacceptable background color after substrate addition. Neutral proteins, such as bovine serum albumin, skim milk powder, casein, hemoglobin and others, at concentrations ranging from 1-10%, have been employed most commonly. Johnson et al. (11) found that a solution of 5% (w/v) nonfat dry milk powder, 0.01% antifoam A, and 0.001% merthiolate in PBS gave consistently good backgrounds in immunoassays on NCM. Excessive amounts of blocking protein can interfere with antigen-antibody reactions and thus reduce assay sensitivity. Low levels of protein (ca. 1%), on the other hand, prevent non-specific binding of antibodies and thereby increase the specificity of the reaction. Non-ionic detergents, such as Nonidet P-40 (NP-40) and Tween

20 or 80, are also excellent blocking agents. In addition to being less expensive, detergents enhance the antibody-antigen reactions. Avoid concentrations in excess of 0.05-0.1% (Tween or NP-40), however, as they can result in desorption of some proteins. The blocking agent of first choice should be 0.05% NP-40 or Tween 20 in TBS.

9. First antibody

Antisera containing the specific antibodies should be diluted with the same material as used for blocking. When assaying for pathogenic components in plant extracts it is advisable to absorb the sera prior to use with healthy plant extract, as plant sap can cause nonspecific precipitation of the substrate regardless of the enzyme system used (20). Affinity purification of the antibodies can also bypass this problem. The diluted antiserum and enzyme conjugate can be reused several times provided that it contains microbial inhibitors and is stored at 4 C.

10. Rinses

Unreacted antibody or conjugate must be removed from the membranes. Two or three brief rinses and one or two 5-10 min soaks in a liberal amount of detergent-containing TBS or PBS buffers are usually sufficient.

11. The enzyme conjugate

Numerous enzyme-conjugated materials are available commercially for detection of the specific antibody. They are identical to those used for conventional ELISA, and their properties and use are discussed in Sections **IV.A.1.** and **IV.A.2.**

12. Quantitation

Quantitation for DIA requires a densitometer which operates in a reflectance mode. The intensity of color formation has been found to be linear within a reasonable span of antigen concentrations and quite satisfactory standard curves can be generated (15). The membranes can be made transparent by dehydrating with absolute alcohols, such as isobutanol or 2-propanol, and then immersing them in xylene. This process does not affect most of the precipitates. Overall, however, DIA is not as suitable for quantitative studies as conventional ELISA.

Applications

Hansel and Matthews (8) used DIA for differentiating *E. coli* colonies from contaminants on agar plates. Lopez-Vidal and Svennerholm (16) and Rhea and Shah (21) used this procedure to screen for production of colonization factor antigen and enterotoxin by *E. coli*. DIA was used successfully to detect and differentiate among *Xanthamonas campestris* pathovars (Fig. 3) (15), isolates of *Xanthamonas campestris* pv. *phaseoli* (17), and *Erwinia amylovora* obtained from infected pear trees (29). Kotani and McGarrity (12,13) found DIA to be a sensitive and reliable technique for detection of *Spiroplasma citri*. Fletcher (6) showed that filter paper was an excellent substitute for nitrocellulose for the detection of this organism. The antigenic relatedness of lipopolysaccharide components of *Chlamydia trachomatis* were compared to that of *Salmonella* and

Acinetobacter by immunoblotting from sodium dodecyl sulfate-polyacrylamide electrophoresis gels (18). Further information on immunoblots and virus detection can be obtained in **Section IV.E.1.**

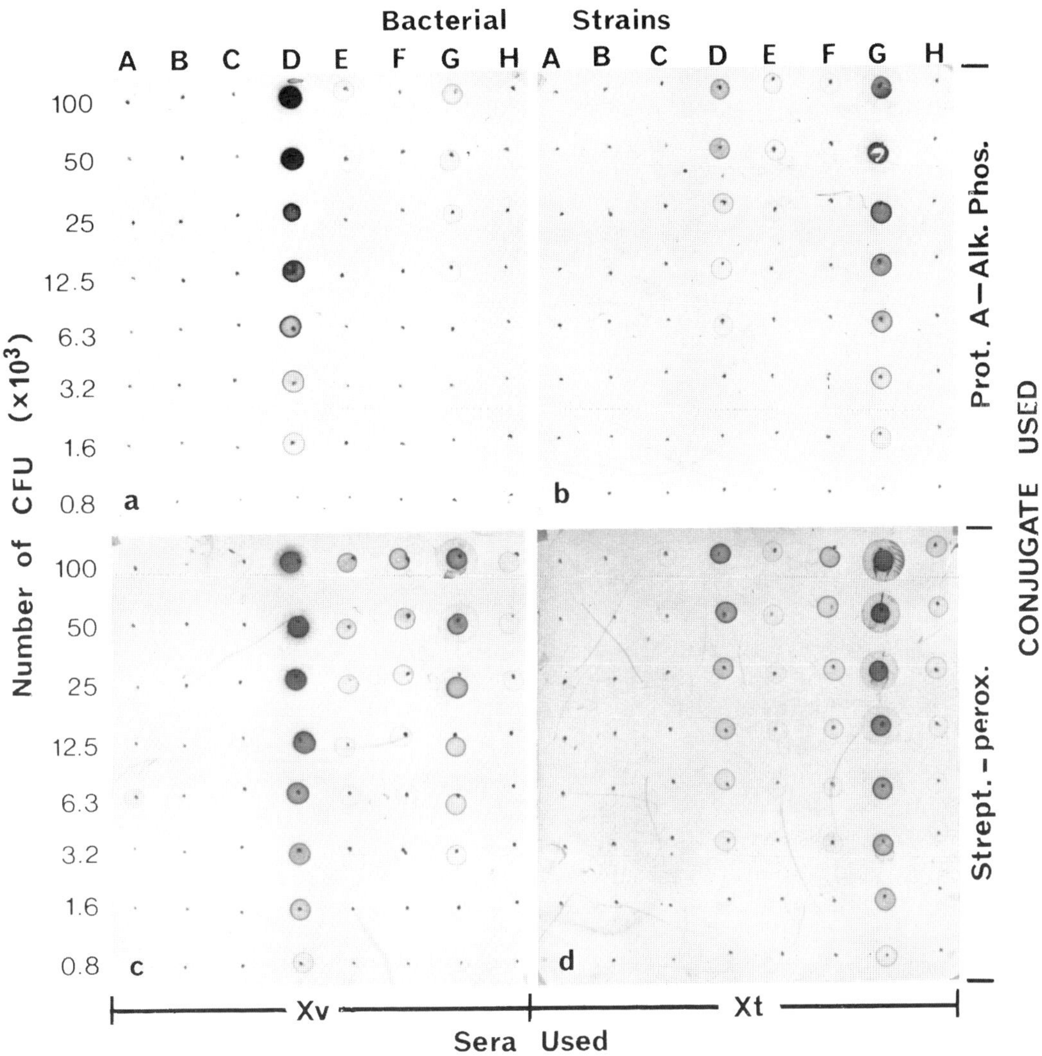

Fig. 3. The reaction of five Xanthamonas campestris pathovars and three unrelated bacteria with two antisera produced against X. campestris pv. vesicatoria (a and c) and pv. translucens (b and d). Membranes a and b were incubated with alkaline phosphatase conjugated Protein A (Prot.A-Alk. Phos.), and c and d with streptavidin-biotinylated peroxidase complex (Strept.-perox.). Key to bacteria strains - lane: A, Erwinia carotovora; B, Corynebacterium fascians; C, Pseudomonas lachrymans; D-H, Xanthamonas campestris pathovars (D) vesicatoria, (E) pelargonium, (F) begoniae, (G) translucens, (H) campestris.

Literature Cited

1. Banttari, E. E., and Goodwin, P. H. 1985. Detection of potato viruses S, X, and Y by enzyme-linked immunosorbent assay on nitrocellulose membranes (Dot-ELISA). Plant Dis. 69:202-205.
2. Blake, M. S., Johnston, K. H., Russell-Jones, G. J. and Gotschlich, E. C. 1984. A rapid, sensitive method for detection of alkaline phosphatase-conjugated anti-antibody on Western blots. Anal. Biochem. 136:175-179.
3. De Blas, A. L., and Cherwinski, H. M. 1983. Detection of antigens on nitrocellulose paper immunoblots with monoclonal antibodies. Anal. Biochem. 133:214-219.
4. Esen, A., Conroy, J. M., and Wang, S. -Z. 1983. A simple and rapid dot immunobinding assay for zein and other prolamins. Anal. Biochem. 132:462-467.
5. Ey, P. L., and Ashman, L. K. 1986. The use of alkaline phosphatase-conjugated anti-immunoglobulin with immunoblots for determining the specificity of monoclonal antibodies to protein mixtures. Methods Enzymol. 121:497-509.
6. Fletcher, J. 1987. Filter paper dot-immunobinding assay for detection of _Spiroplasma citri_. Appl. Environ. Microbiol. 53:183-184.
7. Graddon, D. J., and Randles, J. W. 1986. Single antibody dot immunoassay - a simple technique for rapid detection of a plant virus. J. Virol. Methods 13:63-69.
8. Hansel, T. T., and Matthews, N. 1987. Bacterial identification using nitrocellulose blotting technique that incorporates ELISA to bacterial surface antigen. J. Clin. Pathol. (Lond.) 40:239-241.
9. Hawkes, R. 1982. Identification of concanavalin A-binding proteins after sodium dodecyl sulfate-gel electrophoresis and protein blotting. Anal. Biochem 123:143-146.
10. Hawkes, R., Niday, E., and Gordon, J. 1982. A dot-immunobinding assay for monoclonal and other antibodies. Anal. Biochem. 119:142-147.
11. Johnson, D. A., Gautch, J. W., Sportsman, J. R., and Elder, J. H. 1984. Improved technique utilizing nonfat dry milk for analysis of proteins and nucleic acids transferred to nitrocellulose. Gene Anal. Techn. 1:3-8.
12. Kotani, H., and McGarrity, G. J. 1985. Rapid and simple identification of mycoplasma by immunobinding. J. Immunol. Methods 85:257-267.
13. Kotani, H., and McGarrity, G. J. 1986. Identification of mycoplasma colonies by immunobinding. J. Clin. Microbiol. 23:783-785.
14. Laing, P. 1986. Luminescent visualization of antigens on blots. J. Immunol. Methods 92:161-165.
15. Lazarovits, G., Zutra, D., and Bar-Joseph, M. 1987. Enzyme linked immunosorbent assay on nitrocellulose membranes (dot-ELISA) in the serodiagnosis of plant pathogenic bacteria. Can. J. Microbiol. 33:98-103.
16. Lopez-Vidal, Y., and Svennerholm, A.-M. 1986. Hemadsorption and enzyme-linked immunosorbent assay nitrocellulose replica methods for identification of colonization factor antigen (CFA)-positive _Escherichia coli_ colonies and for isolation of CFA-negative mutants. J. Clin. Microbiol. 24:615-619.
17. Malin, E., Belden, E. L. , and Roth, D. 1985. Evaluation of the radioimmunoassay, indirect enzyme linked immunosorbent assay, and dot blot assay for the identification of _Xanthamonas campestris_ pv. _phaseoli_. Can. J. Plant Pathol. 7:217-222.

18. Nurminen, M., Wahlstrom, E., Kleemola, M., Leinonen, M., Siakku, P., and Mäkelä, P. H. 1984. Immunologically related ketodeoxyoctonate-containing structures in Chlamydia trachomatis, Re mutants of Salmonella species, and Acinetobacter calcoaceticus var. anitratus. Infect. Immun. 44:609-613.
19. Parent, J. -G., Belanger, F., Desjardins, S., and Brisson, J. D. 1985. Dot-immunobinding for detection of tomato mosaic virus and potato virus X infecting greenhouse tomatoes. Phytoprotection 66:53-57.
20. Powell, C. A. 1987. Detection of three plant viruses by dot-immunobinding assay. Phytopathology 77:306-309.
21. Rhea, U. S., and Shah, D. B. 1985. Foodborne enterotoxigenic Escherichia coli: Detection and enumeration on nitrocellulose membrane by enzyme immunoassay. Abstracts of the Annual Meeting of the Amer. Soc. Microbiol. (Abstr. P34). p. 256.
22. Smith, F. D., and Banttari, E. E. 1987. Dot-ELISA on nitrocellulose membranes for detection of potato leafroll virus. Plant Dis. 71:795-799.
23. Swerdlow, P. S., Finley, D., and Varshavsky, A. 1986. Enhancement of immunoblot sensitivity by heating of hydrated filters. Anal. Biochem. 156:147-153.
24. Taketa, K. 1987. A tetrazolium method for peroxidase staining: Application to the antibody-affinity blotting of α-fetoprotein separated by lectin affinity electrophoresis. Electrophoresis 8:409-414.
25. Towbin, H., and Gordon, J. 1984. Immunoblotting and dot immunobinding-current status and outlook. J. Immunol. Methods 72:313-340.
26. Turner, B. M. 1986. Use of alkaline phosphatase-conjugated antibodies for detection of protein antigens on nitrocellulose filters. Methods Enzymol. 121:848-855.
27. Vissing, H., and Madsen, O. D. 1984. Comparison of detection limits for various nitrocellulose binding immunoassays using β_2-microglobulin as a model antigen. Electrophoresis 5:313-314.
28. Walsh, B. J., Sutton, R., Wrigley, C. W., and Baldo, B. A. 1984. Allergen discs prepared from nitrocellulose: Detection of IgE binding to soluble and insoluble allergens. J. Immunol. Methods 73:139-145.
29. Zutra, D., Shabi, E., and Lazarovits, G. 1986. Fire blight on pear: A new disease in Israel. Plant Dis. 70:1071-1073.

Section IV.E.3.

RAPID SCREENING FOR MONOCLONAL ANTIBODIES BY DOT IMMUNOBINDING (DIA)

G. Lazarovits
Agriculture Canada, London Research Centre
London, Ontario, Canada

The usual techniques for screening hybridoma cell culture fluids for monoclonal antibodies (McAbs) are solid phase radioimmunoassay (RIA) and enzyme-linked immunosorbent assay (ELISA). Both procedures are described elsewhere in this volume. If these assays have any shortcomings, it is that they are time consuming, labor intensive and expensive in their requirements for antigens, secondary antibodies and laboratory reagents. Conversely, dot immunobinding assay (DIA) is simple, quick and frugal in its use of antigens and other reagents. For detailed discussion of DIA, please refer to **Sections IV.E.1.** and **IV.E.2**. Briefly, the procedure involves applying a small amount of antigen to a nitrocellulose membrane (NCM) and then incubating this in the hybridoma cell culture fluid. Alternatively, a small amount of hybridoma cell culture fluid can be applied directly over the antigen spot (8). The membrane is then transferred to a solution containing anti-mouse immunoglobulin (IgG) conjugated to horse-radish peroxidase or alkaline phosphatase. After addition of a suitable substrate, a colored precipitate is formed at the site of enzymatic activity.

The simplicity and ease of use of the protocols described in this section will greatly expedite the selection of McAbs in hybridoma cultures. The selection process for specificity is also greatly simplified because the reaction of the McAbs can be assessed against multiple antigens into one step. This assay will be useful to those seeking antibodies that perform well in DIA procedures or that can be used to detect specific components on Western blots (see **Section IV.F.1**.). Antibodies do not perform equally well under all assay conditions; therefore it is advisable to use assay conditions for the initial screenings that most closely resemble the assay system for which antibody use is intended (2).

Materials and Solutions

1. Nitrocellulose: use 0.45 μm pore size or 13 mm disks available from a number of manufacturers such as Schleicher & Schuell and Gelman.
2. Containers which allow for complete coverage of membranes and minimize volume of reactants, i.e. covers of microtiter plates, Petri dishes, plastic baggies, 100 and 500 ml beaker.
3. Distilled or deionized water.
4. Tris buffered saline (TBS): [50 mM tris (hydroxymethyl)-aminomethane, 150 mM sodium chloride, 0.01% Thiomersal, pH 7.4].
5. Blocking solution and antibody diluent (TBS-BSA-Tween): TBS containing 0.05% Tween 20 or Nonidet 40 and 1.0% bovine serum albumen (BSA) or similar protein.
6. 2X Diluent - same as the solution in 5 except double concentration of all components.
7. Rinse buffer: TBS + 0.05% Tween 20 or Nonidet 80.

8. 10 μl Hamilton syringe or similar pipeting device.
9. Antigen solutions for testing specificity of McAbs (0.5-1.0 mg protein ml^{-1} buffer is ideal).
10. Supernatants from hybridomas in 24 or 96 well microtiter plates.
11. Anti-mouse immunoglobulin enzyme conjugate.
12. Substrate buffer (see **Section IV.E.2.**).
13. Substrate (see **Section IV.E.2.**)
14. Platform shaker.

Procedure

Membrane preparation and antigen application

Protocol A

1. Handle nitrocellulose with gloves or forceps.
2. Rule sheets into 12 x 8 grids of 6 x 12 mm sections (see Fig. 1) with a non-smearing ball point pen. A pen designed specifically for marking NCM is available from Schleicher and Schuell. Number sections sequentially from 1-96 at the bottom boundary of each small rectangle. To test supernatant fluid in 24-well microtiter plates, use 13 mm diameter NCM disks.
3. Wash NCM by floating on distilled water and stir for 10 min on a rotary shaker. Dry NCM thoroughly between layers of filter paper.
4. With a 10 μl Hamilton syringe apply as small a volume of antigen solution (0.5-1.0 mg ml^{-1} protein) as possible (0.1-0.5 μl) to a location on the top half portion of each rectangle. Generally, three different drops of antigen can be accommodated in this area per strip when 0.5 μl is used (Fig. 1) and up to six with 0.1 μl. Nine spots of 0.5 μl are readily accommodated on 13 mm diameter NCM disks (Fig. 2). Maintain the same pattern of antigen application for each strip or disk.
5. Allow the membranes to air-dry and then block unreacted sites in blocking solution for 10 min on a rotary shaker. Rinse in a generous amount of distilled water and incubate for another 5-10 min in blocking solution.
6. Place on filter paper, and while still moist, cut out the individual rectangles with a razor blade or a scalpel. Membranes not immediately required can be stored at or below 4 C in the dark for several weeks. It is easiest to cut along strips with a ruler and a razor blade and then to cut out the individual sections with the flat edge of the razor blade.
7. Pipet off 125 μl of the hybridoma cell culture fluids into 96-well microtiter plates and add an equivalent amount of 2X strength blocking solution. Insert the NCM rectangles with the antigen side down into each of the wells, ensuring that well and rectangle numbers coincide. Do not allow membranes to contact each other. Incubate plates at room temperature on a platform shaker for 2 hr.
8. Remove the membranes from the cell culture fluid and rinse for 1-2 min in 250 ml of TBS-T in a beaker. Continue rinsing the membranes in four changes of 50 ml of TBS-T at 3-5 min intervals, on a platform shaker.
9. Incubate strips for 90 min in 10-20 ml of 1:500 - 1:2000 enzyme conjugated anti-mouse immunoglobulin in blocking solution.
10. Rinse with 50 ml TBS-T five times for 3-5 min. The last rinse should be in substrate buffer. The enzyme conjugate can be saved and reused

several times.

11. Apply appropriate substrate solution (see **Section IV.E.2.**) and incubate until color develops to desired intensity. Rinse membrane thoroughly in distilled water and dry between pieces of filter paper. Arrange in order and mount on cardboard.

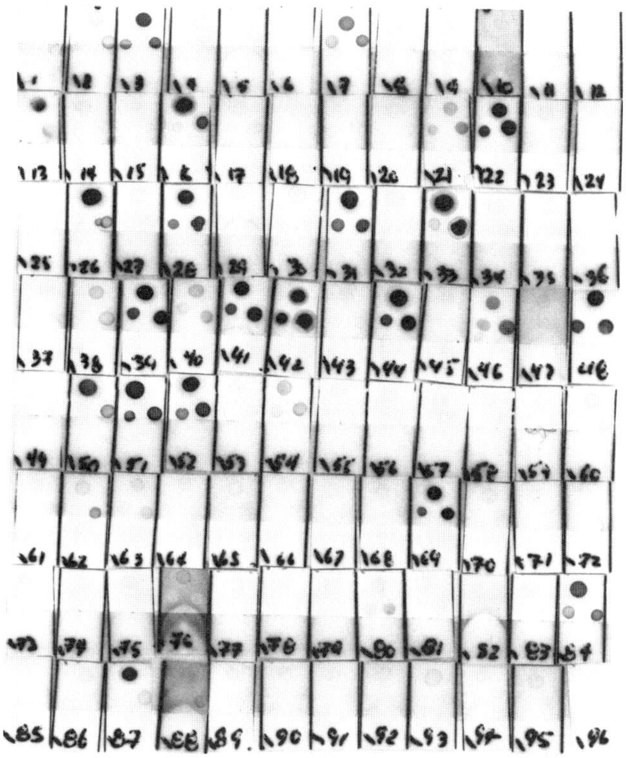

Fig. 1. Dot immunobinding assay used for initial screening of hybridoma culture filtrates. Spleen cells, from a mouse immunized with a crude fungal antigen preparation, were fused with a mouse myeloma cells. Samples (0.5 μl) of the immunizing antigen were spotted on the uppermost section of the NCM strips. Similar antigen preparations from two closely related fungal species were spotted below. The NCM strips were incubated for 2 hr in a 1:1 mixture of hybridoma cell culture fluid and blocking solution in 96-well microtiter plates. They were then rinsed and placed in a 1:1000 dilution of anti-mouse (FAb) alkaline phosphatase conjugate. The enzyme reaction was visualized in naphthol AS-MX phosphate and fast red TR salt. Most cultures had antibodies that reacted with antigens from all three fungi (strip 41) but some had antibodies reacting with antigens from only two fungi (strips 26 and 33) or with only the immunizing antigen (strip 87).

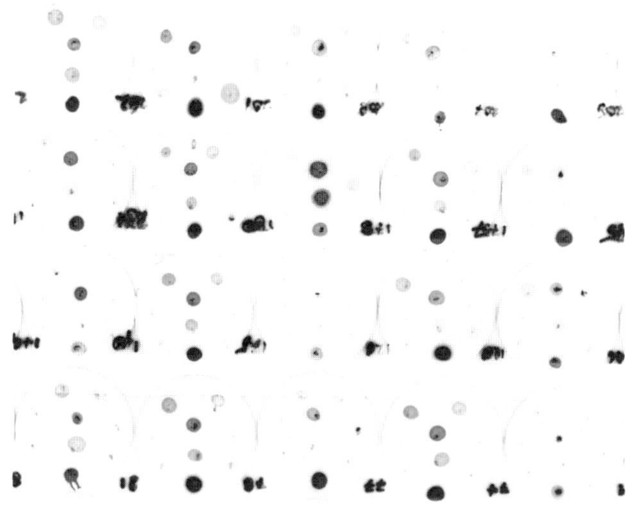

Fig. 2. Dot immunobinding assay used for secondary screening of hybridoma cell culture fluids. NCM disks were spotted with 0.5 μl aliquots of extracts from three unrelated fungi (left row), the immunizing antigen (center row - top), extracts from two closely related fungi (center row - middle and bottom), from non-related antigens (right row - top and middle) and extract from tissue infected with the fungus used for immunization (right - bottom). The NCM were reacted as described in Fig. 1. Note the various reactions of the hybridoma (antibody) lines with 9 different antigen preparations

Protocol B

1. Handle nitrocellulose with gloves and forceps.
2. Rule nitrocellulose sheets into 12 x 8 grids of 9 mm squares.
3. Wash sheet in distilled water for 10 min and allow to air-dry.
4. Apply 1 μl dots of 1 mg ml^{-1} solution of antigen to the center of each square.
5. Air-dry sheet and then block in blocking solution for 10 min on a rotary shaker. Rinse in distilled water and continue blocking for another 10 min.
6. Rinse in TBS and dry.
7. Apply 1.5-2.0 μl samples of hybridoma cell culture fluid supernatant directly onto antigen spots and leave for 5 min.
8. Wash for 30 min in several changes of TBS-Tween.
9. Incubate in solution of enzyme conjugated anti-mouse immunoglobulin in blocking solution for 2 hr.
10. Wash for 30 min with several changes of TBS-Tween with the last wash in substrate buffer.
11. Develop in substrate solution (see **Section IV.E.2.**). After color development rinse NCM in distilled water and allow to dry between sheets of filter paper in a dark place.

Applications

Hawkes et al (4) developed a procedure for screening McAbs which involved incubating small discs of antigen coated NCM in hybridoma cell culture fluids in wells of microtiter plates. They found this procedure highly reliable in diagnostic use for screening human pathological sera for multiple antibodies in one operation. Bode et al (1) used a similar procedure to identify human viral antigens and to select antibodies produced against them. Refinements in these procedures were described by Ey and Ashman (3). The most interesting modification was described by Smith et al (8). Their procedure permitted the screening of hybridoma cell cultures for production of McAbs against a glycoprotein from the alga _Chlamydomonas reinhardii_ using one sheet of NCM. Sherwood et al (7) compared the reactivity of monoclonal and polyclonal Abs for detection of peanut mottle virus. Möller and Borrebaeck (6) developed an assay for detecting single cell lines of antibody producing splenocytes. Cell suspensions, after immunization either in vivo or in vitro, were plated on NCM impregnated with antigen, and antibody production was detected with an anti-immunoglobulin conjugated to peroxidase. Antibody producing cell lines could be identified after 1-3 hr of incubation.

The author extends sincere appreciation to Dr. Z. Eshhar and Ms. T. Waks (Dept. of Chemical Immunology, Weitzmann Inst. of Science) for collaboration in work shown in Fig. 1 and 2.

Literature Cited

1. Bode, L., Beutin, L., and Köhler, H. 1984. Nitrocellulose-enzyme-linked immunosorbent assay (NC-ELISA) - a sensitive technique for the rapid visual detection of both viral antigens and antibodies. J. Virol. Methods 8:111-121.
2. Eshhar, Z. 1985. Monoclonal antibody strategy and techniques. Pages 3-41 in: Hybridoma Technology in the Biosciences and Medicine, T. Springer, ed. Plenum Publishers. 602 pp.
3. Ey, P. L., and Ashman, L. K. 1986. The use of alkaline phosphatase-conjugated anti-immunoglobulin with immunoblots for determining the specificity of monoclonal antibodies to protein mixtures. Methods Enzymol. 121:497-509.
4. Hawkes, R. 1986. The dot immunobinding assay. Methods Enzymol. 121:484-491.
5. Hawkes, R., Niday, E., and Gordon, J. 1982. A dot-immunobinding assay for monoclonal and other antibodies. Anal. Biochem. 119:142-147.
6. Möller, S. A., and Borrebaeck, C. A. K. 1985. A filter immuno-plaque assay for the detection of antibody-secreting cells in vitro. J. Immunol. Methods 79:195-204.
7. Sherwood, J. L., Sanborn, M. R., and Keyser, G. C. 1987. Production of monoclonal antibodies to peanut mottle virus and their use in enzyme-linked immunosorbent assay and dot-immunobinding assay. Phytopathology 77:1158-1161.
8. Smith, E., Roberts, K., Butcher, G. W., and Galfre, G. 1984. Monoclonal antibody screening: Two methods using antigens immobilized on nitrocellulose. Anal. Biochem. 138:119-124.

Section IV.F.1.

WESTERN BLOTTING AND THE USE OF MEMBRANES TO ADSORB ANTISERA AND TO AFFINITY PURIFY ANTIBODIES

J. Hammond
USDA ARS Florist and Nursery Crops Laboratory
Beltsville, MD 20705

Western Blotting

Proteins separated in almost any type of gel system (usually SDS-polyacrylamide gels) can be transferred to a membrane which is subsequently reacted with antibody and a detection system. The prime advantage of Western blotting (2) over dot-blotting is that antibody reactivity can be correlated with proteins or peptides of particular molecular weights, even in complex mixtures.

The simplest and cheapest way of transferring proteins from a gel to a membrane is by passive capillary blotting (6); no specialized equipment is required. However, passive transfer takes longer than electrophoretic transfer, the bands may diffuse slightly, and the gel may start to dry out and stick to the membrane. The necessary separation of the gel from the membrane may then lower the quality of the final result. Higher molecular weight proteins may be less efficiently transferred by passive blotting than by electroblotting.

Electroblotting was introduced by Towbin et al. (10) using an apparatus in which the gel and membrane were sandwiched together in a buffer tank across which an electrical field is applied. Several companies now manufacture similar apparatus; the major differences between units are in density and uniformity of the electric field and volume of buffer required. Transfer time is dependent upon gel thickness and strength of field applied, but is generally from 1 hr to overnight. Generated heat must be dissipated.

More recently a semi-dry blotting technique was developed (4) in which the electrodes are graphite plates instead of platinum grids, and the liquid is reduced to sheets of buffer-saturated filter papers on each side of the gel-membrane sandwich. There are several apparatus models commercially available and transfer times of as little as 30 min are claimed for 0.5-0.75 mm gels. These units generate less heat and require much smaller volumes of buffer. A further advantage is that multiple-gel transfers can be made simultaneously.

I. Passive Blotting

Principle

A gel containing the proteins of interest is sandwiched between filter paper wicks on one side, and a membrane overlayered with an absorbent material such as paper towels on the other (6). Mass flow of liquid through the gel by capillary action results in elution of the proteins from the gel, and the majority of the protein is retained on the membrane. The membrane is then incubated with the antiserum of interest, and the location of the reaction visualized and correlated with molecular weight. The degree of reaction may be compared with the total protein loading by staining a duplicate membrane to reveal total protein.

Materials

1. A dish that will hold at least 1 liter of transfer buffer (a baking dish is suitable and readily available).
2. A plate (such as a Plexiglass sheet) to span the dish, or with a support to raise it above the buffer in the dish.
3. Two filter paper wicks, each long enough to overlap the plate and hang down on one side into the buffer.
4. A stack of filter paper (Whatman 3MM or equivalent) cut just larger than the gel containing proteins to be transferred; the stack should be ca. 2 cm thick.
5. A stack of absorbent paper towels of the same or slightly larger dimensions than the filter papers (high quality absorbent papers are available, but paper towels are much cheaper).
6. A piece of membrane for protein transfer, cut just larger than the gel.
7. Parafilm to lay around the gel, to divert all of the liquid through the gel and membrane.
8. Second dish or tray in which to wet the membrane and the filter papers.
9. A weight (1-2 kg). A reagent bottle filled with water and a plate to distribute the weight evenly, or a brick, make suitable weights.
10. One liter of TBS. TBS = 0.02 M Tris, 0.15 M NaCl, pH 7.5; a 5x TBS stock is useful. To prepare a 5x TBS stock, dissolve 12.1 g Tris base and 43.8 g NaCl in 800 ml distilled water. Adjust the pH to 7.5 with HCl and then bring the volume to 1000 ml.

Procedure

1. Wear gloves. Place the support plate across the reservoir dish, and pour some of the TBS into the reservoir. Pour TBS into the wetting dish.
2. Wet the wicks thoroughly in TBS and place them across the support plate so that they overlap, with the free ends in the buffer on opposite sides of the support (Fig. 1). Be careful to avoid trapping air bubbles between these or any layers. Wet a filter paper just larger than the gel and place on top of the wicks. Carefully lay the gel on wet filter paper, and surround it with Parafilm.
3. Wet the membrane and lay it over the gel, followed by five or six wet filter papers, and then the stack of dry filter papers.
4. Stack the paper towels on top and add the weight. Pour all remaining buffer into the reservoir dish.
5. Completed blot transfer usually requires 48 to 72 hr, during which two changes of paper towels may be necessary.
6. Remove the weight, paper towels, and upper filter papers. Lift the membrane and gel carefully into a shallow dish of TBS, and carefully peel the membrane from the gel. If the gel sticks to the membrane, allow time for rehydration; however, it may be necessary to use a spatula or similar tool to separate the membrane and gel. Use only a very smooth-edged implement, and proceed very carefully to avoid scrape marks on blot, after development. A gloved fingernail may suffice if the adhesion is minor.

7. The membrane can then be dried and stored between two pieces of filter paper or blocked immediately and incubated with antibody.

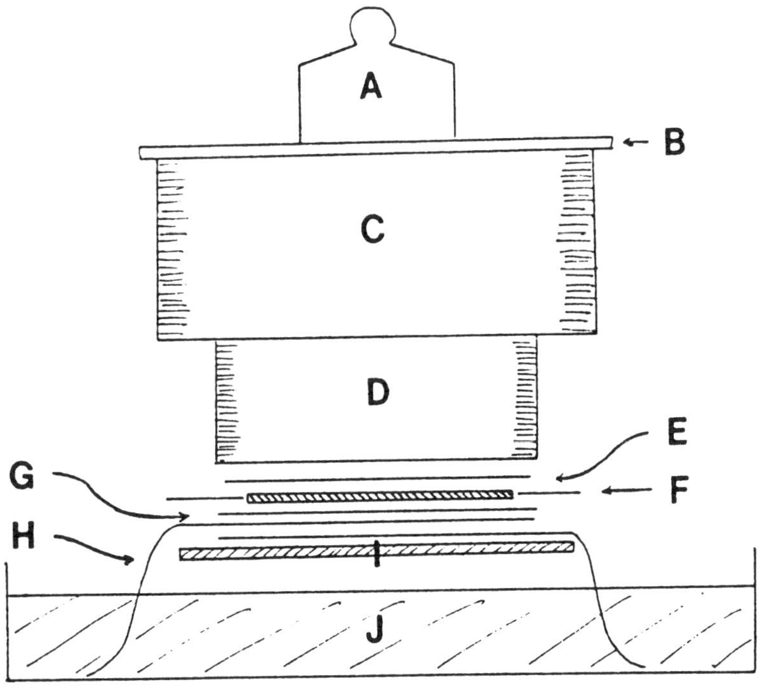

Fig. 1. Schematic diagram of passive blotting. A) weight; B) plate to distribute weight (if necessary); C) stack of absorbent paper towels; D) ca. 2 cm thick stack of filter paper or blot-block paper. The bottom five or six should be wetted in TBS; E) membrane; F) gel surrounded by Parafilm; G) filter paper; H) filter paper wick; I) support plate spanning sides of J) reservoir, filled with TBS.

Comments

1. Gloves should be worn at all times when handling membranes, to prevent deposit of skin oils which would result in background reactions or obscuring of specific reactions.
2. Passive blotting is the simplest and cheapest way to transfer proteins from a gel to a membrane, but not the fastest.
3. Air bubbles trapped between any of the layers of the sandwich will prevent local transfer.
4. Provided that there is absolutely no contact between the absorbent layers and the wicks (ensured by the presence of a non-permeable layer surrounding the gel and membrane), increasing the area of the absorbent

layer may speed the rate of liquid movement, and hence protein transfer.

5. The wetted filter papers above the gel sandwich provide the initial movement of fluid into the absorbent layer; omission results in slower and possibly uneven transfer.

II. Electroblotting

Principle

The gel containing proteins separated by electrophoresis is sandwiched with filter paper on one side, and a membrane and filter paper on the other, and electrical current passed through the sandwich. The proteins are electrophoresed out of the gel onto the membrane. The membrane is then incubated with specific antiserum, and the antibody reaction visualized and compared to the protein pattern in a duplicate gel. "Wet" units (10) usually have a cassette made to compress the gel sandwich gently between foam pads (to support the gel), which is then placed between the electrodes in a buffer tank. "Semi-dry" units (4) reduce the buffer volume to filter papers saturated in buffer placed on both sides of the gel, with the outer filter papers directly in contact with plate electrodes.

A. "Wet Units"

Materials

1. Electroblotting apparatus and constant current power supply.
2. A dish deep enough to submerge the assembled gel cassette.
3. A piece of membrane just larger than the gel.
4. Two filter papers of the same size as the membrane.
5. Sufficient buffer to fill the apparatus (1.5-4.0 l, depending upon the apparatus). A typical transfer buffer is 25 mM Tris, 150 mM glycine, pH 8.3, containing 20% methanol, or the same buffer containing 0.02% SDS instead of methanol. Equilibration of the gel in transfer buffer may be necessary if the transfer buffer ionic strength differs significantly from the original electrophoresis buffer.
6. A sheet of plastic or Parafilm the size of the gel cassette, with a "window" the size of the gel cut out of the center.
7. A magnetic stirrer and a stirring bar that will fit inside the assembled unit.

Procedure

1. Soak the foam pads in the transfer buffer, taking care to remove all air bubbles; keep the pads submerged.
2. Soak the two filter papers and the membrane in the transfer buffer.
3. Place one wet filter paper on top of one foam pad in the dish of transfer buffer.
4. Place the membrane against the gel to be transferred; this is most easily done with the gel still on one of the plates of the electrophoresis apparatus. Avoid trapping any air bubbles.
5. Transfer the gel and membrane into the dish of transfer buffer,

and lay them membrane-side-down over the filter paper and foam pad.

6. Lay the plastic or Parafilm sheet over the gel and cover with the remaining filter paper. Lay the second foam pad on top, ensuring that no air bubbles are trapped in these operations. Close the cassette around the assembled foam pads and gel sandwich with spring clips.
7. Transfer the gel cassette into the buffer-filled apparatus between the electrodes; be certain that the membrane is toward the anode.
8. Cover the tank, place on the stirring plate, and connect the electrodes to the power supply (set to provide constant current).
9. The time necessary for transfer depends on the thickness of the gel, the power supplied and distance between the electrodes, the molecular weight of the proteins of interest, and the buffer system used. As a rough guide, 3 hr at 200 mA or 12 hr at 70 mA should be sufficient for qualitative transfer of proteins of up to 100 kD from a 0.75 mm thick gel in the presence of methanol. If SDS is used and methanol omitted, much shorter times may suffice.
10. Disconnect the power and remove the gel cassette. Disassemble the cassette and remove the membrane from the gel. Rinse the membrane briefly in TBS, and either air-dry or process immediately (see below).

Comments

1. There are several units commercially available; the set-up instructions may be slightly different for each. The basic requirements are to assemble the gel and wetted membrane between buffer-saturated filter papers and sponge supports, without trapping any air bubbles. Bubbles anywhere will prevent local transfer. The easiest way to accomplish the assembly is in a buffer-filled dish next to the apparatus, keeping all of the parts submerged and bubble-free.
2. For an SDS-polyacrylamide gel system the assembled transfer cassette is then placed in the buffer-filled unit with the membrane on the anode side of the gel, and power is applied. If other gel systems are being used, it must be determined whether the protein will migrate toward the anode or the cathode.
3. As with other types of transfer, the area around the gel can and should be screened with Parafilm or non-conductive plastic sheet to direct all of the current through the gel and membrane.
4. Constant current is used rather than constant voltage because the resistance may drop as a result of electrolytes being eluted from the gel; heat build-up might be significant if constant voltage were used. Transfer may require 1 to 15 hr, depending upon gel thickness, buffer composition, protein molecular weight, and the power applied. In all cases heat build-up may be a problem, and transfers can be carried out at 4 C, with a stirrer in the buffer tank to ensure heat dissipation, or with a cooling manifold in the apparatus.
5. Other buffers are summarized by Towbin and Gordon (9). Methanol is included to prevent swelling of the gel during transfer (although it slows the movement of protein out of the gel), and also to improve adsorption of protein to the membrane. SDS

improves mobilization of proteins from the gel, and improves the efficiency of transfer of higher molecular weight proteins, but may inhibit binding to the membrane to some extent.

6. The use of nitrocellulose (NC) membranes of smaller pore size will increase retention of proteins of less than 30 kD, which may be poorly retained on 0.45 μm membranes; the interaction with nylon membranes is less affected by pore size, and more protein will be retained. However, nylon membranes require more stringent blocking than nitrocellulose prior to the antibody reaction.

B. "Semi-dry" Units

Materials

1. Semi-dry blotter unit and constant voltage power supply.
2. Four filter papers cut to the exact size of the gel.
3. A piece of membrane cut to the exact size of the gel.
4. Cathode buffer (25 mM Tris, 40 mM 6-aminohexanoic acid, 0.1% SDS, pH 9.4).
5. Anode buffer #1 (0.3 M Tris, 20% methanol, pH 10.4).
6. Anode buffer #2 (25 mM Tris, 20% methanol, pH 10.4).
7. Distilled water.
8. Dishes or trays for wetting the membrane and filter papers.
9. A glass rod or a pipet to smooth out the membrane and filter papers.

Procedure

1. Wear gloves. Wet the membrane in distilled water.
2. Saturate two pieces of filter paper in cathode buffer, and one in each anode buffer.
3. Drain excess liquid from the filter soaked in anode buffer #1, and lay the filter paper carefully on the anode of the unit. If necessary smooth out any air bubbles or wrinkles by gently rolling the glass rod over the surface.
4. Drain excess buffer from the filter paper in anode buffer #2, and position the filter paper carefully over the first. Smooth carefully with the glass rod.
5. Lay the wetted membrane over the gel (easily done if the gel is still on one of the glass plates of the electrophoresis apparatus). Smooth out any air bubbles.
6. Pick up both the gel and membrane and lay them, membrane side down, over the filters on the anode. Again smooth with the glass rod if necessary.
7. Drain excess buffer from the filters saturated in cathode buffer, and lay each in turn over the gel, smoothing each with the glass rod to remove any air bubbles.
8. Place the cathode plate over the assembled sandwich. If the unit has clamps, these should be tightened only to hold the cathode parallel to the anode, and not to exert any pressure on the sandwich.
9. Attach the unit to the power supply, and set constant voltage such that the initial current is 0.8 mA per cm^2 of gel (or as recommended by the supplier of the unit). The current may drop

slightly over the course of the transfer. Depending upon the thickness of the gel and the molecular weight, transfer may be complete in as little as 30 min. Thicker gels and high molecular weight proteins may take longer. Essentially complete transfer of proteins of up to at least 90 kD from a 1.5 mm thick, 10% acrylamide gel is achieved within an hr.

10. Switch off and disconnect from the power supply, and disassemble. There should be no sign of brown discoloration of the filter papers; this is a sign of incorrect assembly leading to overheating. Separate the membrane from the gel, rinse briefly in TBS, and either air-dry for storage or process immediately.

Comments

1. These units offer the advantage of rapidity with low heat build-up and low buffer requirements. Transfer may be complete in as little as 30 min for proteins of up to 50,000 M_r from gels up to 0.75 mm thick. Higher molecular weight proteins and thicker gels require longer transfer times.
2. Exact fitting of filter paper, membrane and gel prevents shorting across the electrodes. Surrounding with Parafilm or a similar non-conducting film prevents current by-passing the gel.
3. Methanol is used in the anode buffers to ensure protein retention on NC membranes, but can be omitted if nylon membranes are used. As noted above, SDS may aid in mobilization of proteins from the acrylamide gel and methanol may inhibit migration while enhancing retention on the membrane. With the semi-dry units both can be used; Svoboda et al. (8) found that best results were obtained when the cathode buffer contained SDS but not methanol (thus increasing protein mobility) and the anode buffer contained methanol but not SDS (thus increasing retention on the membrane of the proteins electrophoresed from the gel).
4. The recommended initial current depends in large part on the electrode material; the first versions of apparatus used graphite electrodes, which can be damaged by incautious handling and may become pitted if not adequately rinsed after use. Other versions use graphite-loaded polymer electrodes, which are less vulnerable to mechanical damage but lose efficiency over time and must be periodically replaced, or metal plate electrodes which are presumably longer lasting.
5. With this type of unit multiple gels can be transferred at once. For multiple transfers each gel and membrane are sandwiched between one filter paper in anode buffer #2 and one in cathode buffer. Each "transunit" is separated from the next by a sheet of dialysis membrane to prevent movement of protein between transunits. One additional filter paper soaked in Anode buffer #1 (on the bottom) and one in Cathode buffer (on the top) must contact the electrodes (4).

Applications

Related proteins of different origins, precursor-product relationships, and proteolytic degradation are among the topics that can be examined by Western blotting. The occurrence of non-specific antibodies (to host antigens in antisera against plant pathogens, for example) can also be visualized by Western blotting (by comparison of extracts of healthy and infected plants); identification of such host antigens might allow improvement of a purification protocol to eliminate the contamination.

Interactions between proteins and nucleic acids (both DNA and RNA) have been analyzed on protein blots, as have interactions between hormones and receptors. Enzyme subunits have been identified by trapping active enzyme on the free reactive sites of antibody bound to non-reactive subunits blotted from a gel; the subunit localization being by detection of enzyme activity with a further overlay. Whole cell interactions with isolated proteins have also been demonstrated on Western blots. Antigen structure can be examined by reaction of monoclonal antibodies with transfers of proteolytic digests of the antigen; by this means domains of the protein may be separated, and may be localized to a single peptide, perhaps leading to correlation of particular domains with specific biological activities. Identification of electrophoretic variants such as isozymes is also possible. These and other applications are discussed in reviews by Gerhsoni and Palade (3), Towbin and Gordon (9) and Beisiegel (1).

Von Wechmar et al. (11) used Western blotting to confirm the presence of brome mosaic virus in infected seed, and detected a large proportion of partially degraded coat protein. However, Western blotting is more often used as an analytical tool than as a method of virus diagnosis, except in some medical uses.

III. Membrane Development

(For details of membrane development and related references, see **Section IV.E.1.**, and **Section IV.E.2.**).

Use of Membranes to Cross-absorb Antibodies

Principle

Proteins are adsorbed to NC or nylon membranes; once adsorbed, quite rigorous treatment is necessary to desorb the protein. This can be put to good use to remove non-viral-specific antibodies from antisera with a number of advantages over conventional methods of adsorption. Antibodies to plant proteins may be present in antiviral antiserum as a result of incomplete separation of virus from host components prior to immunization. An extract of the host plant can be used to saturate the protein binding sites on a membrane. The membrane is then incubated in blocking solution to assure that binding sites are saturated. Antiserum can then be cross-absorbed by incubation with host-antigen saturated membrane. Anti-host protein antibodies will be bound to the membrane and anti-viral antibodies will remain in solution. The membrane piece can then be discarded, and if necessary the process repeated.

Materials

1. An extract of the host plant concentrated by methods analagous to the virus purification used for antiserum preparation.
2. A piece of membrane (nitrocellulose or nylon).
3. A dish of suitable size for incubation of the membrane, and of a material

(such as glass or polypropylene) that adsorbs insignificant amounts of antibody.
4. Blocking solution: 1% non-fat dried milk, 0.5% BSA in TBS.
5. Wash buffer: TBS.
6. Antiserum to be absorbed.
7. A shaker.

Procedure

1. Incubate the membrane in the host plant extract for 1.5 hr at room temperature on the shaker.
2. Incubate membrane in blocking solution for 1 hr.
3. Dilute antiserum about 1:50 in TBS, and incubate treated membrane in the diluted serum for 1.5 hr. Discard the membrane.
4. Dilute a sample of the cross-absorbed antiserum to standardized concentration and test in application of choice.
5. If necessary repeat absorption with further pieces of membrane.
6. Reconcentrate the absorbed antiserum to the original volume by ammonium (or sodium) sulfate precipitation, if desired.

Comments

This use of the membrane to remove unwanted antibodies is essentially a contaminant affinity purification of the antiserum. By using this means of cross-absorption, unwanted antibodies and the absorbing material are completely removed from the serum without any color change in the serum, with minimal change in serum volume, and without adding to the serum unwanted soluble plant proteins. Diluted serum can easily be reconcentrated by ammonium (or sodium) sulfate precipitation.

Use of Membranes to Affinity Purify Virus-specific Antibodies

Principle

Following transfer of electrophoretically separated proteins to a membrane, the area representing a particular protein may be cut out and reacted with antiserum; alternatively, the whole membrane may be reacted with the antiserum and the desired area excised. Bound specific antibodies can be eluted by acid treatment as described by Olmsted (5), and as used to purify antibodies to a plant virus by Rybicki (7) in the procedure outlined below.

Materials

1. A blotted preparative gel (i.e. a single, full-width loading of the protein mixture, transferred to a membrane).
2. Blocking solution (10 mM Tris, 150 mM NaCl, 2% BSA, 0.05% Nonidet P-40, pH 7.5; BSA in TBS may be equally effective).
3. Conjugated second antibody and substrate (to locate desired area on alignment strips).
4. Diluted antiserum.
5. Wash buffer (150 mM NaCl, 0.05% Nonidet P-40).
6. TBS (TBS alone may substitute for the wash buffer; however, Rybicki (7) found that NP-40 increased the specificity of the reactions).
7. Antibody elution buffer (0.1 M glycine-HCl pH 2.9).

8. Neutralization solution (0.1 M NaOH).
9. Shaker.

Procedure

1. Incubate the whole membrane in blocking solution on the shaker; one hr at room temperature or overnight at 4 C is usually sufficient.
2. Cut "alignment strips" from each side of the membrane, and incubate these in an appropriate dilution of antiserum (this must be empirically determined for each antiserum, and should be the highest concentration that produces minimal background staining on blots) for 90 min.
3. Rinse alignment strips in four changes of wash buffer (about 5 min per rinse).
4. Incubate alignment strips with conjugated second antibody and develop.
5. The strips are then realigned with the remainder of the membrane, and the area corresponding to the specific blot region cut from the main membrane.
6. Incubate the selected region of the unreacted blot in diluted antiserum and wash as above. If the NP-40 wash solution was used, follow by two rinses in TBS.
7. Move the selected membrane piece to a small beaker or shallow trough, and agitate for 10 min in 7.5 ml of antibody elution buffer; decant the buffer containing the eluted antibody to a test-tube and neutralize by addition of 1.4 ml of the 0.1 M NaOH. The elution step should then be repeated, after which the strips can be re-blocked and used again to purify further antibody at least twice more.

Comments

1. Multiple blots, of course, can be processed simultaneously with separate antisera. Likewise, multiple protein bands can be used to bind specific antibodies from antiserum made to a mixture of proteins.
2. The actual yield from this type of affinity purification will depend upon the amount of antigen that can be transferred to the membrane without overlapping any non-specific antigen bands as the specific band broadens. Nylon membranes, with their higher protein-binding capacity than NC may prove superior, although extra care must be taken to ensure complete blocking of the membrane. Rybicki (7) has pointed out that this means may be used to prepare monospecific antibodies without the equipment, expense, and expertise required for monoclonal antibody production, but the antibodies will not be to single epitopes, nor will large quantities ever be available through this procedure.

Literature Cited

1. Beisiegel, U. 1986. Protein blotting. Electrophoresis 7:1-18.
2. Burnette, W. M. 1981. "Western blotting": Electrophoretic transfer of proteins from sodium dodecyl sulfate-polyacrylamide gels to unmodified nitrocellulose and radiographic detection with antibody and radioiodinated protein A. Anal. Biochem. 112:195-203.
3. Gershoni, J. M., and Palade, G. E. 1983. Protein blotting: Principles and application. Anal. Biochem. 131:1-15.
4. Kyhse-Anderson, J. 1984. Electroblotting of multiple gels: A simple apparatus without buffer tank for rapid transfer of proteins from polyacrylamide to nitrocellulose. J. Biochem. Biophys. Meth. 10:203-209.

5. Olmsted, J. B. 1981. Affinity purification of antibodies from diazotized paper blots of heterogeneous protein samples. J. Biol. Chem. 256:11955-11957.
6. Renart, J., Reiser, J., and Stark, G. R. 1979. Transfer of proteins from gels to diazobenzyloxymethyl paper and detection with antisera: A method for studying antibody specificity and antigen structure. Proc. Natl. Acad. Sci. USA 76:3116-3120.
7. Rybicki, E. P. 1986. Affinity purification of specific antibodies from plant virus capsid protein immobilized on nitrocellulose. J. Phytopathol. 116:30-38.
8. Svoboda, M., Meuris, M., Robyn, C., and Christophe, J. 1985. Rapid electrotransfer of proteins from polyacrylamide gel to nitrocellulose membrane using surface conductive glass as anode. Anal. Biochem. 151:16-23.
9. Towbin, H., and Gordon, J. 1984. Immunoblotting and dot immunobinding - Current status and outlook. J. Immunol. Methods 72:313-340.
10. Towbin, H., Staehelin, T, and Gordon, J. 1979. Electrophoretic transfer of proteins from polyacrylamide gels to nitrocellulose sheets: Procedure and some applications. Proc. Natl. Acad. Sci. USA 76:4350-4354.
11. Von Wechmar, M. B., Kaufmann, A., Desmarais, F., and Rybicki, E. P. 1984. Detection of seed-transmitted brome mosaic virus by ELISA, radial immunodiffusion and immunoelectroblotting tests. Phytopathol. Z. 109:341-352.

Section IV.F.2.

WESTERN BLOTTING FOR DETECTION OF BACTERIAL ANTIGENS

S. H. De Boer and L. Ward
Agriculture Canada Research Station
Vancouver, B.C., Canada

Principle

In comparison to viruses, bacteria present a wide array of antigenic components and hence exhibit complex serological relationships. To identify specific antigens and sort out serological relatedness between strains, modern techniques of protein analysis can be used (6). Crude protein [and even lipopolysaccharide, (1)] preparations can be separated by sodium dodecyl sulfate polyacrylamide gel electrophoresis (SDS-PAGE). Subsequently the separated proteins can be transferred to a membrane such as nitrocellulose in an electric field by the blotting process (2,8). The blot can then be probed with an antibody preparation and the site of antibody reaction identified by an enzyme-linked immunoassay (3) (Fig. 1).

Materials

SDS-PAGE

1. Acrylamide/bis (30% and 2.67%, respectively)
2. 1.5 M Tris-HCl, pH 8.8
3. 0.5 M Tris-HCl, pH 6.8
4. 10% (w/v) sodium dodecyl sulfate (SDS)
5. 10% (w/v) ammonium persulfate
6. Temed
7. Sample buffer [4.0 ml distilled water, 1.0 ml 0.5 M Tris-HCl, pH 6.8, 0.8 ml glycerol, 1.6 ml 10% SDS, 0.4 ml 2-mercaptoethanol, 0.2 ml 0.05% (w/v) bromophenol blue]
8. Running buffer (5X), pH 8.3 (9.0 g Tris base, 43.2 g glycine, 3.0 g SDS, 600 ml distilled water)
9. Vertical mini gel apparatus (e.g. Mini Protean, Bio-Rad)

Blotting

1. Nitrocellulose membrane
2. Filter paper
3. Transfer buffer, pH 8.3 (25 mM Tris, 200 mM glycine, 20% methanol)
4. Transblot apparatus (e.g. Mighty Small Transphor, Hoefer)

Immunological probing

1. 0.01 M phosphate buffered (pH 7.2) containing 0.85% NaCl (PBS)
2. BLOTTO (10% skim milk powder, 0.2% sodium azide)

3. Anti-primary antibody/alkaline phosphatase conjugate (Jackson Immunoresearch Labs)
4. Naphthol AS-MX phosphate (Sigma, cat. no. N-5000)
5. Fast red TR salt (Sigma, cat. no. F-1500)
6. 100 mM Tris-HCl, pH 8.0

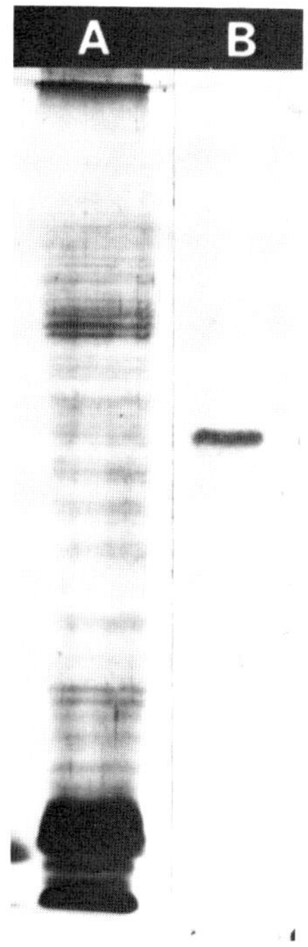

Fig. 1. Separation of total cell proteins of Erwinia carotovora subsp. atroseptica on SDS-PAGE (12% acrylamide) and stained with Coomassie blue (Lane A). The same preparation separated by SDS-PAGE was transferred to nitrocellulose, probed with a monoclonal antibody, and developed with anti-mouse alkaline phosphatase and the naphthol AS-MX/phosphate/Fast red substrate (Lane B). The monoclonal was specific to E. carotovora subsp. atroseptica flagella.

Procedure

1. Separate proteins in antigen preparation by the standard SDS-PAGE procedure of Laemmli (4). Details of the procedure are given in many textbooks and in gel apparatus instruction manuals (e.g. Bio-Rad Mini Protean II Dual Slab Cell Instruction Manual, or Hoefer Scientific Instruments Catalog 1986) (see also **Section II.B.1.**).
2. Remove gel from apparatus and soak in transfer buffer for 45 min.
3. Wet nitrocellulose membrane by floating on transfer buffer for 5 min and then submerging it for 45 min.
4. Assemble transfer-sandwich under buffer. Place the SDS-PAGE gel on the submerged nitrocellulose membrane and sandwich these between filter paper, Scotch Brite pads, and the sandwich holder (Fig. 2).
5. Transfer the transfer-sandwich to the blotting apparatus containing chilled (4 C) transfer buffer. The buffer must cover the sandwich completely and the nitrocellulose must be on the anode side of the gel.
6. The optimum voltage and running time for the transfer varies with antigen of

interest and apparatus being used (5). A standard transfer using a Hoefer model TE22 may involve 75 V for 1 hr followed by 100 V for another hour. Alternatively, 4 hr at 100 V may be adequate for a wide range of proteins in a Bio-Rad Trans-Blot Cell.

7. After transfer, remove the transfer-sandwich and air-dry the nitrocellulose blot. It may be stored between tissue for future use at this stage.

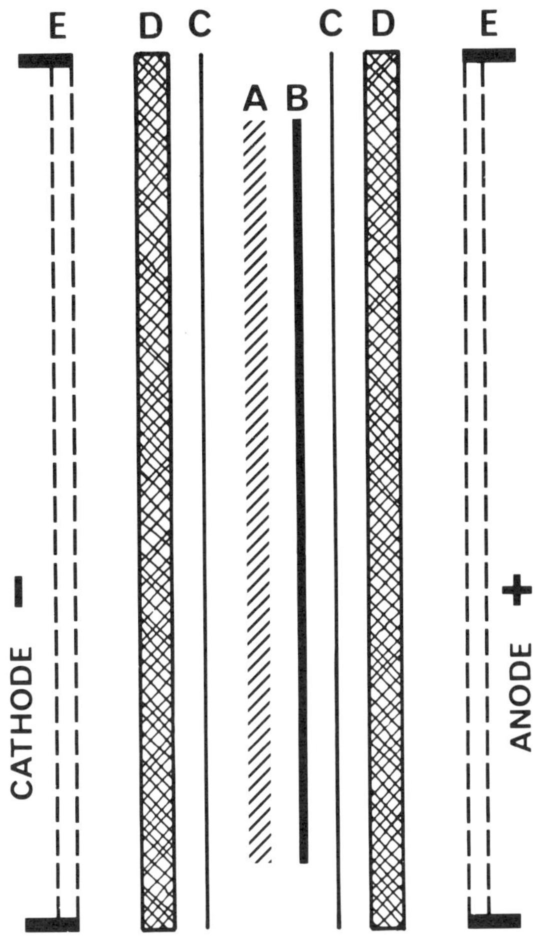

Fig. 2. Schematic representation of "transfer-sandwich" in which proteins from the gel (A) are transferred to the nitrocellulose membrane (B). The gel and nitrocellulose are sandwiched between filter paper (C), Scotch Brite pads (D), and the sandwich holder (E).

8. Block the blot by floating on 2% BLOTTO in PBS for 5 min and submerging it for 1 hr at 37 C with agitation.
9. Incubate the blot in primary antibody diluted in PBS plus 0.5% BLOTTO (PBS/BL) for 2-4 hr at 37 C with agitation.
10. Wash the blot 3-5 times with agitation at room temperature for 15 min in PBS/BL.
11. Incubate the blot in the secondary antibody/alkaline phosphatase conjugate diluted in PBS/BL for 1 hr.
12. Wash blot in PBS/BL as in #10 above and then once in 100 mM Tris-HCL, pH 8.0.
13. Submerge blot in substrate (3.3 mg Naphthol AS-MX phosphate, 50 mg Fast Red

TR salt, 10 ml 100 mM Tris-HCl). Incubate the blot with agitation in the dark for 5-35 min.

14. Reactive antigens are visible as red bands on blots that correspond to antigen bands on the original SDS-PAGE gel.

Modifications

Direct procedure - rather than use a secondary antibody/enzyme conjugate, the primary antibody can be conjugated directly with the enzyme to simplify the staining procedure.

Alternate substrate for alkaline phosphatase - To 10 ml 0.06 M sodium borate add 2.5 mg o-dansidine tetraoxide, 2.5 mg β-naphthyl acid phosphate and 12 mg magnesium sulfate heptahydrate. This precipitating substrate develops a blue color upon enzymatic conversion.

Alternate enzyme for conjugate - Peroxidase conjugate can be used in place of alkaline phosphatase conjugate. The procedure is the same but avoids use of BLOTTO with sodium azide since peroxidase is sensitive to the azide. As substrate use 6 mg 4-chloro-1-naphthol or 3 mg 3,3'-diaminobenzidine and 0.01% hydrogen peroxide (100 μl of a 30% solution) in 10 ml 50 mM Tris-HCl buffer, pH 7.4. The 4-chloro-1-naphthol develops a dark blue-green color and 3,3'-diaminobenzidine a dark brown color upon enzymatic conversion.

Sequential probing - It is possible to detect two different antigens on the same blot by the use of two different antibody preparations with different enzyme conjugates. Deposits of differently colored products permit the two antigens to be differentiated (7). The simplest approach is to use the direct procedure for the two antibody probes. However, the indirect procedure can be used if the primary antibodies are derived from different animal species.

Alternate membranes - Although nitrocellulose membranes have been used most commonly, other membranes such as nylon and polyvinylidene difluoride are now also available (e.g. Millipore). These membranes have a higher mechanical strength and higher protein binding capacity.

Literature Cited

1. De Boer, S. H., and McNaughton, M. E. 1987. Monoclonal antibodies to the lipopolysaccharide of *Erwinia carotovora* subsp. *atroseptica* serogroup I. Phytopathology 77:828-832.
2. Gershoni, J. M., and Palade, G. E. 1983. Protein blotting: principles and applications. Anal. Biochem. 131:1-15.
3. Jurs, M., Peters, H., Timmis, K. N., and Bitter-Suermann, D. 1985. Immunoblotting with monoclonal antibodies: A highly specific system for detection and identification of bacterial outer membrane proteins. Pages 94-102 in: Rapid Methods and Automation in Microbiology and Immunology. K.-O. Habermehl, ed. Springer-Verlag. 780 pp.
4. Laemmli, U. K. 1970. Cleavage of structural proteins during the assembly of the head of bacteriophage T4. Nature (London) 227:680-685.
5. Otter, T., King, S. M., and Witman, G. B. 1987. A two-step procedure for efficient electrotransfer of both high-molecular-weight (>400,000) and low-molecular-weight (<20,000) proteins. Anal. Biochem. 162:370-377.

6. Smyth, C. J. 1985. Immunology of outer membrane proteins of gram-negative bacteria. Pages 177-201 in: Immunology of the Bacterial Cell Envelope. D. E. S. Stewart-Hull and M. Davies, eds. John Wiley & Sons Ltd. 391 pp.
7. Theisen, N., Lohoff, E. S., von Figura, E., and Hasilik, A. 1986. Sequential detection of antigens in western blots with differently colored products. Anal. Biochem. 152:211-214.
8. Towbin, H., Staehelin, T., and Gordon, J. 1979. Electrophoretic transfer of proteins from polyacrylamide gels to nitrocellulose sheets: Procedures and some applications. Proc. Nat. Acad. Sci. USA 76:4350-4354.

Section IV.G.1.

FLUORESCENT ANTIBODY TEST, VIRUSES

R. E. Gingery
USDA ARS, Ohio State University
Wooster, Ohio 44691

Principle

Immunofluorescence, the detection of antigens by antibodies to which fluorochromes have been attached, is used principally for locating antigens in tissues and cells. In the "direct" method, the fluorochrome is attached directly to the antigen-specific antibody (called the primary antibody). In the "indirect" method, the primary antibody is unlabeled and is located after a second reaction using either fluorochrome-conjugated Protein A or fluorochrome-conjugated anti-primary-antibody antibody (called the secondary antibody).

The indirect method is usually more sensitive than the direct method because of the double-reaction amplification effect, resulting in more fluorochrome molecules per antigen molecule during the sequence of applied reagents. Another advantage of the indirect method is that a single fluorochrome-conjugated secondary antibody (or conjugated Protein A) can be used with many primary antibodies, thus eliminating the need to separately conjugate primary antibodies. Conjugated secondary antibodies, and conjugated Protein A are available commercially.

Materials

1. Buffers
   a) PBS - 0.14 M NaCl, 0.02 M sodium phosphate, pH 7.2
   b) 0.4 M sodium carbonate/bicarbonate, pH 9.5.
2. Fluorochromes
   FITC (fluorescein isothiocyanate)
   TRITC (tetramethylrhodamine isothiocyanate)
   See Goding (9) for others
3. Conjugated fluorochromes for indirect staining:
   Fluorochrome-conjugated anti-immunoglobulin
   Fluorochrome-conjugated Protein A
   These reagents are widely available commercially, but only affinity-purified grades should be used.
4. Chemicals: ethanol, methanol, acetone, glycerol, n-propyl gallate, ammonium sulfate
5. Bovine serum albumin (BSA)
6. Mayer's Albumin Fixative (20% v/v egg whites, 80% v/v glycerol)
7. Sephadex G-25
8. Photomicroscope with ultraviolet light source, appropriate filters, and film
9. Microscope slides and coverslips
10. Dialysis tubing
11. UV-VIS Spectrophotometer

12. Humidification chamber

Procedures

A. Purification of antibodies by ammonium sulfate precipitation

1. Place antiserum in beaker on a magnetic stirrer. Stir slowly because frothing denatures proteins.
2. Dropwise, add saturated ammonium sulfate, v/v, until it constitutes 40-45% of its mixture with antiserum
3. Continue slow stirring for 30 min to 1 hr.
4. Recover precipitated immunoglobulin by low-speed centrifugation (2,000 to 10,000 x g).
5. Wash pellet two or three times with 40%-saturated ammonium sulfate (i.e., v/v of saturated ammonium sulfate in distilled water).
6. Dissolve final pellet in PBS and dialyze against PBS to remove residual ammonium sulfate.

   Other methods for antibody purification (ion exchange or affinity chromatography, Rivanol precipitation, etc.) may be used, but ammonium sulfate precipitation is easy, gentle, and almost always yields a product suitable for fluorochrome conjugation.

B. Preparation and purification of FITC-conjugated antibodies

1. Dialyze purified antibody preparation (1-20 mg/l) overnight against 0.4 M sodium carbonate/bicarbonate, pH 9.5. Determine the protein concentration by measuring the A_{280} across a 1 cm light path and dividing by 1.4 to give mg protein/ml.
2. Dropwise, add a fresh stock of FITC (10 mg/ml in dimethyl sulfoxide) to the purified antibodies. Add about 10 μg FITC per mg of protein if the immunoglobulin concentration is greater than 10 mg/ml, or 100 μg FITC/mg protein if the antibody concentration is 1-2 mg/ml.
3. Shield the reaction from light and allow to proceed at room temperature for 1-2 hr (2-3 hr at low antibody concentrations).
4. Separate unreacted or hydrolyzed fluorochrome from conjugated protein by chromatography on Sephadex G-25 (or equivalent) columns equilibrated in PBS containing 0.1% NaN_3.
5. Check the fluorochrome/protein (F/P) ratio by the equation:

$$\text{F/P ratio} = \frac{2.87 \times A_{495}}{A_{280} - 0.35 \times A_{495}}$$

   Ideal ratios are 2-3 if the antibodies are to be used with fixed cells and 4-6 for intact living cells. High ratios are favored at high protein and fluorochrome concentrations and with long reaction times.
6. Conjugated antibodies are best stored at 4 C or at -20 C in 50% glycerol and protected from light.

C. Preparation of tissue sections for immunofluorescence

1. Cut 4- to 6-micron-thick sections in cryostat and transfer to clean glass slides. Air drying for 20 min may fix the samples well enough. If not, dipping the slides in acetone, ethanol, or methanol may suffice. The optimal fixation procedure must be determined for each

antigen. Alternatively, paraffin-embedded tissue may be sectioned, but there is a risk of epitope inactivation during the embedding procedure.

D. Preparation of protoplasts for immunofluorescence

1. Place a drop of protoplast suspension on a clean glass slide previously coated with a thin layer of Mayer's albumin fixative. Allow the drops to air dry or dry them with a blow dryer. Mark "front slide" of slide.
2. Fix the protoplasts by dipping the slide into 95% ethanol for 15 min at room temperature.

E. Staining

1. Place slides with fixed tissue sections or fixed protoplasts into a rack and wash for 1 hr or longer in 2 or 3 changes of PBS + 1% BSA with gentle stirring. This step should block most sites of nonspecific protein binding.
2. Remove the slides and allow them to drain. Remove excess buffer and place the slides flat with samples on top in a humidification chamber. Cover each sample with 10-20 μl of conjugated virus-specific antibody or with unconjugated virus-specific antibody if the indirect method is used. Incubate at room temperature for 30 min or longer. Be sure that the solutions covering the samples do not evaporate to dryness in this and subsequent steps.
3. Wash off excess antibody in a stream of PBS from a wash bottle. Direct the stream at a point above the section and let the stream flow over the section. Repeat the PBS-BSA wash as described in step 1.
4. If the indirect method is being done, repeat steps 2 and 3 with the conjugated secondary antibody (e.g., goat anti-rabbit IgG) or conjugated Protein A.
5. Add a drop of mounting medium and a cover slip. Equal parts of glycerol and PBS may be used as mounting medium, but 5% w/v n-propyl gallate in glycerol will greatly reduce photobleaching if FITC or TRITC is used.
6. Seal the edges of the cover slip with paraffin or nail polish.

F. Visualization and photography of stained specimens

Procedures for visualization and photography of fluorescing images vary greatly depending on the available microscope and illumination system. The comments below, however, may be helpful.

Comments

For optimal visualization of stained specimens, epi-illumination systems are far superior to those using transmitted light because of higher illumination intensities which produce brighter images of better quality. Also, epi-illumination allows rapid switching between visible and fluorescent optics. High-pressure mercury (most common), high-pressure xenon, or quartz-halogen lamps are suitable for fluorescence work. However, because different fluorochromes have different optical properties, check to be sure that the lamp used emits at the appropriate wavelength. The highest image intensity is achieved with oil-immersion objectives because of the higher numerical aperture compared to air. Because fluorescence intensity decreases exponentially with magnification, using the lowest magnification necessary for the desired resolution will yield the brightest images.

The excitation and barrier filters depend on the fluorochrome used. All fluorochromes emit light of a longer wavelength than the excitation wavelength. For example, FITC is maximally excited at 495 nm and emits at 500-580 nm with a peak at 515 nm. Thus, excitation through a band-pass filter that transmits light from 450-490 nm and visualization after a barrier filter that transmits all wavelengths longer that 510 or 520 nm should give suitable fluorescence with minimal background. Be sure that the barrier filter is in place before viewing to prevent eye injury.

Photography is generally straightforward. The main difficulties are fading of images due to photobleaching and color changes during intense irradiation of the sample. Usually, fast color or black-and-white films are used and exposure times are determined empirically or by using an exposure meter with an automatic system. If photobleaching occurs and the fluorescence intensity is so low that exposure times are needed whose length exceeds the time after which photobleaching has occurred, it may not be possible to photograph the image very well. Either the photobleaching must be reduced (see E. Staining, above) or a faster film must be used.

The main problems encountered in immunofluorescence are nonspecific fluorescence or lack of fluorescence. Nonspecific fluorescence has many causes and it is important that appropriate negative controls be included in each experiment. The more common causes of nonspecificity include residual unconjugated fluorochrome, high F/P ratios of the antibodies, inadequate washing of samples during staining, and autofluorescence of tissues and cells. In general, nonspecificity is minimized if highest quality reagents are used. Primary antibodies should be purified as much as possible and steps to prevent antibody aggregation should be taken. In some cases, Fab or $F(ab')_2$ fragments may be needed because of Fc binding sites in the sample. To conjugate fluorochromes and antibodies, the unconjugated fluorochrome should always be separated from the conjugated antibodies by gel filtration, never dialysis, making sure that the column is large enough to give good separation of free and conjugated molecules.

To minimize nonspecific fluorescence in indirect staining, the quality of the anti-immunoglobulin or Protein A is of paramount concern. Only affinity-purified reagents should be used.

Negative controls should include nonstained tissue sections or cells from infected plants to assess autofluorescence and stained sections from healthy plants to assess nonspecific fluorescence. Two other controls are useful: a) treatment of sample with non-conjugated antibody before direct staining which should preclude staining, and b) no treatment with primary antibody (or treatment with an unrelated primary antibody) before staining with labeled anti-immunoglobulin or Protein A, which should also preclude staining. The optics should routinely be checked with a blank slide to determine the degree of light scattering.

FITC is usually the fluorochrome of choice because it is inexpensive, fairly hydrophilic, and has a high quantum efficiency. FITC fluoresces an apple-green color. The next most-used fluorochrome is TRITC, which has a red fluorescence, is easily distinguishable from FITC, and so may be used in double-labeling procedures. However, TRITC is somewhat hydrophobic and tends to bind nonspecifically to antibodies and other proteins. Slow release of such loosely bound fluorochrome can cause background fluorescence. There are a number of other fluorochromes available for use and Goding (9) may be consulted for a discussion of their properties and applications.

Applications

Fluorescent antibodies have been used to locate plant viruses in plants (3,13,19,34,35,38), in insect vectors and vector-cell monolayers (15,16,29, 30,31), in callus tissue (21), and to measure the rate and progress of infection of plant protoplasts by viruses (7,10,20,22,23,27,37). Other uses have included: measuring virus spread in infected plants both by determining the numbers of infected cells isolated at various times after infection (39) and by directly staining intact leaf tissue (17,18); selecting appropriate cells for further study by electron microscopy (26); detecting virus adsorbed to uredospores of a suspected fungal vector (5); and studying interference between virus strains in protoplasts (1).

Viral antigens detected with fluorescent antibodies include the coat protein (above references), viral inclusion proteins (2,12,25) and a protein required for insect transmission of plant viruses, the so-called helper protein (11).

Although immunofluorescence-based viral assays have been developed (32,8) enzyme-linked immunosorbent assay (ELISA) and nucleic acid hybridization assays are generally more sensitive and specific. However, Siitari and Kurppa (28) have developed a time-resolved fluoroimmunoassay reported to be 50-100 times more sensitive than ELISA.

Recent reviews of immunofluorescence include Deluca (4), Feteanu (6), Goding (9), Kawamura (14), Peters and Coons (24), Sternberger (33) and van Regenmortel (36).

Literature Cited

1. Barker, H., and Harrison, B. D. 1978. Double infection, interference and superinfection in protoplasts exposed to two strains of raspberry ringspot virus. J. Gen. Virol. 40:647-658.
2. Chang, C. A., Purcifull, D. E., Hiebert, E., and Edwardson, J. R. 1986. Immunofluorescence evidence for the origin of nuclear inclusion and cytoplasmic crystals induced by bean yellow mosaic virus. (Abstract) Phytopathology 76:1061.
3. Chiu, R.-J., and Black, L. M. 1969. Assay of wound tumor virus by the fluorescent cell counting technique. Virology 37:667-677.
4. Deluca, D. 1982. Immunofluorescence analysis. Pages 189-231 in: Antibody as a Tool. J. J. Marchalonis and G. W. Warr, eds. John Wiley, New York. 568 pp.
5. Erasmus, D. S., Rybicki, E. P., and Von Wechmar, M. B. 1983. The association of brome mosaic virus and wheat rusts. II. Detection of BMV in/on uredospores of wheat stem rust. Phytopathol. Z. 108:34-40.
6. Feteanu, A. 1978. Labelled Antibodies in Biology and Medicine. McGraw-Hill, New York. 411 pp.
7. Furusawa, I., Yamaoka, N., Okuno, T., Yamamoto, M., Kohno, M., and Kunoh, H. 1980. Infection of turnip protoplasts with cauliflower mosaic virus. J. Gen. Virol. 48:431-435.
8. Gingery, R. E. 1978. An immunofluorescence test for maize chlorotic dwarf virus. Phytopathology 68:1526-1529.
9. Goding, J. W. 1983. Immunofluorescence. Pages 208-249 in: Monoclonal Antibodies: Principles and Practice: Production and application of monoclonal antibodies in cell biology, biochemistry and immunology. B. F. Haynes and G. S. Eisenbath, eds. Academic Press, New York. 276 pp.
10. Hibi, T., Rezelman, G., and van Kammen, A. 1975. Infection of cowpea mesophyll protoplasts with cowpea mosaic virus. Virology 64:308-318.

11. Hiebert, E., de Mejia, M. V. C., and Purcifull, D. E. 1985. Intracellular visualization of potyviral antigens by immunofluorescence microscopy. (Abstract) Phytopathology 75:1177.
12. Hiebert, E., Purcifull, D. E., and Christie, R. G. 1984. Purification and immunological analyses of plant viral inclusion bodies. Pages 225-280 in: Methods in Virology, Vol. VIII. K. Maramorosch and H. Koprowski, eds. Academic Press, New York. 396 pp.
13. Hosokawa, D., and Mori, K. 1982. Studies on multiplication and distribution of viruses in plants by use of fluorescent antibody technique. III. Multiplication and distribution of tobacco mosaic virus in floral parts of petunia, Petunia hybrida. Ann. Phytopath. Soc. Japan. 48:534-537.
14. Kawamura, A. 1977. Fluorescent Antibody Techniques and their Applications, 2nd Ed. Univ. of Tokyo Press, Tokyo and Univ. Park Press, Baltimore, Maryland. 292 pp.
15. Kimura, I. 1986. A study of rice dwarf virus in vector cell monolayers by fluorescent antibody focus counting. J. Gen. Virol. 67:2119-2124.
16. Kimura, T., and Black, L. M. 1971. Some factors affecting infectivity assays of wound-tumor virus on cell monolayers from an insect vector. Virology 46:266-276.
17. Lei, J. D., and Agrios, G. N. 1986. Detection of virus infection and spread by immunofluorescent staining of enzyme-treated leaves. Phytopathology 76:1031-1033.
18. Lei, J. D., and Agrios, G. N. 1986. Mechanisms of resistance in corn to maize dwarf mosaic virus. Phytopathology 76:1034-1040.
19. Nagaraj, A. N. 1965. Immunofluorescence studies on synthesis and distribution of tobacco mosaic virus antigen in tobacco. Virology 25:133-142.
20. Nishiguchi, M., Langridge, W. H. R., Sqalay, A. A., and Zaitlin, M. 1986. Electroporation-mediated infection of tobacco leaf protoplasts with tobacco mosaic virus RNA and cucumber mosaic virus RNA. Plant Cell Reports 5:57-60.
21. Omura, T. 1978. Localization of tobacco mosaic virus in tobacco callus tissue. Ann. Phytopath. Soc. Japan. 44:151-158.
22. Otsuki, Y., and Takebe, I. 1969. Fluorescent antibody staining of tobacco mosaic virus antigen in tobacco mesophyll protoplasts. Virology 38:497-499.
23. Pappu, H. R., and Hiruki, C. 1986. Infection of mesophyll protoplasts of Nicotiana clevelandii with sweet clover necrotic mosaic virus. (Abstract) Phytopathology 76:1121.
24. Peters, D., and Coons, A. H. 1976. Fluorescent antibody as specific cytochemical reagents. Methods Immunol. Immunochem. 5:424-444.
25. Rao, D. V., Shukla, P., and Hiruki, C. 1978. In situ reaction of clover yellow mosaic virus (CYMV) inclusion bodies with fluorescent antibodies to CYMV. Phytopathology 68:1156-1159.
26. Rieder, C. L., and Bowser, S. S. 1985. Correlative immunofluorescence and electron microscopy on the same section of Epon-embedded material. J. Histochem. Cytochem. 33:165-171.
27. Shalla, T. A., and Petersen, L. J. 1973. Infection of isolated plant protoplasts with potato virus X. Phytopathology 63:1125-1130.
28. Siitari, H., and Kurppa, A. 1987. Time-resolved fluoroimmunoassay in the detection of plant viruses. J. Gen. Virol. 68:1423-1428.
29. Sinha, R. C. 1965. Sequential infection and distribution of wound-tumor virus in the internal organs of a vector after ingestion of virus. Virology 26:673-686.

30. Sinha, R. C., and Black, L. M. 1962. Studies of the smear technique for detecting virus antigens in an insect vector by use of fluorescent antibodies. Virology 17:582-587.
31. Sinha R, C., and Black, L. M. 1963. Wound-tumor virus antigens in the internal organs of an insect vector. Virology 21:183-187.
32. Sinha, R. C., and Thottappilly, G. 1974. Sensitivity of three serological tests for detecting wheat striate mosaic virus purified from infected plants. Phytopathol. Z. 81:124-132.
33. Sternberger, L. A. 1986. Immunocytochemistry, 3rd Ed. John Wiley, New York. 524 pp.
34. Thornley, W. R., and Mumford, D. L. 1979. Intracellular location of beet curly top virus antigen as revealed by fluorescent antibody staining. Phytopathology 69:738-740.
35. Urban, L. A., Klomparens, K., and Ramsdell, D. C. 1986. Preliminary studies on location of blueberry shoestring virus (BBSSV) in highbush blueberry cv. 'Jersey' using indirect fluorescent antibody and dot-blot immunoassay techniques. (Abstract) Phytopathology 76:1074.
36. Van Regenmortel, M. H. V. 1982. Serology and Immunochemistry of Plant Viruses. Academic Press, New York. 302 pp.
37. Watanabe, Y., Ohno, T., and Okada, Y. 1982. Virus multiplication in tobacco protoplasts inoculated with tobacco mosaic virus RNA encapsulated in large unilamellar vesicle liposomes. Virology 120:478-480.
38. Worley, J. J., and Schneider, I. R. 1963. Progressive distribution of southern bean mosaic virus antigen in bean leaves determined with a fluorescent antibody stain. Phytopathology 53:1255-1257.
39. Wyatt, S. D., and Wilkinson, T. C. 1984. Increase and spread of cowpea chlorotic mottle virus in resistant and fully susceptible cowpeas. Physiol. Plant Path. 24:339-345.

Section IV.G.2.

IMMUNOFLUORESCENCE FOR BACTERIA

S. H. De Boer
Agriculture Canada Research Station
Vancouver, B.C., Canada

Introduction

The immunofluorescence procedure is useful for identifying bacterial strains, diagnosing the etiology of bacterial plant diseases, and detecting bacterial pathogens in seed, vegetative propagules, etc. The advantage of this technique over other serological procedures is that it allows one to actually see the bacterial cells. Thus, rather than simply observing a precipitin band or measuring an enzyme reaction, bacterial cell morphology and the size of the cell population can be seen. The additional information is valuable for assessing the validity of serological reactions.

The procedure is one of the most sensitive, permitting detection in the range of 10^3 cells/ml. Because of its sensitivity, the possibility of detecting a cross-reacting microorganism is of concern, but this concern has been greatly diminished with the advent of monoclonal antibodies (3).

Principle

In the immunofluorescence procedure the antibody is tagged with a fluorescent dye which can be visualized on microscope preparations with an appropriately equipped microscope. The technique can be carried out in the direct mode in which the fluorescent dye is conjugated with the primary antibody, or in the indirect mode in which the fluorescent dye is conjugated to a secondary antibody which is specific for the primary antibody. The indirect procedure is advantageous in that one commercially available conjugate can be used for any primary antibody preparation and is, therefore, most practical in many laboratories.

The indirect procedure involves fixing an appropriately diluted sample onto a microscope slide, treating the preparation with an antibody specific for the bacterium of interest, and then reacting the first antibody with a secondary antibody to which the fluorescent dye, usually fluorescein, has been conjugated (2) (Fig. 1).

The prepared slides are observed with a microscope equipped with filters specified for the fluorescent dye being used, and a light source which provides adequate illumination in the near ultraviolet range. Epi-illumination, in which the illumination is provided through the objective lens, is recommended. Positive bacteria fluoresce a bright yellow-green color (with fluorescein) against a dark background.

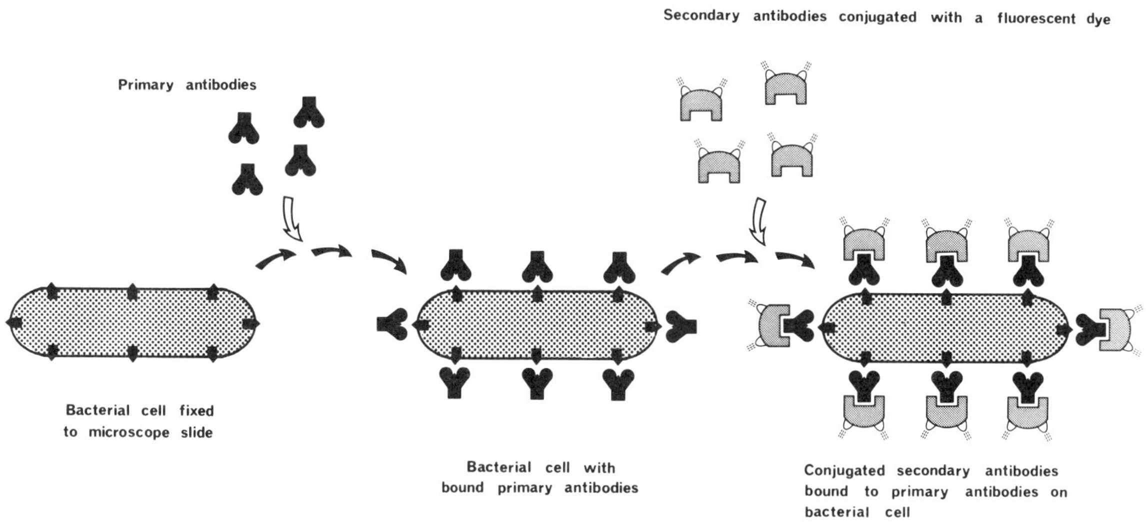

Fig. 1. Schematic representation of indirect immunofluorescence staining procedure.

Materials

1. Multi-well microscope slides (e.g., toxoplasmosis slides from Bellco Glass, cat. no. 5638-01940) and 24 x 60 mm No. 1 coverslips.
2. Fluorescein conjugated anti-primary antibody (anti-rabbit for primary rabbit antisera; anti-mouse for primary mouse monoclonal antibodies) available from many sources (e.g., Jackson Immunoresearch Labs).
3. Phosphate buffered saline (PBS) 0.01 M, pH 7.2.
   Solution A: Potassium phosphate, dibasic 69.67 g/l distilled water.
   Solution B: Potassium phosphate, monobasic 54.44 g/l distilled water.
   Stock solution = 250 ml solution A + 100 ml solution B (check pH = 7.2)
   PBS = 25 ml stock solution + 975 ml distilled H_2O + 8.5 g NaCl.
4. Blocking agent - Prepare 10% skim milk powder stock solution and add 0.2% sodium azide as preservative. Use at 20 μl/ml in diluted antibody preparations.
5. Mounting fluid - Mix 90 ml glycerol with 10 ml PBS. To prevent fading of fluorescein during microscopic observation, add p-phenylenediamine (PPD, Sigma, cat. no. P-6001) to mounting fluid at 1 mg/ml. To dissolve PPD add to glycerol/PBS and keep at -20 C a few days in the dark. Also store it under these conditions since PPD in solution turns brown at room temperature.
6. Microscope - Most standard light microscopes can be equipped for fluorescence work. A high intensity light source such as a high pressure mercury or xenon burner set up for epi-illumination and an appropriate filter-set for the fluorescent dye being used are required. A 40X Neofluar objective is often sufficient but a 100X oil objective is preferred for small cells. A phase contrast condenser and objectives are highly recommended to aid in focusing preparations in which little fluorescing material is present. Phase contrast at low light levels can be used simultaneously with epi-illumination when scanning preparations.

Indirect IF procedure

1. Prepare three 10-fold dilutions of sample in distilled water (presence of salts may interfere with adhesion of cells to the glass slide).
2. Apply 20 μl of undiluted and the three diluted samples to separate wells (5 mm dia.) of multiwell slides. Air-dry the preparations on a slide warmer set at 50 C or under a hair dryer.
3. Fix preparations by soaking slides in acetone for 10 min followed by air-drying, or fix by flaming. Flaming must be done carefully to avoid destroying the antigenic structure of bacterial cells.
4. Dilute primary antibody in PBS plus blocking agent and apply 20 μl to each well. Use the lowest antibody concentration which still gives good staining when there are up to 100 positive cells/microscope field.
5. Incubate slides 30-60 min in a moist chamber at 37 C.
6. Rinse slides with a gentle stream of distilled H_2O but avoid directing the stream directly on the preparation; air-dry.
7. Dilute secondary antibody conjugate as in 4. The optimum dilution will vary with each batch of conjugated antibody and therefore must be determined by trying a range of dilutions (usually 1:20-1:200 of commercial preparations) on a known positive sample.
8. Incubate slides 30-60 min in a dark, moist chamber at 37 C.
9. Rinse slides with distilled H_2O as in 6; air-dry.
10. Mount coverslip with mounting fluid.

Related procedures and modifications

Direct staining - The immunofluorescence staining procedure can be simplified by conjugating the fluorescent dye directly to the primary antibody and thereby obviating the need for the secondary antibody conjugate. To conjugate antibody with an isothiocyanate fluorochrome, 5 mg of purified immunoglobulin in 1 ml of 0.1 M carbonate buffer pH 9.0 is incubated with 25 μg of fluorochrome dissolved in dimethylsulfoxide for 3 hr at room temperature or 8 hr at 4 C (4). The conjugated antibody is separated from unconjugated dye on a PD-10 or Sephadex G-25 column.

Double staining - The specificity of immunofluorescence staining can be improved by staining with two different antibodies that react with different antigens on the same bacterium permitting individual cells to be characterized on the basis of two rather than one antigen. This approach is especially applicable with two monoclonal antibodies of known specificity. The two antibodies are tagged with different fluorescent dyes such as fluorescein and rhodamine. Reaction of a bacterium with one or both monoclonals can then be ascertained by microscopic viewing with the appropriate filters.

Immunosorbent fixation - To selectively increase the number of bacterial cells in a preparation, microscope slides can be coated with an antiserum to trap bacteria on the slide. Trapped bacteria are subsequently stained by a standard procedure. Antibodies can be fixed on the slide with ethanol, Kirkpatrick fixative, or on a pre-applied formvar layer (7).

Membrane filters - To detect plant pathogenic bacteria in dilute suspensions, the bacteria can be deposited on polycarbonate membranes by filtration under vacuum and stained for immunofluorescence directly on the membranes (1). Alternatively, the bacteria can be stained in the liquid phase prior to deposition onto the membranes (5). Staining in the liquid phase is done simply by suspending the pelleted preparation in diluted antibody and after incubation removing excess antibody by centrifugation and resuspension.

Immunogold silver staining - As an alternative to fluorochrome conjugates as the secondary antibody, antibodies labelled with colloidal gold (Janssen Life Sciences Products) can be used. The procedure is the same as for immunofluorescence except for the change in secondary antibody preparation. After staining, the gold beads require further enhancement by a silver staining procedure in which the diameter of the gold particles is increased as much as ten-fold. The advantage of this approach is that preparations can be observed with a normal brightfield light microscope (6). The disadvantage is that the reagents are very costly.

Literature Cited

1. De Boer, S. H. 1984. Enumeration of two competing Erwinia carotovora populations in potato tubers by a membrane filter - immunofluorescence procedure. J. Appl. Bacteriol. 57:517-522.
2. De Boer, S. H., and McNaughton, M. E. 1986. Evaluation of immunofluorescence with monoclonal antibodies for detecting latent bacterial ring rot infections. Am. Potato J. 63:533-543.
3. De Boer, S. H., and Wieczorek, A. 1984. Production of monoclonal antibodies to Corynebacterium sepedonicum. Phytopathology 74:1431-1434.
4. Haaijman, J. J. 1983. Labelling of protein with fluorescent dyes: Quantitative aspects of immunofluorescence. Pages 47-85 in: Immunohistochemistry. A. C. Cuello, ed. John Wiley & Sons. 501 pp.
5. Lin, C. P., Chen, T. A., Wells, J. M., and van der Zwet, T. 1987. Identification and detection of Erwinia amylovora with monoclonal antibodies. Phytopathology 77:376-380.
6. Van Laere, O., De Wael, L., and De May, J. 1985. Immuno gold staining (IGS) and immuno gold silver staining (IGSS) for the identification of the plant pathogenic bacterium Erwinia amylovora (Burrill) Winslow et al. Histochemistry 83:397-399.
7. Van Vuurde, J. W. L., and Van Henten, C. 1983. Immunosorbent immunofluorescence microscopy (ISIF) and immunosorbent dilution - plating (ISDP): New methods for the detection of plant pathogenic bacteria. Seed Sci. & Technol. 11:523-533.

Section IV.G.3.

IMMUNOFLUORESCENCE COLONY STAINING

J. W. L. van Vuurde
Research Institute for Plant Protection
Wageningen, The Netherlands

Principle

Immunofluorescence colony staining (IFC) was developed to improve the sensitivity of detection and isolation and to combine isolation directly with serological colony identification (4). The procedure for IFC is diagramed in Fig. 1. IFC enables the fluorescent colonies of the target bacterium to be distinguished from those of other microorganisms in the agar with an UV-microscope at low (10 to 60 times) magnifications. Results are available after one to three days and preparations can be interpreted more easily by less-experienced technicians than immunofluoresce (IF) cell staining. Isolation from IF-positive colonies can be made for confirmation of identity, pathogenicity testing, collection of cross-reacting organisms for research on antiserum specificity, and further improvement of the assay procedure. Illustrations of IFC are presented for macroscopically visible colonies (Fig. 1 and 2) and for microcolonies (Fig. 3) in a miniaturized assay.

Materials

1. Polystyrene Petri dishes (6 cm)
2. Microscope slides
3. Agar medium
4. Warm air dryer
5. Microscope with incident blue light (490 nm)
6. Fluorescein isothiocyanate (FITC)-conjugated antiserum
7. Phosphate buffer, pH 7.0, contining 0.85% (w/v) NaCl (PBS) and mounting medium (e.g. 'Aqua Mount', Lerner Laboratories)

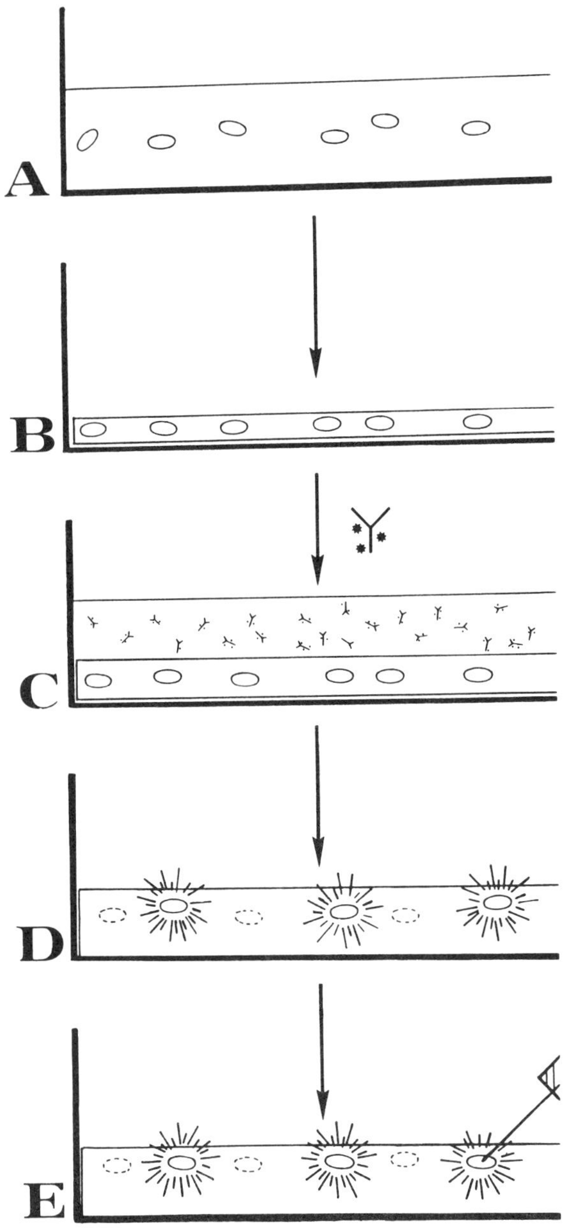

Fig. 1. Diagram of immunofluorescence colony staining. A. Growth of colonies in nutrient agar. B. Drying of agar medium with warm air blower. C. Incubation with FITC-conjugated antibodies. D. Observation with incident blue light at low magnification. E. Isolation of bacteria from a serologically positive reacting colony.

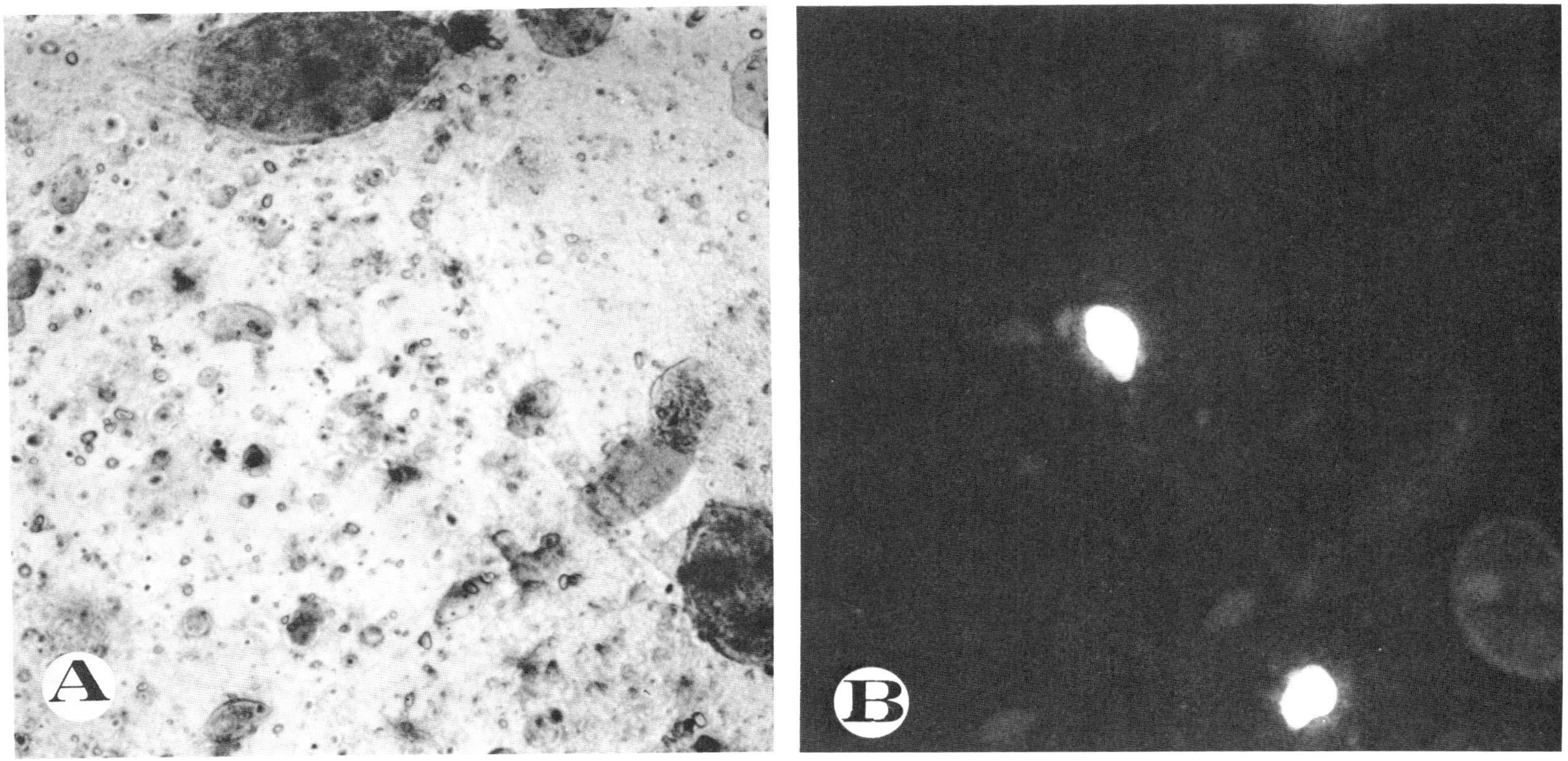

Fig. 2. IFC-positive colonies of _Erwinia chrysanthemi_ (Ech) among saprophyte colonies in tuber peel extract. FITC-conjugate dilution 1:200, detection level 10^2 cfu per ml for 16 mm Y wells of tissue culture plates. Microscope magnification 16 times. A. Agarfilm stained with anti-Ech FITC conjugate observed with normal light. B. Same preparation with incident blue light showing two IFC-positive colonies of Ech.

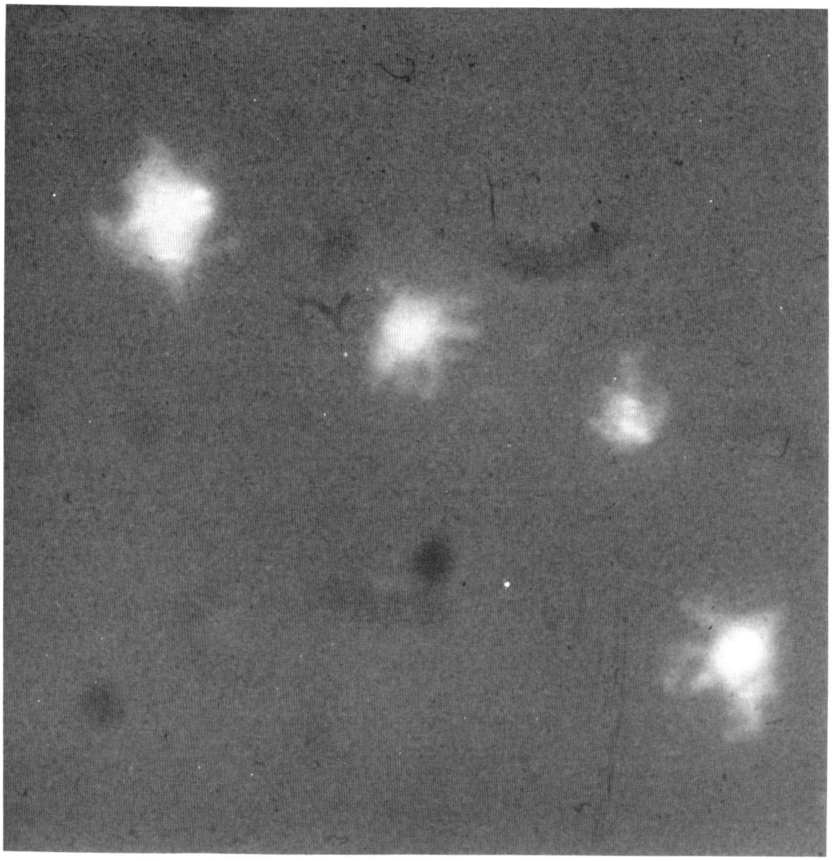

Fig. 3. Microcolonies of Clavibacter michiganense pv. sepedonicum (10-50 cells) stained using IFC after 48 hr growth in YPG-medium (traditional isolation on YPG takes 6-8 da). FITC-conjugate dilution 1:100, microscope magnification 500 times (N. J. M. Roozen and J. W. L. van Vuurde, unpublished results).

A. Macro colony assay

Procedure

1. Add the test sample (max. 2 ml) and sufficient agar medium at 45 C to a 6 cm diameter polystyrene Petri dish to obtain a final volume of ca. 6 ml. Swirl the plate gently clockwise and counter-clockwise for a homogeneous distribution of the bacteria in the agar. Prepare controls with extracts of known healthy and contaminated material.
2. Incubate at optimal conditions for the target bacterium. Stop the incubation when pinhead size colonies are formed (1 or 2 da for most bacteria).
3. Open the Petri dish and dry the agar into a thin (< 1 mm thick) film by blowing warm air (ca. 40 C) over the agar surface.
4. Add 5 ml of diluted FITC-conjugated antiserum to the Petri dish and incubate the agar film overnight at room temperature. The conjugate is diluted with 0.01 M PBS, pH 7.4, usually at a two-fold higher concentration than used for staining homologous bacterial cells in

direct IF.

5. Remove the non-bound conjugate by two 5 min washes in PBS. Remove excess PBS from the Petri dish with a pipet.
6. Place the Petri dish with the stained agar film under a microscope with incident blue light and 3.5 to 10x objective and 4 to 6.3x ocular magnification. Inspect the plate for colonies with green fluorescence and colony morphology comparable with the positive control.
7. Isolate from IF-positive colonies, if needed, by puncturing the colony under the microscope with a fine needle or glass capillary tube (by hand or with a micromanipulator) and plating on a medium.

Comments

1. The format of the test (step #1) can easily be adapted for routine test application at a somewhat lower sensitivity level by using the wells of a 6, 12, or 24-well tissue culture plate instead of Petri dishes. A final agar concentration of 0.8 to 1.0% after mixing with the sample is recommended. The temperature of the sample should be close to that of the agar medium if large sample volumes are used.
2. Dried agar film with colonies (step #3) can be stored several weeks in a container with silica gel at 4 C before IFC staining is done.
3. Depending on experimental conditions, incubation with stain reagents (step #4) can sometimes be reduced to 2 hr.
4. Since blue light is used (step #6), the objectives of a standard light microscope can replace the more expensive special UV objectives. The highest fluorescence intensity is obtained with a relatively high objective and relatively low ocular magnification. The expensive high pressure Mercury or Xenon UV-light source can be replaced by a cheaper and more constant Tungsten light source (minimum 50 W, 12 V) when the somewhat lower light intensity does not affect the detection of IF-positive colonies.
5. Interpretation of test results (step #7) should be based on comparison with those of the positive and the negative control, which can be prepared in advance and can be stored according to comment 3. Part of the agar film of these controls can be stained simultaneously with each test series as a standard reference. Fading of IF-stained colonies (observed with 10x objective) is much less noticeable than that of IF-stained cells (observed with 40-100x objective).

B. Microcolony assay

Procedure

1. Mix the sample and the agar, and prepare a layer ca. 1 mm thick on a microscope slide, in a well of a multi-well microscope slide (Cel-line), in a chamber of a tissue culture chamber/slide (Lab-Tek, Miles Scientific), or in a well of a 24-well tissue culture plate or a 96-well polystyrene flat-bottom microplate.
2. Incubate at optimal conditions for the target bacterium to form micro-colonies (5 to 50 cells per colony). For most plant pathogenic bacteria an 8 hr or overnight incubation is suitable. This can be checked microscopically
3. Dry and stain the agar layer as described above. Wash carefully to prevent washing away the agar film.

4. Mount the stained agar film under a cover glass and observe the slide at ca. 400x magnification, or observe the preparation in the well of a tissue culture plate or microplate through the bottom of the inverted plate with a 40x long distance objective (e.g., Olympus LWDCDPL 40/0.55 with 1.9 mm working distance, suitable for all modern microscope types).

Comments

1. The diagnostic sensitivity (3) of a test involving isolation of a target bacterium will be determined mainly by the interference of saprophytic microorganisms and the capacity of the test system for developing individual colonies. Semi-selective media (2) decrease the interference of saprophytes but may also affect the plating efficiency for various strains of the target organism (1). The widely used agar surface plating technique is 100 to 1,000 times less sensitive, compared with the technique in which the sample is mixed with the agar ('pour plates'). This is mainly due to stronger inhibition among colonies of microorganisms on the agar surface than among those in the agar (colonies stay much smaller and are 3-dimensionally distributed). With 'pour plates' in 9 cm diameter Petri dishes, sample quantities up to 5 ml can be used per plate and the capacity for separate colonies in the agar is 10^5 to 10^6 per plate. Because of poor visual colony differentiation, 'pour plates' are now mainly used to determine total numbers of colonies. The combination of 'pour plate' and serological identification with IFC provides an assay with a high diagnostic sensitivity and specificity (3), particularly when IF-positive colonies are checked by subsequent isolation and identification or pathogenicity tests. Other colony identification methods like 'pour agar plating' with antiserum simultaneously added with the test sample (4) and immuno-enzyme assays using precipitating substrate are possible but are less sensitive and specific, respectively (van Vuurde, unpublished results).
2. The major critical aspects in the procedure are the suitability of the agar medium for development of colonies of the target organism in the agar even at high saprophyte to pathogen ratios, and the specificity of the antiserum against cell wall-linked antigens. The best antisera for IFC have high titers in IF-cell staining. The addition of Tween 20 to the conjugate (0.1%) may help reduce background fluorescence due to non-specific binding of the conjugate to the agar.
3. Since polyclonal antisera contain antibodies against both soluble and cell wall-linked antigens, bacterial strains can give (a) fluorescent colonies, (b) fluorescent colonies with a fluorescent precipitation halo (due to a reaction with soluble antigens diffused from the colony) or (c) unstained colonies with a fluorescent halo. The immunostaining pattern of cross-reacting organisms can provide information on the possible interference of these organisms with other assays (e.g. a fluorescent halo but no colony staining indicates a higher risk for cross-reaction in ELISA than in IF). If necessary, IFC can be further improved (or used to improve other serological methods) by isolating the cross-reacting microorganisms and using them for modifying the isolation medium to prevent growth of these organisms or for selecting more specific mono- or polyclonal antisera.

4. The slow growth of *C. michiganense*. subsp. *sepedonicum* colonies (4 da for pinhead size colonies) (4 da for pinhead size colonies) necessitated a 100 to 1,000 times greater dilution of the sample than for *Erwinia* spp., to reduce the antibiosis of saprophytes (N. J. M. Roozen, J. W. L. van Vuurde, unpublished results). IFC was also tested with the indirect immunofluorescence staining, first using unlabeled rabbit antibodies against *E. chrysanthemi*, followed by FITC-conjugated goat anti-rabbit antibodies as second antibodies. Indirect IFC gave a good staining of the *E. chrysanthemi* colonies, however some cross reactivity with *E. carotovora*. subsp. *atroseptica* colonies and increased background fluorescence was observed compared with direct IFC of *E. chrysanthemi* colonies.

Applications

Experiments with the macrocolony assay for various plant pathogenic bacteria, using the wells (ca. 16 mm diameter) of 24-well tissue culture plates, have demonstrated the value of IFC in detecting them in complex natural substrates. Representatives of plant pathogenic *Pseudomonas*, *Xanthomonas*, *Clavibacter*, and *Erwinia* spp. showed good growth in agar. *E. chrysanthemi* and *E. carotovora* subsp. *atroseptica* colonies could be detected at a concentration of 10^2 cfu per ml in potato peel extract and cattle manure slurry (J. W. L. van Vuurde, M. W. Bos, N. J. M. Roozen, unpublished results). *Clavibacter michiganense* subsp. *sepedonicum* was detected in cattle manure slurry artificially inoculated with 10^4 to 10^5 cfu per ml.

Literature Cited

1. Chun, W. W. C., and Alvarez, A. M. 1983. A starch-methionine medium for isolating *Xanthomonas campestris* pv. *campestris* from plant debris in soil. Plant Dis. 67:632-635.
2. Fahy, P. C., and Hayward, A. C. 1983. Media and methods for isolation and diagnostic tests. Pages 337-378 in: Plant Bacterial Diseases. P. C. Fahy and G. J. Persley, eds. Academic Press, Sidney. 393 pp.
3. Sheppard, J. W., Wright, P. F., and Desavigny, D. H. 1986. Methods for the evaluation of EIA tests for use in the detection of seed-borne diseases. Seed Sci. and Techn. 14:49-60.
4. van Vuurde, J. W. L. 1987. New approach in detecting phytopathogenic bacteria by combined immunoisolation and immunoidentification assays. EPPO Bulletin 17:139-148.

Section IV.H.

LATEX AGGLUTINATION FOR VIRUSES AND BACTERIA

S. A. Slack
Cornell University
Ithaca, NY 14583

E. M. Ball
University of Nebraska
Lincoln, NE 68583

Principle

Uniform latex particles (spherical polystyrene beads) are coated with antibodies via adsorption or covalent binding protocols. The "sensitized" latex particles are mixed with antigen (viral or bacterial), agglutination occurs and is detectable within minutes (Fig. 1) The test is most useful when a combination of speed and sensitivity are desired in an assay. An assay based on the commonly used adsorption protocol is described.

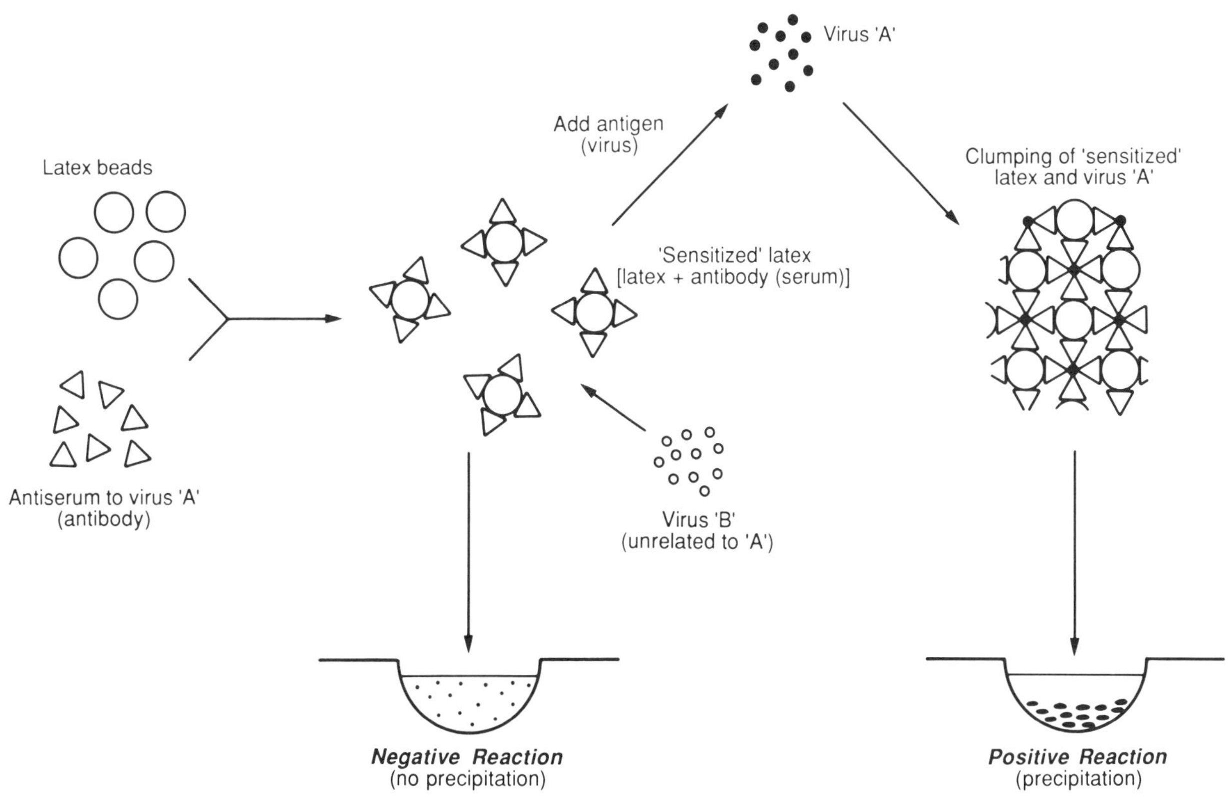

Fig. 1. Components and principle of the latex agglutination test.

Materials

1. Chemicals:
   a. Tris (hydroxymethyl) aminomethane (0.1 M Tris, pH 7.4)
   b. NaCl (0.85%)
   c. Sodium azide (2% stock solution)
   d. Polyvinyl pyrrolidone, M_r 40,000 (PVP)
   e. Bovine serum albumin - crystallized and lyophilized (Sigma, Catalog #A-4378)
2. 0.8 µm diam. polystyrene latex beads (Sigma, Catalog #LB-8)
3. Centrifuge, Sorvall SS-34 rotor or equivalent
4. Whatman filter paper #2
5. 100 Lambda micropipets (Clay-Adams, Catalog #4625, 1 mm x 127 mm). Inside diameter not to exceed 1 mm. (Alternatively, reactions can be carried out using 96-well tissue culture plate, test tubes, or ringed slides.)
6. Mortar and pestle or motor-driven leaf crusher
7. Small vials or tubes for samples
8. Rotator (small electric motor with plexiglass arm for end-over-end rotation of micropipets <u>or</u> variable speed platform rotator for shaking tubes or plates)
9. Dissecting microscope (10-100x)

Procedure

1. <u>Latex conjugation</u>

   a. Use immunoglobulin G (IgG) fraction of serum in 0.1 M Tris, pH 7.4, and prepare several twofold dilutions (e.g., 1:50-1:1600). Note: 1:200-1:800 dilutions of IgG fraction have worked best for us with rabbit polyclonal antibodies to several different antigens. A microprecipitin titer $\geq$512 is desired.
   b. To each antibody dilution (5 ml), add an equal volume of latex beads diluted 1/15 (v:v) in 0.85% NaCl. (Gently shake latex beads to disperse before making dilutions. Latex beads should be stored at 4 C and should be usable indefinitely unless aggregated.)
   c. Mix and let stand at room temperature (22 C) for 1 hr. Shake several times.
   d. Centrifuge for 30 min at 4,340 g. The pellet will be very soft; pipet off supernatant carefully, but quickly.
   e. Stabilize the sensitized latex particles by washing the pellet twice with 0.02% PVP in 0.85% NaCl. Centrifuge at 4,340 g for 20 min and resuspend pellets in 5 ml PVP-NaCl.
   f. For antisera to viruses, resuspend the final pellet in 5 ml of 0.1 M Tris, pH 7.4, containing 0.02% sodium azide. The addition of 1.0% bovine serum albumin to the sensitized latex preparation was found to minimize or eliminate nonspecific agglutination in a bacterial assay.
   g. Filter preparations through Whatman filter paper #2. Use a Buchner funnel and gentle vacuum. (Vacuum, ca. 15 psi, is required to filter preparation.)
   h. Store in refrigerator (4 C). Do not freeze! Stability varies, but preparations usually can be used for 8-12 mo.

2. Antigen preparation and testing

a. Dilute plant samples 10x in 0.1 M Tris, pH 7.4, containing 0.85% NaCl or other suitable buffer. A tenfold dilution of the initial extract (100x final) also is prepared to enhance reactions when the antigen occurs in high concentrations.
b. Include appropriate positive and negative controls.
c. Draw 2 cm (10 μl) sensitized latex by capillary action into 100-lambda micropipets. Wipe the pipet tip with a disposable wipe and then draw another 2 cm (20 μl total) antigen into pipet.
d. Select at least two dilutions of sensitized latex and mix each with the two antigen dilutions to optimize reactant concentrations.
e. Attach pipets to arm of rotator and rotate end-over-end for 15 min at 7 rpm. Tape may be used to attach and label micropipets (micropipets must be clean or reactants will not mix thoroughly). Alternatively, droplets can be mixed in ceramic depression plates, microtiter plates, or test tubes and oscillated at 100-200x/min on a platform rotator for 20-30 min.
f. Score the reactions using a dissecting microscope with dark field illumination (10-100x) (Fig. 2).
g. Positive reactions are characterized by distinct clumping or agglutination and a clearing of the suspension background. The suspension remains uniformly milky in a negative test. A subjective scale (e.g., 0-5) may be used to score reactions; however, agglutination can be quantified by turbidimetric or nephelometric measurement.

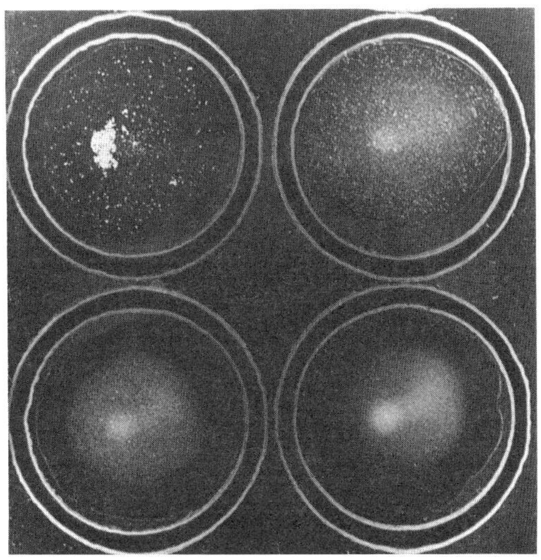

Fig. 2. Application of latex agglutination test to purified potato virus X. Pictured are readings of 4-upper left, 2-upper right, 1-lower left, negative-lower right of 60, 15, 8, and 0 mg/ml of antigen, respectively. Conditions: latex particles - Sigma LB 8 coated with 60 μg of potato virus X IgG; buffer for antigen dilutions 0.01 M neutral phosphate; 10 μl each reactant/ring; rotated 30 min, 100 rpm, room temperature and photographed.

Comments

1. Depending upon the virus being investigated, sap extracts from virus-infected plants have remained reactive for several months by adding 0.1% sodium azide and storing at 4 C. Bacterial suspensions (0.1-0.5 OD at 600 nm) in 0.85% NaCl with 0.1% sodium azide have also remained reactive for several months at 4 C.
2. Assay sensitivity is about 100 ng/ml for viruses and 10^7 cfu/ml for bacteria. Sensitivity of bacterial systems can be enhanced significantly by solubilizing reactive antigen.
3. Some plant species, especially those containing latex, cannot be tested directly. Some method of sample clarification is required. In some cases, addition of bovine serum albumin (1-3%) or other blocking agents will minimize or eliminate background problems.
4. In cases where greater stability of the sensitized latex is required, covalent linkage of the antibody to the latex particle may be desired. There is an extensive array of commercially prepared latex products or kits by which investigators may perform their own linkage (e.g., Polysciences). "Derivatized" polystyrene latex is available as either carboxyl or amino reactive microparticles. Carbodiimide or glutaraldehyde protocols are usually followed for linkage. Latex particles with spacer arms to move the ligand away from the latex particle surface and colored or fluorescent latex particles can also be purchased. The latex particles can be purchased in a range of sizes, about 50 nm to 1 mm diameter, depending on need and ingenuity of the investigator. Perhaps the newest variation is superparamagnetic particles (i.e., particles that respond to weak magnetic fields, but do not themselves become magnetized) (Advanced Magnetics, Polysciences, Seradyn). Note: Adsorption of antibodies produced in chickens has not worked in our limited trials, whereas covalent linkage to superparamagnetic particles with glutaraldehyde has worked with the same antibody preparation.
5. In some cases, the antibody titer may be too low to be usable for latex agglutination. The use of Protein A (PALLAS technique) has been shown to enhance low-titered antiserum to a usable level (11). Assay sensitivity, however, is usually not enhanced. Note: Protein A will not bind chicken IgG.

Applications

This is a rapid, sensitive assay that can be used for serodiagnosis or for monitoring purification or separation steps. Generally, the assay is about 10x more sensitive than comparable radial diffusion assays and 0.1x as sensitive as ELISA. In the medical field, an array of commercial kits is available to test for viruses, bacteria and bacterial products, drugs, hormones, etc. (2). Plant pathologists have primarily used latex tests for studies on plant viruses (3,4,7,9,10), but test utility has been demonstrated for procaryotes (8,14). In the latter systems, test sensitivity is about 10^7 cfu/ml, but soluble antigen (extracellular polysaccharide) significantly enhances actual detection of *Corynebacterium sepedonicum*. Latex tests also have been used to detect plant pathogens in insect vectors (8,10).

Tanned Red Blood Cell Agglutination

This test is a passive hemagglutination procedure since receptors for plant viruses are not present on blood cell surfaces (1,6,12,13). Blood cells, usually from sheep, are treated with a dilute solution of tannic acid (0.0005% final concentration) thereby changing the surface structure so they will adsorb protein. Antigen or IgG can be adsorbed to the cells forming a "sensitized" product which readily agglutinates when mixed with the homologous counterpart. Endpoint titrations are in the microgram range.

Other Adsorption - Bentonite Flocculation

Purified IgG or antigens can be adsorbed to bentonite, a finely ground earth product widely used in virus purification (3,5). Coated bentonite particles combine with the second reactant to produce a flocculation easily read under 100x magnification. The addition of methylene blue dye aids in reading the degree of flocculation, but is not required for a good test. The test is sensitive giving dilution endpoints of less than a microgram: tobacco mosaic virus 0.02-0.05 μg/ml, southern bean mosaic virus 0.4 μg/ml, wheat streak mosaic virus 0.10-0.20 μg/ml.

Literature Cited

1. Abu Salih, H. S., Murant, A. F., and Daft, M. J. 1968. Comparison of the passive hemagglutination and bentonite flocculation tests for serological work with plant viruses. J. Gen. Virol. 23:155-166.
2. Bangs, L. B. 1987. Uniform latex particles. Biotechnology Laboratory 5(3):10-16.
3. Bercks, R. 1967. Methodische untersuchungen uber den serologischen Nachweis pflanzenpathogener viren mit dem bentonit-flockungtest, dem latex-test und dem bariumsulfat-test. Phytopath. Z. 58:1-17.
4. Bercks, R., and Querfurth, G. 1969. Weitere methodische untersuchungen Uber den latex test zum serologischen nachweis pflanzenpathogener viren. Phytopath. Z. 65:243-256.
5. Bozicevich, J., Scott, H. A., and Vincent, M. M. 1963. The bentonite flocculation test for detection of plant viruses and titration of antibody. Proc. Soc. Exp. Biol. Med. 114:794-798.
6. Brown, R. A., and Ellis, J. 1970. The use of red blood stroma as an immunoabsorbent and analytical tool. I. Studies with immune sera. Immunochemistry 6:63-79.
7. Demski, J. W., Bays, D. C., and Khan, M. A. 1986. Simple latex agglutination test for detecting flexuous rod-shaped viruses in forage legumes. Plant Dis. 70:777-779.
8. Fletcher, J., and Slack, S. A. 1986. Latex agglutination as a rapid detection assay for _Spiroplasma citri_. Plant Dis. 70:754-756.
9. Khan, M. A., and Slack, S. A. 1978. Studies on the sensitivity of a latex agglutination test for the serological detection of potato virus S and potato virus X in Wisconsin. Am. Potato J. 55:627-637.
10. Omura, T., Hibino, H., Usugi, T., Inoue, H., Morinka, T., Tsurumachi, S., Ong, C. A., Putta, M., Tsuchizaki, T., and Saito, Y. 1984. Detection of rice viruses in plants and individual insect vectors by latex flocculation test. Plant Dis. 68:374-378.
11. Querfurth, G., and Paul, H. L. 1979. Protein-A coated latex-linked antisera (PALLAS): New reagents for a sensitive test permitting the use of antisera unsuitable for the latex test. Phytopath. Z. 94:282-285.
12. Rajeshwari, R., Reddy, D. V. R., and Iizuka, N. 1981. Improvement in the passive hemagglutination technique for serological detection of plant viruses. Phytopathology 71:306-308.

13. Saito, Y., and Iwata, Y. 1964. Hemagglutination test for titration of plant virus. Virology 22:426-428.
14. Slack, S. A., Sanford, H. A., and Manzer, F. E. 1979. The latex agglutination test as a rapid serological assay for Corynebacterium sepedonicum. Am. Potato J. 56:441-446.

Section IV.I.

SEROLOGICALLY SPECIFIC ELECTRON MICROSCOPY (SSEM)

K. S. Derrick
Citrus Research and Education Center
Lake Alfred, Florida 33850

Principle

The leaf-dip technique is a procedure for the detection of virus particles where a freshly cut leaf surface is dipped into a drop of water which is subsequently dried on a filmed electron microscope grid, which is then stained and examined by transmission electron microscopy (TEM) (2,3). An extension and improvement of the technique was addition of the virus-containing drop to appropriately diluted antiserum already on the grid. This causes the virus particles to be clumped together and coated with antibody molecules. This technique, named leaf-dip serology, gives information on both the morphological and serological properties of the virus in one easy experiment (1).

Leaf-dip serology was modified by adsorbing specific antibodies to filmed electron microscope grids rather than mixing it with the virus-containing extract. When serum coated grids are floated on extracts, virus particles are specifically attached to the grids while cellular debris, salts, etc. are removed by washing prior to staining. This technique was named serologically specific electron microscopy (SSEM) (4) to distinguish it from leaf-dip procedures. The principles of SSEM are similar to other methods, e.g. ELISA and RIA, that involve adsorption of antiserum to plastic surfaces, except that the virus particles are detected with an electron microscope instead of using antibodies linked to enzymes or radioisotopes.

Materials and Equipment

1. Electron microscope grids.
2. Fine tip forceps, regular and anticapillary.
3. Formvar (0.5% in ethylene dichloride).
4. Unfractionated antisera diluted 1/200 with 0.05 M Tris:HCl, pH 7.2 (Tris buffer).
5. Mortars and pestles, 1.5 ml microcentrifuge tubes, plastic Petri plates, filter paper.
6. Extraction buffer, 0.05 M Tris:HCl, pH 8.4, containing 0.4 M sucrose and 0.15 M NaCl (TSS).
7. Protein A (0.02 mg/ml in Tris buffer).
8. Uranyl acetate, 2.0% in distilled water, prepare just before use.
9. Microcentrifuge.
10. Vacuum evaporator, equipped for carbon and metal evaporation.
11. Electron microscope.

Procedures

1. Grids. Spread electron microscope grids, as received from the supplier, on filter paper, treat with a solution of 0.5% Formvar in ethylene dichloride and air dry.
2. Preparation of filmed grids. Place one drop of 0.5% Formvar in ethylene dichloride on water in a plastic Petri dish. Allow the "cast" film to dry for about five minutes. View with indirect light, and place grids, shiny side down, on the surface of the film. Tap each grid with a toothpick, and float a piece of Parafilm over the filmed grids. Lift the Parafilm from the water (the filmed grids should stick to the Parafilm) and air dry. Deposit a visible layer of carbon on the filmed grids using a vacuum evaporator. This can be observed in the bell jar of the evaporator by turning up a corner of the Parafilm. Deposit carbon until a visible "shadow" of the corner is observed on filter paper placed under the Parafilm.
3. Adsorption of sera to grids. Float filmed grids, shiny side down, on small drops of diluted sera in plastic Petri dishes for about 30 min (1 grid/drop). Wash grids to remove unadsorbed serum proteins by touching to a series of about 10 drops of Tris buffer. When Protein A is used, float grids on a solution of 0.02 mg/ml of Protein A in Tris buffer for 30 min, wash in a series of 10 drops of Tris buffer and then place on drops of antiserum.
4. Attachment of virus particles to grids. Float serum coated grids on drops of extracts in plastic Petri dishes for 1 to 4 hr. Put a piece of moistened filter paper in the dish and close the lid to prevent evaporation of the extracts. Reactions are usually done at room temperature, but unstable viruses can be assayed at 4 C. Stop the reactions by touching the grids to a series of 3 drops of TSS. Wash the grids in a firm stream of distilled water for a minimum of 20 sec.
5. Staining. Float the washed grids on drops of freshly prepared 2% aqueous uranyl acetate for about 10 min. Remove the stain by blotting dry for a positive stain or partially blot, leaving the grid slightly wet, for a negative stain.
6. Assays for double-stranded RNA. By using antisera specific for double-stranded RNA (dsRNA), SSEM can be used to assay tissue extracts for dsRNA (6). The procedure is similar to that used for virus particles up to the staining step. After the final wash, prior to staining, dip dsRNA grids in 95% ethyl alcohol for 30 sec, rinse in distilled water, and place on drops of 0.1 mg/ml of cytochrome C in water for 15 min. Wash grids on 3 drops of distilled water and place on drops of 0.01 phosphate, pH 7.2, for 15 min. Hold the grids in a stream of drops of 1 nM aqueous uranyl acetate delivered from a syringe equipped with a 0.2 μm filter. Touch the grids to a drop of distilled water and blot dry.

Comments

1. Copper, 400 mesh, tabbed grids are recommended for SSEM assays, but grids with larger openings and without tabs can be used. Anticapillary forceps are used to transfer grids between drops of antiserum or virus extracts.
2. Antisera can be used at extremely high dilutions (5), but a 1 to 200 dilution works well (50 μl, or 1 drop from a glass pipet, in 10 ml of 0.05 M Tris:HCl, pH 7.2). This rather concentrated solution has worked well and will withstand repeated freeze-thawing without loss of serological activity. Extremely high dilutions of antiserum may leave unblocked areas

on the grid surface resulting in nonspecific attachment of virus particles.

3. Treating grids with Protein A before applying antiserum has been shown to increase the sensitivity of the technique (9). When an antiserum has a sufficient titer, the use of Protein A may have little effect on the number of particles attached. But with a low titer serum, the use of Protein A can make the difference between no virus particles attached and a number sufficient for a positive assay. The use of Protein A is recommended when a low titer serum must be used and in testing trial bleedings.
4. A variety of extraction buffers have been used for SSEM. Use TSS and grind 1 weight of tissue with two volumes of extraction buffer with a mortar and pestle. The extract is clarified in a microcentrifuge prior to assay. Depending upon the titer of the virus being assayed, extracts can be prepared using larger volumes of buffer, in which case the microcentrifugation step can be omitted. If there is some question as to which extraction buffer should be used for a particularly unstable virus, buffers that have been used in purification of the virus may also be used for SSEM.
5. DsRNA should be readily detected at a magnification of about 10,000 when viewed with a well-aligned, high resolution electron microscope. Depending on the performance of the microscope, it may be necessary to rotary shadow the dsRNA with metal prior to examination.
6. SSEM assays can be completed in a few hours, but there is a limit to the number of grids that can be processed. Laboratories assaying a small number of samples of several different viruses are more likely to favor SSEM. Conversely, when a large number of assays of a single virus are necessary, ELISA or RIA are usually preferred. Testing a given antiserum by SSEM to establish that it reacts with virus particles of the correct morphology before using it for other serologically procedures is desirable. In addition, antisera that react strongly with host antigens can be used for SSEM assays. The host antigens, if they are detected at all, are readily distinguished from virus particles.
7. To use SSEM as a routine diagnostic tool, it is necessary to be able to consistently prepare useable filmed grids. Floating films off glass slides is the commonly used procedure for producing high quality, uniform thickness films. There are numerous difficulties with this technique, and different laboratories offer various suggestions for success (4,8). While cast films are not as strong as those produced on glass slides, they are much easier to prepare and are recommended for routine SSEM. Some workers have suggested preparing filmed grids and storing them with and without antiserum adsorbed to be used at a later date. Since the procedure for casting films is easy, quick and reliable, it is recommended that the grids be prepared and used the same day.

Applications

Some typical results of analysis of citrus infected with citrus tristeza virus (CTV) and St. Augustinegrass infected with St. Augustinegrass decline virus (SADV) are shown in Fig. 1-3. In each case, assays were made using crude extracts prepared with one weight of leaf tissue and two volumes of extracting buffer. The CTV grids were given a positive stain with 2% uranyl acetate in water. A similar uranyl acetate solution was used to give the SADV particles a negative stain. The dsRNA molecules were stained as described herein.

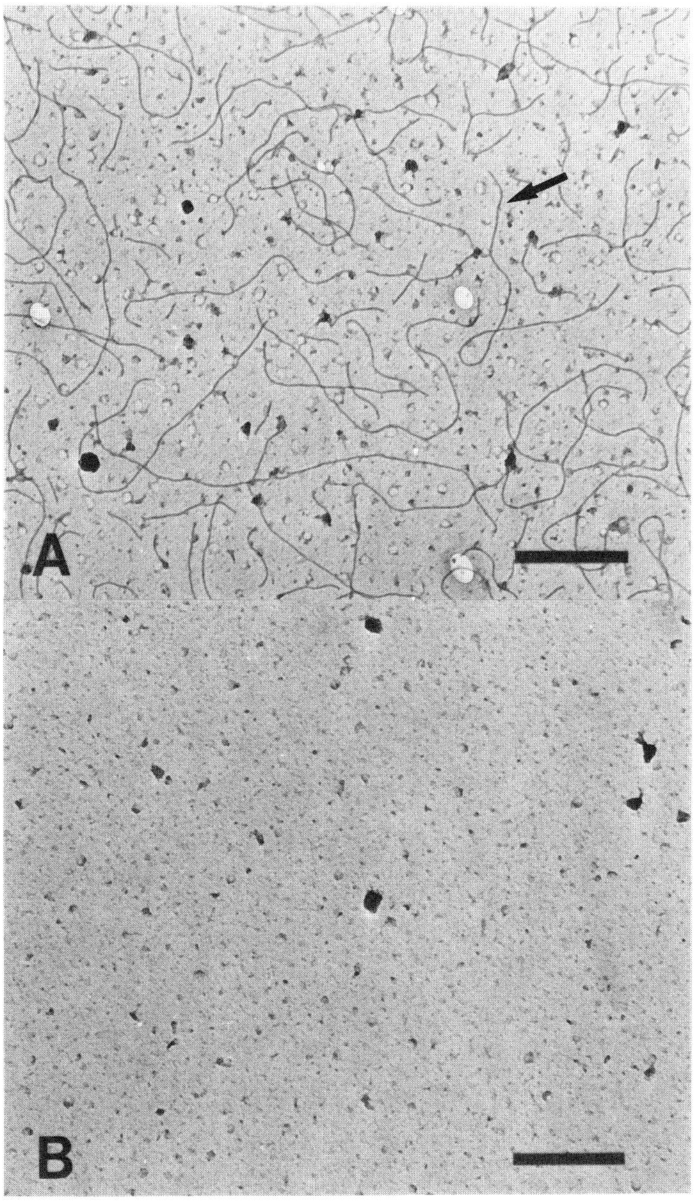

Fig. 1. SSEM assays of CTV-infected citrus on grids prepared using CTV antiserum (A) and normal serum (B). Scale bars represent 500 nm. Electron micrographs taken by Beverly Hewitt. Intact CTV particle, arrow.

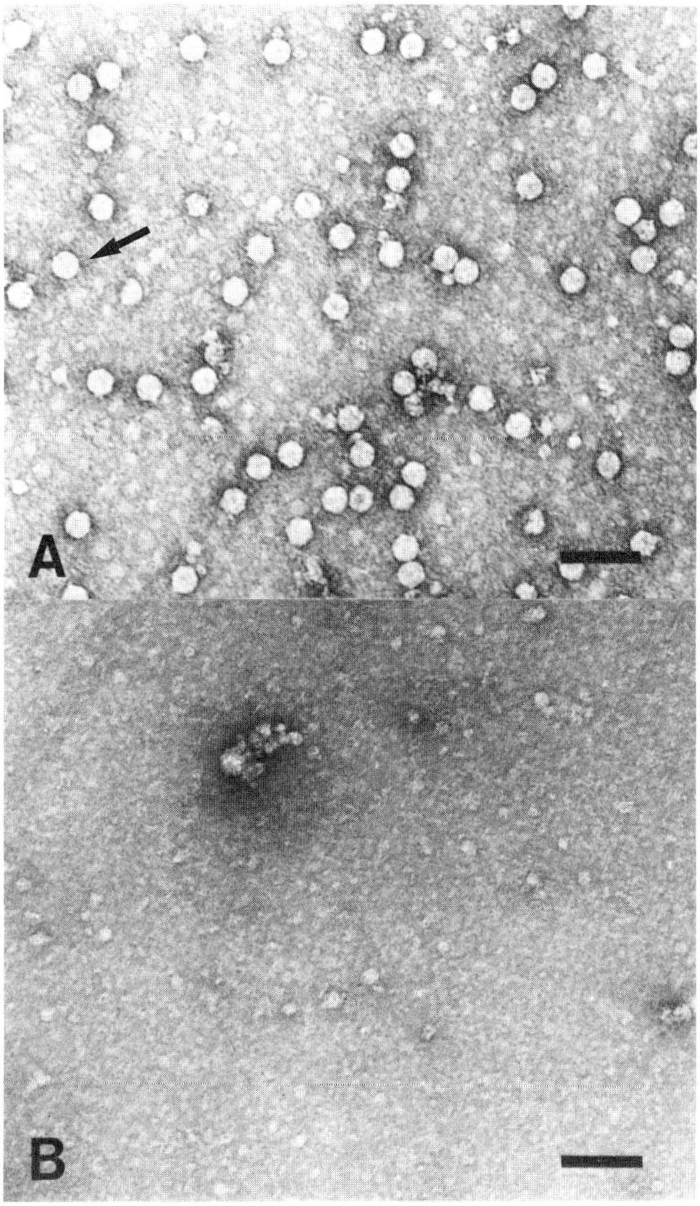

Fig. 2. SSEM assays of SADV-infected St. Augustinegrass on grids prepared using SADV antiserum (A) and normal serum (B). SADV particle, arrow. Scale bars represent 100 nm. Electron micrographs taken by Zoe Liu and G. E. Holcomb.

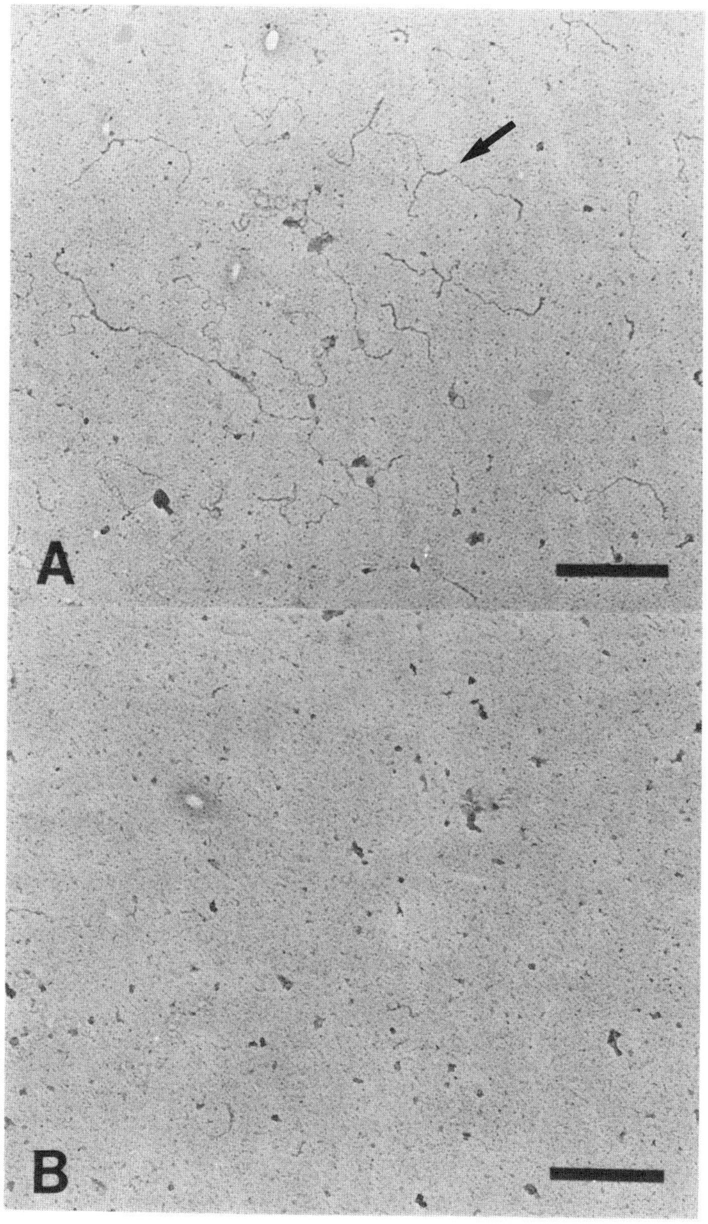

Fig. 3. SSEM assays of SADV-infected (A) and healthy (B) ST. Augustinegrass on grids prepared using dsRNA antiserum. SADV dsRNA, arrow. Scale bars represent 500 nm.

Leaf-dip serology and SSEM have been extensively used by plant virologists, and there are numerous reviews and descriptions of the techniques. Both techniques will obviously work using a wide variety of buffers, reaction times, staining methods, etc. SSEM was recently used to determine the particle morphology of citrus ringspot virus (CRSV) (7). An antiserum to a partially purified preparation of CRSV was used to detect a mixture of short and long, extremely flexible, filamentous particles in extracts of infected tissue.

Literature Cited

1. Ball, E. M., and Brakke, M. K. 1968. Leaf-dip serology for electron microscopic identification of plant viruses. Virology 36:152-155.
2. Brandes, J. 1957. Eine elektronenmikrospopische Schnelle-Methode zum Nachweis faden-und stabchenformiger Viren, inbesondere in Kartoffeldunkelkeimen. Nachrichtembl. Dtsche. Planzenschutsdientes (Braunschweig) 9:151-152.
3. Christie, S. R., Purcifull, D. E., Crawford, W. E., and Ahmend, N. A. 1987. Electron microscopy of negatively stained clarified viral concentrates obtained from small tissue samples with appendices on negative staining techniques. Bulletin 872. Institute of Food and Agricultural Sciences, University of Florida, Gainesville. 45 pp.
4. Derrick, K. S. 1973. Quantitative assay for plant viruses using serologically specific electron microscopy. Virology 56:652-653.
5. Derrick, K. S., and Brlansky, R. H. 1976. Assay for viruses and mycoplasmas using serologically specific electron microscopy. Phytopathology 66:815-820.
6. Derrick, K. S., French, R. C., Clark, C. A., and Gabriel, C. J. 1984. Detection of double-stranded RNA by serologically specific electron microscopy. J. Virol. Methods 9:293-299.
7. Derrick, K. S., Brlansky, R. H., da Graca, J. V., Lee, R. F., Timmer, L. W., and Nguyen, T. K. 1988. Partial characterization of a virus associated with citrus ringspot. Phytopathology 78:1298-1301.
8. Pease, D. C. 1964. Histological Techniques for Electron Microscopy, 2nd edition. Academic Press, New York. 381 pp.
9. Shukla, D. D., and Gough, K. H. 1979. The use of protein A from *Staphylococcus aureus* in immune electron microscopy for detecting plant virus particles. J. Gen. Virol. 45:533-536.

Section IV.J.

COLLOIDAL GOLD, ITS GENERATION AND USE IN IMMUNO-ELECTRON MICROSCOPY

W. G. Langenberg
USDA ARS and University of Nebraska
Lincoln, NE 68583

Principle

The colloidal gold labeling technique for transmission and scanning electron microscopy is simple, rapid and relatively inexpensive when reagents are prepared in one's own laboratory. Although lectins, antibodies, Protein A, biotin, or avidin labeled with colloidal gold of various sizes and of excellent quality are available from a number of suppliers, those who intend to use gold labeling regularly in their research may find it advantageous to prepare their own gold-labeled reagents. Successful results can be obtained without much practice when some of the precautions listed below are taken into consideration.

Colloidal gold particles are highly negatively charged. Proteins, such as IgG, bind to colloidal gold particles by electrostatic charges only. It is, therefore, important to know the isoelectric point (pI) of proteins to be labeled. Best binding occurs just above the pI where the protein has a slightly negative charge but less so than colloidal gold particles.

Materials

1. Gold chloride or chloroauric acid ($HAuCl_4$, $AuCl_3$). Other gold formulations such as $AuBr_3$, AuI_3, or mono or divalent halide salts of Au cannot be used as starting material for colloidal gold. Although pure gold content of $AuCl_3$ is often listed and may vary from 49% to 63%, this seems to have no practical significance in the production of colloidal gold. The source of colloidal gold is important. Some suppliers (Fisher, Fluka AG) have consistently had gold chloride from which colloidal gold could be made successfully, which was less true for others. Gold chloride must be kept in a tightly stoppered brown bottle over $CaCl_2$ since it is very hygroscopic and light sensitive. Weigh out required amounts rapidly in subdued light.
2. Sodium citrate. Highest quality, analytical grade. 4% sterile solution.
3. Potassium carbonate. Highest quality, analytical grade.
4. Saturated Uranyl acetate solution.
5. Immunoglobulin G (IgG). IgG is poorly soluble to insoluble in distilled water. Once denatured it cannot be renatured. Goat anti-rabbit (or -mouse or -rat) IgG is purchased freeze dried, desalted and fluorescein-labeled (U. S. Biochemicals). Fluorescein label increases water solubility and probably retards denaturation in triple distilled water (3xDW) sufficiently for the time required to complete the labeling process. When salt (> 1 mM) is present, desalt IgG preparations before labeling with colloidal gold.
6. K_2HPO_4 - citric acid buffer, pH 7.2-7.4, 1 M stock solution.
7. Polyvinylpyrrolidone (PVP), 10,000 M_r.

8. Polyethylene glycol. 20,000 M_r, 1% stock solution in 2x distilled water and 0.05% sodium azide.
9. Sodium azide. 10% solution.
10. Normal goat serum.
11. Pasteur pipets.
12. Test tubes. 10 x 75 mm.
13. 500 ml distilling flask, with stoppered side arm, acid cleaned. Do not use detergents for cleaning. Triple distilled water (3xDW) final rinse. Drip-dry in dust free surrounding.
14. Condenser. Water cooled jacket, cleaned as for the flask.
15. Triple, glass distilled water.
16. Magnetic stirring bars. Assorted sizes, acid cleaned, 3xDW final rinse.
17. NaCl. 10% solution in double glass distilled water, store at room temperature, stable.

A. Procedure to generate colloidal gold of plus/minus 20 nm

1. Rapidly weigh out 20 mg $AuCl_3$ on weighing paper and rinse into boiling flask with a small amount of 3xDW. Bring volume to 240-250 ml with 3xDW (~ .01% $AuCl_3$). The gold chloride concentration cannot be increased much beyond 0.01% without risking spontaneous flocculation and precipitation of colloidal gold as it forms.
2. Bring the clear pale yellow solution to a rolling boil and reflux with side arm closed.
3. Add nucleation promotor: plus/minus 1 mg uranyl acetate (=20 μl of a fresh, saturated UAc solution).
4. Inject 2 ml sodium citrate (from a sterile 4% stock solution).
5. Reflux for 1/2 hr, solution should turn from clear pale yellow to colorless, to pink, to light purple, to wine red, and finally orange red. When arrested at any other color, or when turning brownish purple, the gold source, the 3xDW or the cleanliness of the distilling flask should be suspected as the reason for failure to generate colloidal gold.
6. Cool the colloidal gold solution to room temperature (may be done in an ice bath).

B. Labeling of goat anti-rabbit IgG with colloidal gold

1. Dissolve approximately 0.2-.3 g of K_2CO_3 in 40-50 ml 3xDW.
2. Add 0.3 ml of 1% PEG 20,000 to 10 ml colloidal gold solution in a clean 30 ml beaker and stir.
3. Insert pH electrodes (pH should read from 5.5-6.0) and raise the pH with 20 μl drops of the fresh K_2CO_3 solution to pH 7.4-7.6. Count the number of drops required. Note: failure to add PEG to the gold solution may irreversibly clog the electrode with precipitated gold.
4. Discard the 10 ml gold and rinse the container with 3xDW. To another 10 ml colloidal gold solution add the known number of 20 μl drops of the K_2CO_3 stock required to bring the pH to 7.4-7.6 (do not measure with the pH meter this time): stock colloidal gold.
5. Add 1 ml stock colloidal gold to each of five 10 x 75 mm tubes
6. Add 3 ml of 3xDW to approximately 2 mg dry fluorescein-labeled goat anti-rabbit IgG and mix in a vortex to dissolve.
7. To one tube containing 1 ml of stock colloidal gold add 5 μl goat IgG, mix in a vortex, wait 30 sec, then add 200 μl 10% NaCl. Development of

a blue color (flocculated gold) indicates that not enough IgG was present to stabilize the gold, i.e., more IgG is required. No color change (orange-red) indicates an excess of IgG. An intermediate purple color indicates saturation of gold with IgG. The ideal preparation contains a 10% excess of IgG, indicated by one increment of IgG beyond that necessary to cause development of purple. Excess IgG is removed in subsequent steps.

8. Measure the number of ml of colloidal gold in a clean 250 ml graduate cylinder (there should be ~ 230 ml), decant in a very clean Erlenmeier flask or beaker and stir. Add the required number of drops of K_2CO_3 previously determined for 10 ml (step 3) to raise the pH of the 230 ml to 7.4-7.6 (do not measure with pH meter).
9. Calculate the amount of IgG to be added to the colloidal gold solution (= number of μl/ml x number ml colloidal gold left).
10. Inject the total amount of IgG calculated into the rapidly stirring, but not aerating, colloidal gold solution. Wait 1 min. Discard any excess IgG.
11. Centrifuge the labeled colloidal gold in 50 ml polyethylene tubes in an HB4 swinging bucket rotor in a Sorvall RC2-B refrigerated low speed centrifuge at 10,000 RPM for 45 min (or calculate the equivalent time and RPM for other swinging bucket type rotors and centrifuges).
12. Carefully aspirate (save!) the red supernatant fluid to just above the liquid red-black pellet (save) and subject the supernatant fluid to a second centrifugation, 12,000 RPM for 1 hr.
13. Dissolve the liquid pellets of the 1st and 2nd centrifugation in 5-10 ml protection buffer (0.05 M phosphate-citrate buffer, 0.05 M tris-HCl, pH 7.5, containing 40 mg/ml polyvinylpyrrolidone, 0.2 mg/ml PEG, 0.5 mg/ml NaN_3 (1,12) and centrifuge 10,000 RPM 10 min to pellet large aggregates. Store at 4 C in the centrifugation vial without removing the pellet.

C. Use of colloidal gold-labeled IgG in an indirect labeling procedure

Labeling procedure. The following procedure is based on labeling of ultra-thin sections of L.R. White embedded plant tissue. All reactions are carried out in a petri dish with the bottom covered by clean Parafilm.

1. Float grids with silver to gold sections, supported on carbon-backed formvar membranes, section side down on drops of 3-4% normal goat serum in 0.1 M phosphate-citrate buffer (pH 7.2-7.4) containing 0.05% NaN_3 for 15 min, drain by touching to a filter paper strip.
2. Float grids on a 1/200-1/400 dilution of rabbit primary anti-serum in 1% normal goat serum in 0.1 M phosphate-citrate buffer, containing 0.05% NaN_3 for 15 min.
3. Wash grids on 6 drops of 1% normal goat serum in phosphate-citrate buffer by flotation for several seconds and drain by touching to a filter paper strip.
4. Float grids, section side down, on drops of goat anti-rabbit IgG labeled with colloidal gold for 15 min.
5. Wash on 6 drops of 1% normal goat serum in phosphate-citrate buffer for several seconds each.
6. Wash on 3 drops of 3xDW and place on drops of saturated uranyl acetate for 15 min to 30 min.
7. Wash on 3 drops of freshly boiled and cooled 3xDW.

8. Contrast for 1 min on drops of lead citrate.
9. Wash on 3 drops of distilled water, drain, and view in the electron microscope when convenient.

Comments

1. Do not save left over K_2CO_3 solutions but prepare a small fresh batch each time just before use.
2. Do not subject labeled colloidal gold to repeated centrifugation to remove larger aggregates since this will accelerate the inevitable gradual aggregation and settling out of labeled colloidal gold. Labeled solutions have varied in usefulness after preparation from 8 weeks to 4 years.
3. As nucleation promotors, almost any heavy metal salt (Cl^-) will suffice at 1 mg/20 mg gold chloride. However, the use of nucleation promotors is not necessary with a good source of gold chloride.
4. Colloidal gold can be made in a range of sizes from 16 nm to 147 nm by varying the amount of Na-citrate added (7). For smaller gold particles, tannic acid is added just before the sodium citrate solution injection in amounts inversely related to the desired particle size. Experimentally determine the amount required, starting at approximately 0.45 mg tannic acid/mg $AuCl_3$ for gold of 5-6 nm (20). Ether, saturated with white phosphorus, was used initially to make smaller colloidal gold particles to label IgG (6). The tannic acid method is much less hazardous for laboratory use. Probably any reducing agent can be used in amounts determined experimentally for the gold size desired. For example also Na-borohydride has been used (17).
5. The IgG molecule is about 4 by 5 by 8 nm. Gold colloid exceeding 15 nm diameter can be stabilized with Protein A; however, the Protein A loses most of its reactivity after conjugation with gold of this size or larger.
6. Except for special requirements, use of colloidal gold less than 15 nm diameter is not recommended for immunolabeling because very high EM magnification is thereby required to resolve particles and the spatial relation of gold label to cell organelles is lost. Journal space for EM micrographs is mostly limited to about 15 cm width and full page micrographs are usually discouraged. In Fig. 1, the location of wheat germ agglutinin (WGA) is shown in sorghum root tip cells. If gold of 5 nm had been used to label goat antibody to rabbit anti-WGA IgG, the magnification of this micrograph would have required a full page for the label to be visible to the unaided eye. Instead, gold of approximately 20 nm was used. Larger gold particles may obscure detail and do not improve resolution of the site or locality of the antigen to be labeled.

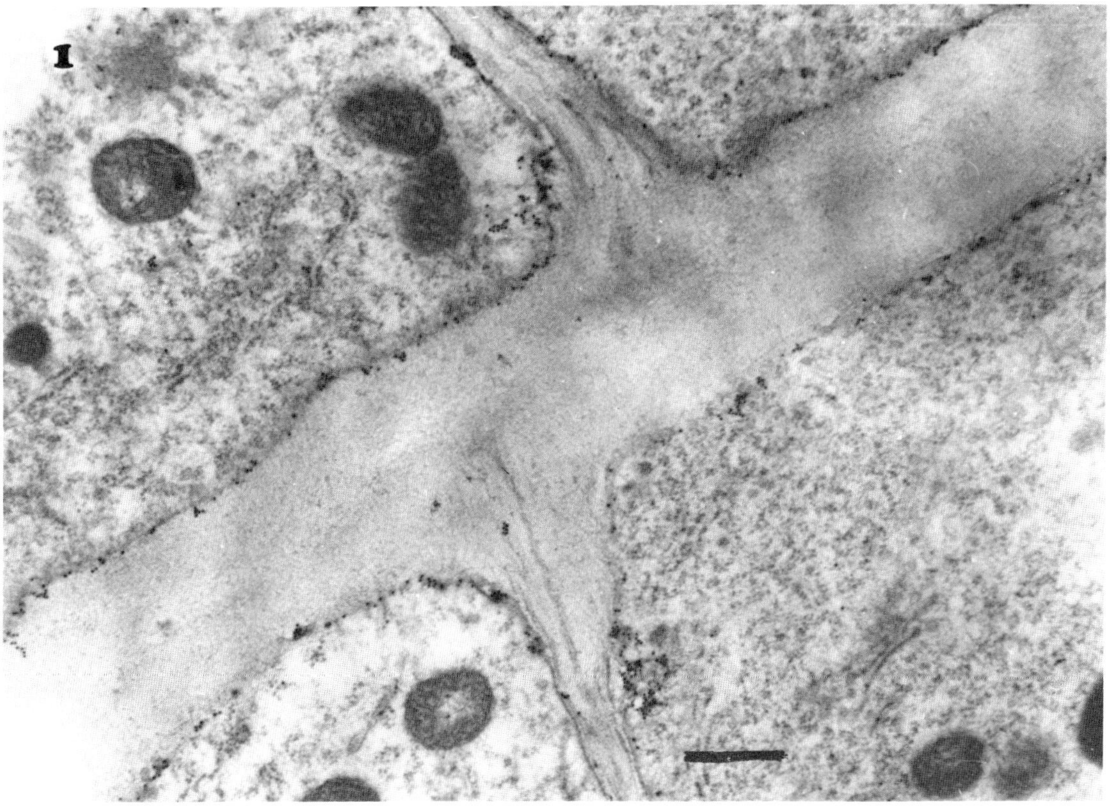

Fig. 1. Sorghum root tip cells labeled with rabbit antibody to wheat germ agglutinin as the primary label. Rabbit antibody bound to the section was detected with goat anti-rabbit IgG bound to colloidal gold. Wheat germ agglutinin is located along or in the plasmalemma and is not distributed within the cell walls.

7. Colloidal gold labeled with IgG can be freeze dried and reconstituted with some loss of activity. Freeze drying is not necessary, though, when sodium azide has been added to the protection buffer and repeated low speed centrifugation is avoided.
8. Proteins can only be attached to colloidal gold at a pH slightly above their isoelectric point when the protein has a slight negative charge. If the pI is unknown, this has to be determined before labeling is attempted. It has not been possible to label colloidal gold directly with highly positively or negatively charged compounds. New techniques to do so await development.
9. Ultraviolet (UV) polymerized acrylic-based resins are preferable to epoxy-based resins for the localization of substances at the EM level. Acrylic-based resins can be polymerized in the cold with resulting lower reduction in antigenicity than with epoxy-based resins. Thin sections of acrylic-based resins do not have to be etched with alcoholic NaOH like epoxy-based resins to enhance label but can be labeled directly after sectioning. Since acrylic-based resins are UV polymerized, OsO_4 should not be used as a post-fixative. The user must get used to the negative image of the membrane system in cells.
10. Additional background information, including a list of isoelectric points for some commonly used lectins, proteins or enzymes, (5,11) may be helpful.

Applications

The low speed pellet of the first centrifugation (gold labeling procedure, steps 11, 12) is useful for:

1. Immunolabeling of Western blots according to Hsu (16), or
2. Enhanced rapid identification of viruses in leaf dips after antibody decoration (18,21). For example, Fig. 2 shows a leaf-dip of a plant doubly infected by wheat streak mosaic virus (WSMV) and barley stripe mosaic virus (BSMV). The grid has been labeled with antiserum to BSMV with the procedure of Lin (18). Only BSMV labels positively; WSMV virions are not labeled. Lin's procedure can rapidly (~ 10 min) identify viruses in crude sap dips by electron microscopy.
3. Light microscope immunofluorescent techniques when the goat IgG to rabbit IgG is labeled with fluorescein.
4. Silver enhancement of gold label in sections (0.5-2.0 μ) for light microscopy according to Danscher (4) and Holgate et al. (10). Silver ions are enzymatically deposited on gold label on the section. Colloidal gold acts as a catalyst in the reaction.

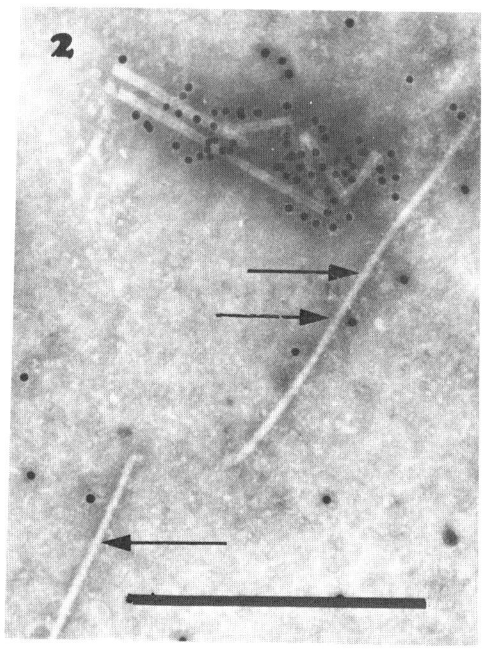

Fig. 2. Rabbit antibody to BSMV labels BSMV only in a leaf-dip from wheat tissue known to be infected with BSMV and WSMV. Arrows point to WSMV virions. Gold labeled goat anti-rabbit IgG clearly shows the presence of specific rabbit anti-BSMV antibody on BSMV virions although the rabbit IgG attachment to BSMV virions is not resolved. The leaf-dip was prepared with the procedures of Lin (18).

The procedures for making colloidal gold have been known for many years. However, gold colloid had not been used as a label for immunological purposes before Faulk and Taylor (6) used it to label bacterial cells before embedding and sectioning. Recently, the procedure has been widely applied to virtually any protein to which antibody can be made by locating that protein precisely at the EM level. An example is given in Fig. 3 for localization of wheat streak mosaic virus in wheat root tip cells. Virions themselves and their entry into protoplasts were traced by immunogold label (22). Also virus encoded non-structural proteins (3), lectins (8,9) and double-stranded viral RNA (19) are pinpointed precisely in ultrathin sections. Fig. 4 shows the location of wheat germ agglutinin in sorghum root tip cells. By labeling colloidal gold with enzymes, their substrate can be localized in the cell (2), or certain sugars in membranes by attaching lectins to gold (13,15).

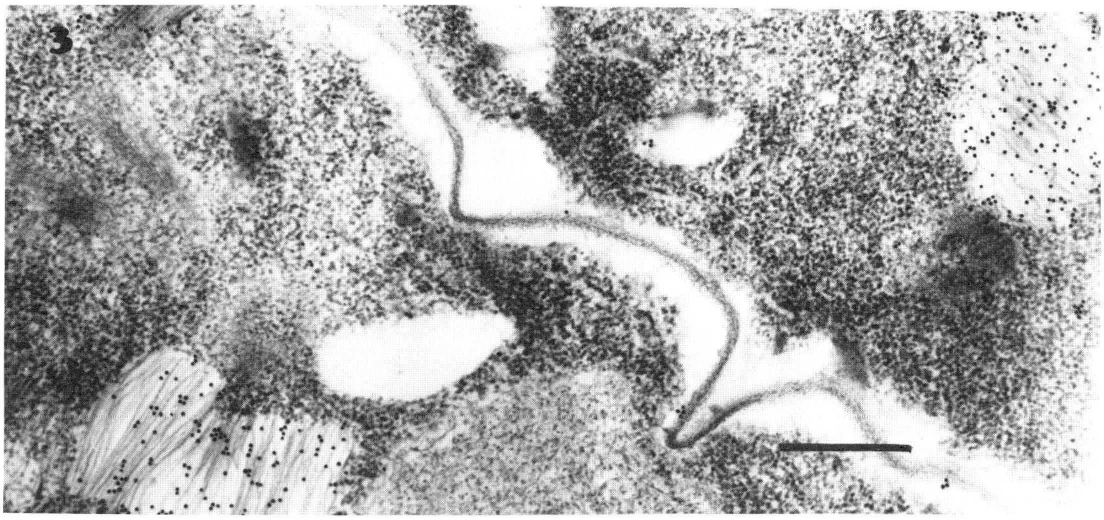

Fig. 3. Section of a wheat root tip infected with WSMV, reacted with rabbit antibody to WSMV and then with gold-labeled goat anti-rabbit IgG. Aggregates of WSMV virions are labeled. The rest of the cells show very little background label.

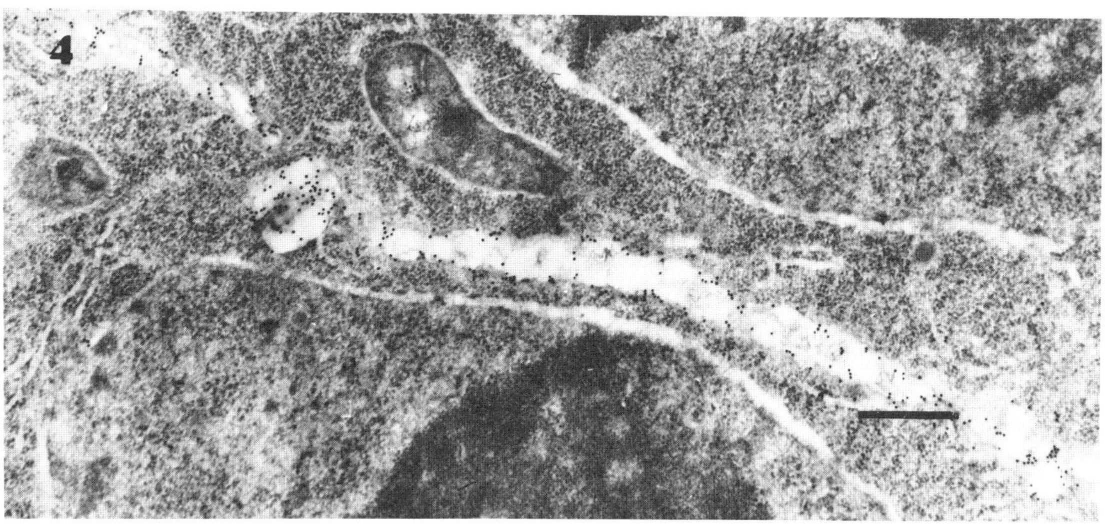

Fig. 4. Sorghum root tip section labeled with rabbit antibody to wheat germ agglutinin and then colloidal gold labeled goat anti-rabbit antibody. Wheat germ agglutinin is present during the cell wall deposition stages along the newly forming cell wall only, not being present elsewhere in the cell at concentrations sufficient to be detected with this method.

Double labeling for the simultaneous detection of two targets in the same cell has been achieved with labeled colloidal gold of different sizes (14). With an increase in the use of colloidal gold as an electron-dense marker, the number of applications and methods will undoubtedly increase exponentially.

Literature Cited

1. Ackerman, G. A., Walken, K. W., and Gelder, F. G. 1980. Surface distribution of monoganglioside G_{M1} on human blood cells and the effect of exogenous G_{M1} and neuraminidase on cholera toxin surface labeling. A quantitative immunocytochemical study. J. Histochem. Cytochem. 28:1100-1112.
2. Bendayan, M. 1981. Ultrastructural localization of nucleic acids by the use of enzyme-gold complexes. J. Histochem. Cytochem. 29:531-541.
3. Brakke, M. K., Ball, E. M., Hsu, Y. H., and Langenberg, W. G. 1987. Wheat streak mosaic virus cylindrical inclusion body protein. J. Gen. Virol. 68:281-287.
4. Danscher, G. 1981. Localization of gold in biological tissue. A photochemical method for light and electron microscopy. Histochemistry 71:81-88.
5. DeMey, J. 1984. Colloidal gold as marker and tracer in light and electron microscopy. Pages 54-66 in: Bulletin of the Electron Microscopy Society of America, Vol. 14(1). R. Anderson ed. Hopewell Jct., New York. 128 pp.
6. Faulk, W. P., and Taylor, G. M. 1971. An immunocolloid method for the electron microscope. Immunochemistry 8:1081-1083.
7. Frens, G. 1973. Controlled nucleation for the regulation of particle size in monodisperse gold suspensions. Nature Phys. Sci. 241:20-22.
8. Geoghegan, W. D., and Ackerman, G. A. 1977. Adsorption of horseradish peroxidase, ovomucoid and anti-immunoglobulin to colloidal gold for the indirect detection of concanavalin A, wheat germ agglutinin and goat anti-human immunoglobulin G on cell surfaces at the electron microscopic level: a new method, theory, and application. J. Histochem. Cytochem. 25:1187-1200.
9. Giunchedi, L., and Langenberg, W. G. 1982. Efficacy of colloidal gold-labeled antibody as measured in a barley stripe mosaic virus-lectin-anti-lectin system. Phytopathology 72:645-647.
10. Holgate, C. S., Jackson, P., Cowen, P. N., and Bird, C. C. 1983. Immunogold-silver staining: A new method of immuno-staining with enhanced sensitivity. J. Histochem. Cytochem. 31:938-944.
11. Horisberger, M. 1979. Evaluation of colloidal gold as a cytochemical marker for transmission and scanning electron microscopy. Biol. Cellulaire 36:253-258.
12. Horisberger, M., Rosset, J., and Bauer, H. 1975. Colloidal gold granules as markers for cell surface receptors in the scanning electron microscope. Experientia 31:1147-1148.
13. Horisberger, M., Rosset, J., and Bauer, H. 1976. Localization of mannan at the surface of yeast protoplasts by scanning electron microscopy. Arch. Microbiol. 109:9-14.
14. Horisberger, M., and Von Lanthen, M. 1978. Simultaneous localization of an hepatic binding protein specific for galactose and galactose-containing receptors on rat hepatocytes. J. Histochem. Cytochem. 26:960-966.
15. Horisberger, M., and VonLanthen, M. 1979. Multiple marking of cell surface receptors by gold granules: simultaneous localization of three lectin receptors on human erythrocytes. J. Microscopy 115:97-102.
16. Hsu, Y.-H. 1984. Immunogold for detection of antigen on nitrocellulose paper. Anal. Biochem. 142:221-225.

17. Keller, G.-A., Tokuyasu, K. T., Dutton, A. H., and Singer, S. J. 1984. An improved procedure for immunoelectron microscopy: Ultrathin plastic embedding of immunolabeled ultrathin frozen sections. Proc. Natl. Acad. Sci. 81:5744-5747.
18. Lin, N.-S. 1984. Gold-IgG complexes improve the detection and identification of viruses in leaf dip preparations. J. Virol. Methods 8:181-190.
19. Lin, N.-S., and Langenberg, W. G. 1985. Peripheral vesicles in proplastids of barley stripe mosaic virus-infected wheat cells contain double-stranded RNA. Virology 142:291-298.
20. Muhlpfordt, M. 1982. The preparation of colloidal gold particles using tannic acid as an additional reducing agent. Experientia 38:1127-1128.
21. Pares, R. D., and Whitecross, M. I. 1982. Gold-labeled antibody decoration (GLAD) in the diagnosis of plant viruses by immunoelectron microscopy. J. Immunol. Methods 51:23-28.
22. Roenhorst, J. W., van Lent, J. W. M., and Verduin, B. T. M. 1988. Binding of cowpea chlorotic mottle virus to cowpea protoplasts and relation of binding to virus entry and infection. Virology 164:91-98.

Section V.A.

IMMUNO-ISOLATION

J. W. L. van Vuurde
Research Institute for Plant Protection
Wageningen, The Netherlands

Principle

Various types of solid phase with antibody binding properties can be used in various formats for immuno-isolation, e.g., rods, bottoms of Petri dishes, tissue culture plates or microplates, microscope slides, columns with macro beads (5,7). Many test parameters affect the efficiency and specificity of an immuno-isolation test. The Petri dish immunosorption evaluation test was designed as a basic procedure for optimizing steps for the various formats. This includes testing the suitability of various antisera and optimizing processes of antiserum coating, sample incubation and agar plating for specific pathogen and substrate combinations.

Immunosorbent dilution plating was developed as an alternative to isolation on selective media for the detection of seed-borne pathogenic bacteria in seed samples with high numbers of saprophytic bacteria (6) By selectively trapping the target bacterium from a contaminated sample onto a solid phase covered with homologous antibodies, non-bound microorganisms can be removed by washing prior to plating. Therefore, plating can be done with media optimized for plating efficiency and recognition of the colonies of the target organism.

A. Petri dish immunosorption evaluation test

Materials

1. Polystyrene disposable Petri dishes (9 cm)
2. Template for the preparation of the test areas (Fig. 1A)
3. Marker (e.g., diamond pen)
4. Pipets
5. Ethyl alcohol (70%)
6. Suitable nutrient agar medium
7. Sterile cotton
8. Homologous antiserum (various preparations and bleedings, if possible)
9. Heterologous antiserum
10. Preimmune serum or serum of similar non-immunized animal (i.e., normal serum)
11. Coating buffer (0.05 M carbonate buffer, pH 9.6, as used for ELISA)
12. Ringer Tween (RT) washing and sample dilution buffer. RT = NaCl 2 g, KCl 0.05 g, $CaCl_2$ 0.05 g, $NaHCO_3$ 0.25 g, KH_2PO_4 1.25 g per 1 of distilled water, adjusted to pH to 7.4 with 1 N NaOH; add 1 ml Tween 20 per liter. Note: this formulation is one-fourth strength of that originally published.

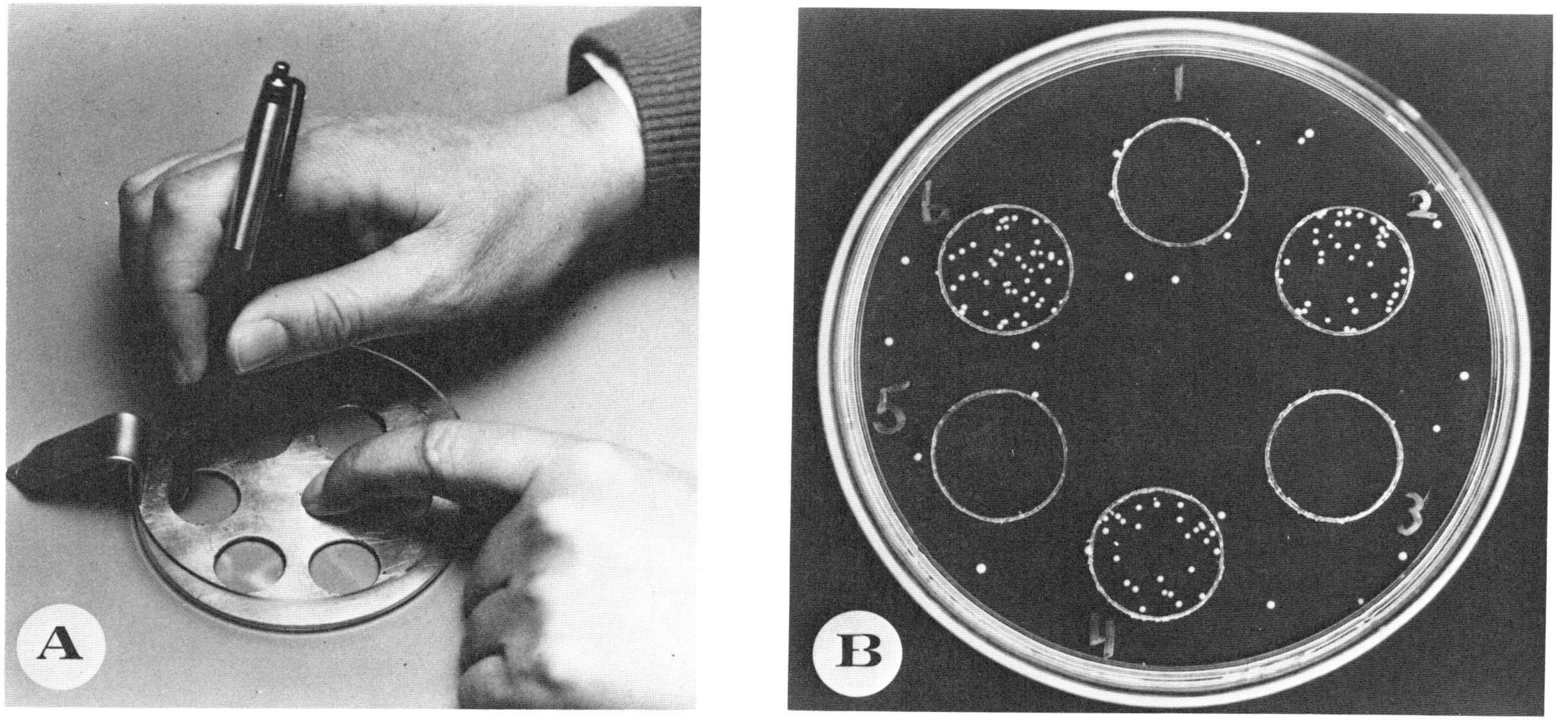

Fig. 1. A. Template for delineating test areas on a Petri dish for the immunosorption procedure. B. Erwinia carotovora subsp. atroseptica (Eca) colonies produced from cells selectively trapped in the circles 2, 4 and 6 coated with anti-Eca serum diluted (1:1000) after 6 ml of an Eca suspension had been incubated in the plate. Circles 1 and 3 were coated with diluted (1:1000) preimmune serum. Circle 5 was coated with buffer only. (Photo 2B, courtesy of R. L. Forster).

Procedure

1. Mark the test circles (e.g., 17 mm diam.) on the bottom inside of the Petri dish (Fig. 1B) and coat the test areas with a droplet of antiserum diluted in coating buffer (Fig. 2A). Prepare negative controls by treating an area with diluted preimmune serum and one with coating buffer alone.
2. Incubate for 1 hr at 27 C in a moist chamber.
3. Remove droplets with a pipet (Fig. 2B).
4. Wash 2 times at 1 min intervals with 6 ml RT by gently pipeting the solution into the middle of the plate.
5. Disinfest by incubating the plate for 10 min with 70% ethanol. Empty the plate and dry in a laminar flow cabinet.
6. Fill the Petri dish with 6 ml of the test suspension diluted in RT (Fig. 2C).
7. Incubate for 1 hr at 27 C or overnight at ca. 4 C.
8. Wash the plate with suitable high molarity and/or low pH elution buffer (Fig. 2D).
9. Clean the inner edge of the Petri dish with a sterile cotton plug.
10. Fill the plate with 10 ml of molten nutrient agar (ca. 45 C) (Fig. 2E).
11. Incubate the plate at suitable conditions for the target bacterium until colonies have been formed (Fig. 2F).
12. Count colonies in the marked areas with a stereo microscope at 5 to 10x total magnification (Fig. 1B).

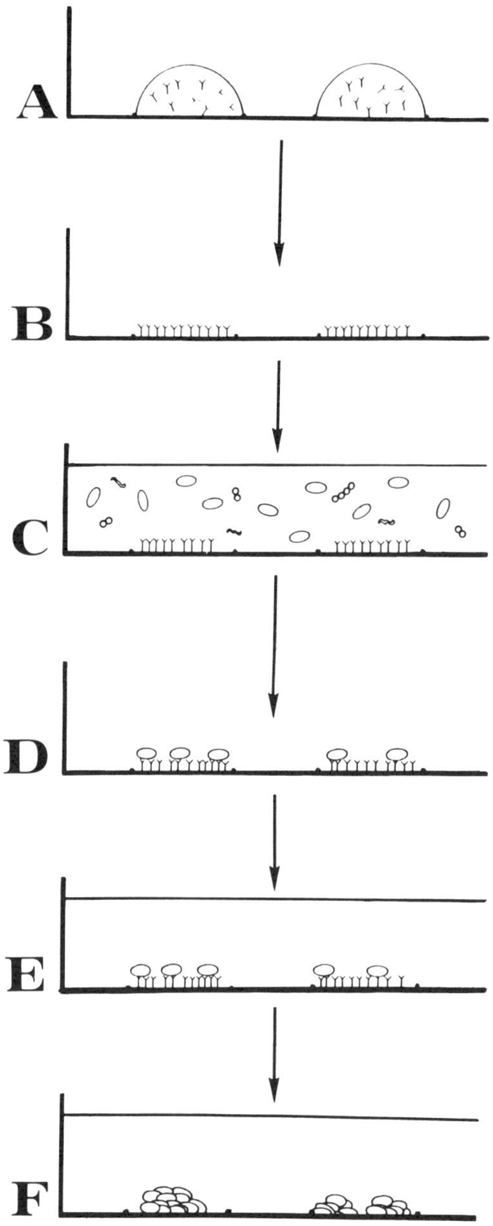

Fig. 2. Petri dish immunosorption evaluation test. A. Coating of test areas with antibodies. B. Coated test areas after washing with Ringer (quarter strength) Tween buffer. C. Incubation of test sample. D. Immunotrapped bacteria after washing away unbound particles. E. Addition of nutrient agar at 45 C. F. Intra-agar growth of colonies from immunotrapped bacteria.

Comments

1. Contamination of antiserum and solvents (including ethanol) with microorganisms can be checked by performing the procedure without the test suspension.
2. Coating antibodies will strongly bind to the plastic and cannot easily be washed away. Part of the coating occurs within seconds giving a background of immunotrapped colonies around the marked area with homologous antiserum when the coating suspension is not removed by pipetting.
3. The washing steps are very critical in the procedure. RT also functions to block the solid phase and antibody coating against non-specific binding and to maintain viability of the bacteria.
4. Vigorous washing can remove almost all immunotrapped bacterial cells. The effect of the washing procedure on the immunotrapped bacteria can be checked by coating the whole bottom of the plate with antibodies and inspecting the plate for uneven distribution of colonies (normally decreasing detachment of bacteria from the center to a radius of 1 5 cm).
5. Whole antisera as well as purified antibodies can be directly tested (step #1) starting with, e.g., 1:100 and 1:1,000 dilutions.
6. Suspensions of a pure culture of the target bacterium (step #6) are diluted with RT. Twenty-four hr cultures on slants may be assumed to have about 10^{10} cells which subsequently may be determined accurately by standard dilution plating. A concentration of 10^3 colony forming units (cfu) is recommended for test systems with a trapping efficiency between 20 and 100%. The specificity can best be investigated starting with a suspension of the target bacterium mixed with one or more strains of saprophytic bacteria with a clearly different colony type.
7. Replacement of washing with RT after the sample incubation (Procedure, step 8) by washing with another solution (e.g., a high molarity buffer such as 0.5 to 1.0 M phosphate buffered saline, chaotropic ions such as 0.5 to 1.0 M $MgCl_2$, and/or a lower pH) may decrease non-specific adsorption and weak cross-reactions (1,2). Such measures should first be tested for deleterious effects on the viability of the target bacterium in the test system.
8. Background contamination of colonies on or in the agar can easily be distinguished from immunotrapped colonies (Procedure, step 12) which develop at the interface of the antibody-coated solid phase and the bottom layer of the agar. However, slow cooling of the agar in the plate may cause release of a part of the immunotrapped bacteria which will form colonies in the agar outside the test area.

B. Immunosorbent dilution plating

Materials

1. Glass rods coated with cellulose nitrate (see Comment 1)
2. Tubes (1 ml) and pipets for diluting sera and test samples
3. Homologous antiserum
4. Preimmune serum
5. Suitable agar nutrient medium
6. Coating buffer (see Materials, Petri dish immunosorption)
7. Ringer Tween (RT) (see Materials, Petri dish immunosorption)

Procedure

1. Prepare solid-phase rods (inoculation stick, Fig. 3) of suitable size and material to fit into a 1-ml sample tube. Polystyrene rods can be coated directly with antibodies, whereas glass rods must first be coated with cellulose nitrate (see Comment 1; 'carrier layer', Fig. 3A). Place rod into a tube containing diluted (e.g., 1:1000) antiserum (carbonate coating buffer). Prepare two negative controls by placing rods into tubes of diluted preimmune serum and coating buffer, respectively.
2. Rinse the rods with Ringer Tween (RT), followed by two incubations in RT for 1 min.
3. Prepare a suspension of the sample in RT. Incubate the antibody-coated rods in the sample tubes for 1 hr at 27 C or overnight at 4 C, for binding of target bacterium (Fig. 3B).
4. Place the rod in a vertical position above a sink and wash gently by dripping RT on the rod ca. 2 cm above the antibody-coated area for 20 sec. Large numbers of rods can be treated simultaneously with a spraying device at low air pressure.
5. Streak the rod on agar surface of one or more nutrient agar plates (Fig. 3 C) to prepare a dilution of the trapped bacteria. Increase the streaking pressure in subsequent streaks.
6. Incubate at suitable conditions to allow growth of the colonies.

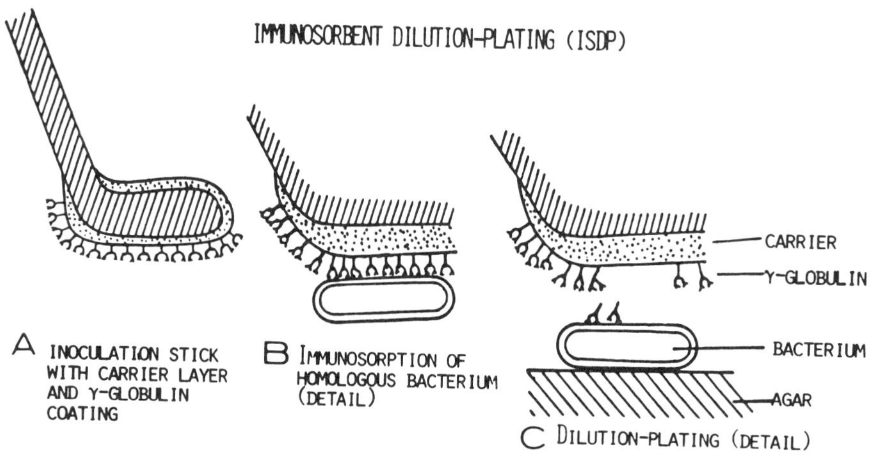

Fig. 3. Scheme of immunosorbent dilution plating. A. Coating of solid phase with antibodies; wash stick (rod) after coating to remove unbound antibodies. B. Incubating antibody-coated rod in sample for immunotrapping of target bacterium; gently wash after incubation to remove unbound microorganisms. C. Releasing immunotrapped bacteria by rubbing the rod on the agar surface of several plates to obtain isolated colonies. Reproduced with permission from van Vuurde and van Henten (6).

Comment

1. Before use in immunosorbent dilution plating, glass rods are first treated (Fig. 3A) with a layer to enhance protein binding, e.g. cellulose nitrate (most types of nail polish contain 90% cellulose nitrate and can be used in a 1:3 dilution with acetone) or polylysine L (4). Coated rods can be stored for several weeks under dry conditions at ca. 4 C before use. Care should be taken that the cellulose nitrate or comparable protein-binding layer is not released from the rod during the procedure.
2. Non-specific binding of bacteria can be reduced by adding blocking compounds (e.g. bovine serum albumin, crude egg albumin, or skimmed milk powder) to the RT.

Applications

Immuno-isolation, applied in various formats has proven useful for selective isolation of various species of Xanthamonas, Pseudomonas, Erwinia, and Clavibacter (3,5). Non-specific binding remains a problem especially for test suspensions from complex natural substrates like soil. The immunotrapping of cross-reacting microorganisms may reduce the sensitivity of immuno-isolation. However, like immunofluorescence colony staining (see **Section VII.B.**) it provides an efficient tool for selective isolation of these organisms, especially when both methods are used in combination. A collection of cross-reacting organisms would enhance research on antiserum specificity, characterization of antigens of the target bacterium and of a test-system reliability. Further, research is needed to improve the binding strength between the antibody-coated solid phase and the target cells, on effective solid phase blocking, and on selective elution procedures to remove non-specific and weakly cross-reacting organisms more efficiently.

Literature Cited

1. Pharmacia Fine Chemicals. 1988. Affinity chromatography, principles and methods. Pharmacia Fine Chemicals, Uppsala, Sweden. 112 pp.
2. Ruissen, M. A., Helderman, C. A. J., Schipper, J., and van Vuurde, J. W. L. 1987. Selective isolation and concentration of phytopathogenic bacteria on immunoaffinity columns. Page 882 in: Plant Pathogenic Bacteria, Proc. 6th Int. Conf. Plant Path. Bact., Maryland, USA. E. L. Civerolo, A. Collmer, R. E. Davis, and A. G. Gillespie, eds. Nijhoff, Dordrecht, The Netherlands. 1050 pp.
3. Stead, D. E., Chauveau, J. F., Janse, J. D., Ruissen, M. A., van Vaerenbergh, J., and van Vuurde, J. W. L. 1987. Immuno-isolation techniques for the detection and isolation of plant pathogenic bacteria. Pages 189-193 in: Immunological Techniques in Microbiology. Soc. for Appl. Bacteriol. 271 pp.
4. Tijssen, P. 1985. Practice and Theory of Enzyme Immunoassays. Elsevier, Amsterdam. 549 pp.
5. van Vuurde, J. W. L. 1987. New approach in detecting phytopathogenic bacteria by combined immunoisolation and immunoidentification assays. EPPO Bulletin 17:139-148.

6. van Vuurde, J. W. L., and van Henten, C. 1983. Immunosorbent immunofluorescence microscopy (ISIF) and immunosorbent dilution-plating (ISDP): New methods for the detection of plant pathogenic bacteria. Seed Sci. & Techn. 11:523-533.
7. van Vuurde, J. W. L., Ruissen, M. A., and Vruggink, H. 1987. Principles and prospects of new serological techniques including immunosorbent immunofluorescence, immunoaffinity isolation and immunosorbent enrichment for sensitive detection of phytopathogenic bacteria. Pages 835-842 in: Plant Pathogenic Bacteria, Proc. 6th Int. Conf. Plant Path. Bact., Maryland, U.S.A. E. L. Civerolo, A. Collmer, R. E. Davis, and A. G. Gillespie, eds. Nijhoff, Dordrecht, The Netherlands. 1050 pp.

Section V.B.

INFECTIVITY NEUTRALIZATION

J. E. Duffus
USDA ARS
Salinas, California 93905

Principle

Infectivity neutralization consists of incubation of virus with antiserum, and then testing the infectivity of the mixture (or a density gradient centrifugation portion of the mixture). The technique is useful for certain plant viruses that are transmitted by insects in the circulative (persistent) manner, that are not mechanically transmissible, and that can be acquired by vectors feeding through membranes into virus preparations. This same procedure may be applied to mechanically transmitted viruses but the mechanical transmission of reactants gives even greater nonspecific inhibition and the reactants must be centrifuged before mechanical inoculation (5).

Materials

1. Cages for confining insects to membranes. Simple feeding chambers can be made from plastic vials (30 mm dia, 60 mm long) (Fig. 1). The bottoms of the vials are cut off and the edges polished with sandpaper. The cut ends of the vials are covered with a piece of Parafilm-M stretched about fourfold in each dimension.

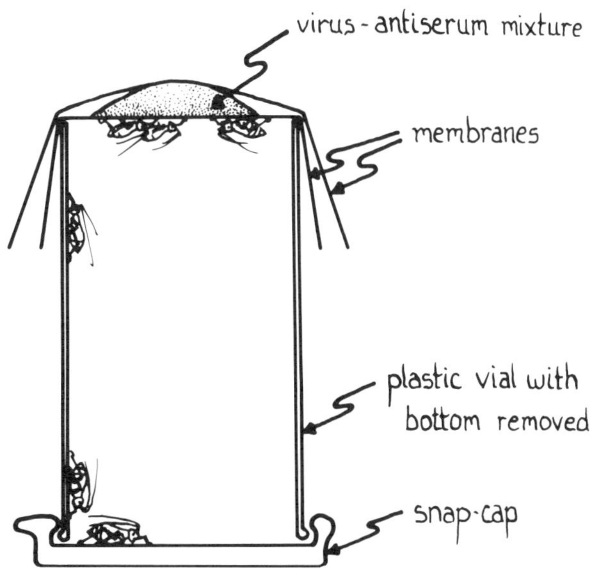

Fig. 1. Feeding cage for insects. Description in **Materials**.

2. Sucrose (60%) dissolved in neutral 0.1 M potassium phosphate or other appropriate buffer.
3. PBS
4. Normal serum.
5. Refractometer.
6. Small test tubes.
7. Incubator or water bath at 37 C.
8. Dropping pipets.
9. Squares of Parafilm-M.
10. Wooden block or rack to hold glass tubes during preparation of feeding solutions.
11. Yellow food coloring
12. Virus free insects.
13. Brush or aspirator tube for handling insects.
14. Appropriate young test plants and cages for confining insects to plants during inoculation test feeding period. If feeding chambers are made large enough, these may be used directly as inoculation cages.
15. Pot labels to identify treatment.
16. Lighted area to permit insects to feed on caged test plants for several days.
17. Fumigation chamber or insecticidal spray to remove insects from plants at end of test feeding period.
18. Greenhouse or growth chamber to permit development of symptoms.

Procedure

1. Place 0.2 ml of each virus preparation (containing an amount of virus previously found to produce infection of most test plants) in each of several test tubes.
2. Mix 0.2 ml of each antiserum to be tested (diluted ca. 1:64) with a virus sample.
3. Add 0.2 ml of PBS, or normal serum to one sample of virus for a control.
4. Mix virus and antiserum; incubate at 37 C for 30 min.
5. Adjust each virus-antiserum mixture and control mixture to 20% sucrose by addition of 60% sucrose.
6. Add a drop of yellow food coloring to the mixtures.
7. Add several drops of virus-antiserum sucrose-dye mixture to Parafilm membranes previously stretched on feeding chambers (Fig.1).
8. Stretch a second clean piece of Parafilm and apply carefully over drops of virus-antiserum preparation on end of feeding chamber to form a layer of liquid between the two membranes.
9. Place 10-30 insects in each of the feeding chamber cages and cap.
10. Keep some insects from each colony for use as controls.
11. Allow the insects to feed from 6 to 18 hr (depending upon virus-insect combination) through the stretched membrane into the virus mixture.
12. Remove insects from feeding chambers and transfer to inoculation cages (or use feeding chambers directly as inoculation cages) and place on test plants.
13. Transfer insects from control mixture to test plants to be sure insects were free of virus at start of test.
14. Allow insects to feed on caged test plants for two days.
15. Remove cages and kill insects by fumigation or spraying.
16. Incubate plants in greenhouse or growth chamber and make frequent inspections for the development of symptoms.

Comments

This relatively simple procedure works well with the luteoviruses and the aphid species that transmit them, the beet curly top virus (geminivirus) and its leafhopper vector and also the whitefly transmitted gemini viruses (1,2,3,4,6,7). Each insect species and/or virus combination may require modification in feeding conditions and caging. With most combinations tested, the insects feed much better on sucrose-gradient fractions rather than directly on the virus-antiserum mixtures.

Low antiserum dilutions (1:4) tend to show group relationships. For example, most all luteoviruses show relationships to beet western yellows viruses at antiserum dilutions of 1:4, but most of the viruses can readily be distinguished from each other at dilutions of 1:64.

Literature Cited

1. Duffus, James E. 1977. Serological relationships among beet western yellows, barley yellow dwarf, and soybean dwarf viruses. Phytopathology 67:1197-1201.
2. Duffus, James E., and Gold, A. H. 1969. Membrane feeding and infectivity neutralization used in a serological comparison of potato leaf roll and beet western yellows viruses. Virology 37:150-153.
3. Duffus, James E., and Gold, A. H. 1973. Infectivity neutralization used in serological tests with partially purified beet curly top virus. Phytopathology 63:1107-1110.
4. Duffus, J. E., and Russell, G. E. 1970. Serological and host range evidence for the occurrence of beet western yellows virus in Europe. Phytopathology 60:1199-1202.
5. Gold, A. H., and Duffus, J. E. 1967. Infectivity neutralization - a serological method as applied to persistent viruses of beets. Virology 31:308-313.
6. Rochow, W. F., Aapola, A. I. E., Brakke, M. K., and Carmichael, L. E. 1971. Purification and antigenicity of three isolates of barley yellow dwarf virus. Virology 46:117-126.
7. Rochow, W. F., and Ball, E. M. 1967. Serological blocking of aphid transmission of barley yellow dwarf virus. Virology 33:359-362.

Section V.C.

TRANSLATION PRODUCT IMMUNOPRECIPITATION

E. M. Ball
University of Nebraska
Lincoln, NE 68583

Introduction

The immunoprecipitation technique, modified from that described by Kessler (29), has been applied to the analysis of viral-RNA products produced in rabbit reticulocyte lysate or in the wheat germ systems. Antisera are prepared to viral products, e.g. virions, coat protein, inclusion bodies or helper component. The translation products, synthesized in the presence of radioactive amino acids, are separated from polysomes, then mixed with antiserum, and the complex precipitated with an immunoadsorbent, Cowan I strain of Staphylococcus aureus cells (IgGSorb) or Protein A sepharose. The precipitate is collected and the contents dissociated and fractionated by sodium dodecyl sulfate-polyacrylamide gel electrophoresis (SDS-PAGE). The antibody-precipitated, radioactive proteins are visualized by fluorography (Fig. 1).

Precipitation of an in vitro translation product of viral RNA by an antiserum is evidence that the protein used to prepare the antiserum is viral-coded. The precipitated protein should be the same molecular weight as the antibody-inducing protein, but if it is of different molecular weight, large numbers of amino acid sequences must be the same as the immunogen. A control with preimmune serum substituted for antiserum must be included in the experiments.

The identification of viral encoded, nonstructural proteins of potyviruses by immunoprecipitation of RNA-translation products and the role of these products in genome organization and function is reviewed by Hiebert (19), Hiebert et al (22), and Dougherty and Carrington (7).

Examples of Immunoprecipitation Techniques

A. Identification of a distinct strain of virus (33)

Materials

1. Staphylococcus aureus Cowan I strain or IgGSorb.
2. Dissociation buffer: 62.5 mM tris-HCl, pH 6.8, containing 2% SDS, 5% 2-mercaptoethanol, 10% glycerol.
3. NET buffer: 150 mM NaCl, 5 mM EDTA, 50 mM tris-HCl, pH 7.4.
4. Nonidet P-40 (NP 40, Sigma N6507).
5. Materials and procedure for centrifugation, SDS-PAGE and fluorography.

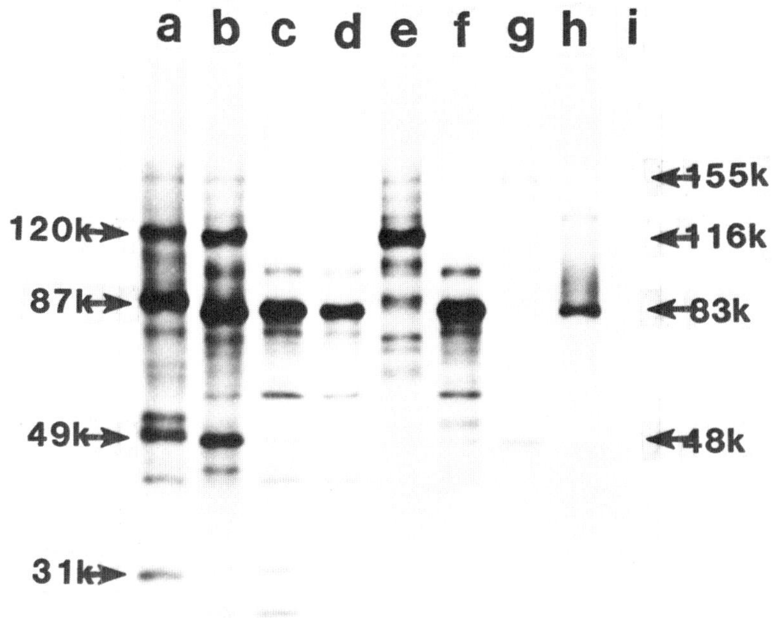

Fig. 1. Immunoprecipitation analysis of the tobacco vein mottling virus (TVMV) RNA cell-free translation products. The [^{35}S] methionine labeled products, dissociated by sodium dodecyl sulfate, were separated on a linear 7.5 to 15% polyacrylamide gel gradient by electrophoresis (SDS-PAGE) and detected by fluorography. (lane a) Total products of tobacco etch virus (TEV) RNA translation; (lane b) total products of TVMV translation; products immunoprecipitated with antiserum to TVMV capsid protein (lane c), to TEV capsid protein (lane d), to pepper mottle virus cylindrical inclusion protein (lane e), to TEV 54K nuclear inclusion protein (lane f), to TEV 49K nuclear inclusion protein (lane g), and to TVMV helper component (HC) protein (lane h); (lane i) products immunoprecipitated with normal serum. The molecular weights of some of the translation products are given on the left for TEV and on the right for TVMV (E. Hiebert, T. Pirone, and D. Thornbury, unpublished).

Procedure

1. The in vitro translation is carried out by standard procedure with purified viral RNA in the presence of a radioactive amino acid. The translation mixture is dissociated in 3 vol. of dissociation buffer and heated at 90 C for 3 min.
2. Forty μl of the dissociated products are added to 80 μl of NET buffer containing 0.5% NP-40, 1 mg/ml of ovalbumin and 2 mM methionine. Twenty μl of antiserum are added, and the mixture is incubated 1 hr at room temperature.
3. The mix is immunoprecipitated by adding 20 μl of *Staphylococcus aureus* Cowan I strain (10% w/v) and incubation is continued for 30 min more.
4. The precipitate is collected by centrifugation at 2000 x g for 5 min,

and the pellet is resuspended in NET buffer + 0.05% NP-40. The centrifugation-resuspension is done 3 times.

5. The final precipitate is resuspended in 50 μl of dissociation buffer and heated at 90 C for 5 min.
6. The resuspended precipitate is layered on gels for SDS-PAGE.
7. Bands representing precipitated, specific polypeptides are visualized by fluorography.

Conclusions

When translation products of a PRSV-like isolate designated Q_{10} were analyzed by SDS-PAGE, a major product had an estimated size of 114 Kd. This was slightly larger than major translation products of watermelon virus-2, papaya ringspot virus-watermelon strain (PRSV-W), or zucchini yellow mosaic virus. Translation products of isolate Q_{10} were immunoprecipitated with antisera to potyvirus-specified proteins: antisera to PRSV-W amorphous inclusion, to PRSV-W cylindrical inclusion, and to Q_{10} capsid protein precipitated the 114 Kd major product, the 70 Kd product, and the 36 Kd product, respectively. With supporting evidence from SDS-immunodiffusion, ELISA, and western blot procedures, the Q_{10} isolate was established as a strain of PRSV.

B. The use of competitor proteins in immunoprecipitation (17)

Materials

1. PBS buffer: 10 mM sodium phosphate buffer, pH 7.5, 15 mM NaCl.
2. Triton X-100, sodium deoxycholate, acetone.
3. IgGSorb: Protein A fixed to *S. aureus* cells (Enzyme Center. Boston, MA)
4. Gel sample buffer: 62.5 mM tris-HCl, pH 6.8, containing 2.3% SDS, 5% β-mercaptoethanol, 10% glycerol.

Procedure

1. Antiserum is diluted to 32 A_{280} units/ml in PBS.
2. Three μl samples of cell free translation mixture are incubated with 2 μl of antiserum in a total volume of 175 μl of PBS containing + 1.5% Triton X-100 and 1.5% sodium deoxycholate.
3. When competition studies are done, 2 μg of non-labeled competitor protein are added directly to a sample. Thus, comparisons of precipitated products with and without competitor indicate whether the competitor protein and the translation products are serologically related.
4. The mixture is incubated 30 min at 25 C with shaking. Then, 15 μl of IgGSorb (90 mg/ml) are added in PBS and incubation is continued for 30 min.
5. The precipitate is collected by sedimentation through a sucrose pad. The precipitate is washed with acetone (80%) and resuspended in gel sample buffer. Prior to electrophoresis, the IgGSorb cells are removed by centrifugation.
6. For sequential immunoprecipitation experiments, translation products are first immunoprecipitated with various sera before immunoprecipitation with the antiserum being tested. In this study, for example,

translation products were subjected to immunoprecipitation with 1) no serum, 2) anti-virus coat protein serum, 3) anti-virus cylindrical inclusion protein serum etc., then the remaining products were tested with antiserum as described in step 2. The procedure is continued as in step 4 and 5. From the alterations of the immunoprecipitated product patterns compared to the control, it was concluded that certain polypeptides were recognized by more than one antiserum.

Conclusions

Viral-RNA translation products of tobacco vein mottling virus were characterized by using antisera of five different proteins: coat protein, cylindrical inclusion, helper component of TVMV and the 49 Kd and 54 Kd nuclear inclusion proteins of tobacco etch virus. When the respective antigens were used as competitor proteins, little or no effect was observed on the immunoprecipitation pattern except with the protein homologous to the antiserum. Likewise, sequentially precipitating with 2 different sera in "preabsorption" experiments established the same relationship, i.e. a change in pattern of precipitated proteins only with the protein homologous to the antiserum. Based on translation products, the 5 sera defined 5 different non-overlapping regions of the virus genome. Presumably the 5 regions code for 5 distinct proteins.

C. Antiserum to a synthetic polypeptide for identification of a translation product (31)

Materials

1. NET buffer: 150 mM NaCl, 5 mM EDTA, 50 mM Tris-HCl, pH 7.4.
2. Gel sample buffer: 10% glycerol, 2.5% 2-mercaptoethanol, 2% SDS, 0.001% bromophenol blue, 61 mM Tris-HCl, pH 6.8.
3. Nonidet P-40 (NP-40, Sigma N 6507).
4. Cowan I cells (*S. aureus*) or Protein A sepharose CL-4B (Pharmacia).

Procedure

1. Mix 2.5-5.0 μl of translation product with 3 μl of preimmune serum and incubate at 37 C for 1 hr and at 4 C for 16 hr.
2. Add a 10% suspension of Cowan I cells (*S. aureus*) or Protein A-Sepharose CL-4B (Pharmacia) to the mixture and incubate at 4 C for 1 hr. Centrifuge to pellet cells and any non-specific reacting material.
3. To the supernatant fluid, add 3 μl of antiserum and process as described in step 1 and 2.
4. Centrifuge and wash the precipitate twice with 0.05% Nonidet P-40 in NET buffer.
5. Suspend the precipitate in gel sample buffer and boil for 3 min.
6. Layer samples on SDS-10% polyacrylamide gels, electrophorese and visualize bands by fluorography on Kodak XAR-5 films at -70 C.

Conclusions

The 30 Kd protein encoded from the tobacco mosaic virus (TMV) genome may have a role in virus transport but had not been identified or proved to be expressed in vivo. From the known sequence of the gene, a polypeptide corresponding to the deduced carboxy-terminal part of the 30 Kd protein was synthesized chemically. An antiserum was prepared to the polypeptide conjugated to bovine serum albumin and was used in immunoprecipitation studies.

Infected protoplasts were processed and in vivo translation products pulse labeled. The 30 Kd polypeptide was detected by immunoprecipitation from cell lysates. When the synthetic peptide (unlabeled) was added in competition studies, the immunoprecipitation was inhibited. Therefore, the procedure positively identified the 30 Kd protein from TMV infection. Additional studies have been reported (30) on immunoprecipitation of TMV-RNA translation products using an antibody to a synthetic peptide containing 11 C-terminal amino acids of 30 Kd protein.

D. Comments on improving specificity of immunoprecipitation (1-5 excerpts from Anderson and Blobel, 7)

1. The amount of antibody added depends upon titer and affinity and the amount of cell-free product. The exact amount used must be determined by titration. Although preincubation for 1 hr at 37 C is not necessary, incubation of the complex for 12 hr at 4 C helps bind the antigen and antibody tighter.
2. An immunoadsorbent is used instead of a second antibody since the complex is easier to wash. The antigen-antibody complex binds to the surface of _S. aureus_ cells that are rich in Protein A. Protein A-sepharose CL-4B beads as adsorbents have the advantage of being very resistant to denaturants and can be recycled. Also, the beads can be washed more thoroughly than _S. aureus_ cells.
3. The major source of contamination in immunoprecipitation is nonspecific adsorption of material to the solid-phase immunoadsorbent. Selecting the appropriate buffer is a compromise between conditions stringent enough to remove adsorbed contaminants but not elute the antigen-antibody complex.
4. To demonstrate specificity of the reaction, run parallel incubation with nonimmune or preimmune sera. Identical serum can be used in competition with excess unlabeled antigen. Finally, peptide mapping or partial radio sequence analysis can prove the identity of the immunoprecipitate.
5. Several procedures that would contribute to cleaner immunoprecipitates are: 1) treat the solubilized lysate reaction mix with immunoadsorbent prior to the addition of antisera (especially effective when using antibody-sepharose); 2) clean up the antiserum by preparing IgG or affinity purified antibody; 3) reduce the amount of antiserum added, thereby reducing the background; 4) reduce the volume of the original synthesis reaction, since the degree of nonspecific contamination is inversely proportional to the specific activity of protein labeling in the cell-free system; and 5) make new antibody because contamination may depend upon the serum used.
6. Sodium dodecyl sulfate at a final concentration of about 0.5% can also be used to reduce or eliminate nonspecific serological reactions during immunoprecipitation (23).

E. Application

1. Variation in parameters for immunoprecipitation

Since the science of analysis of RNA translation products is relatively recent, most of the references to the application of immunoprecipitation have appeared since 1980. Of these procedures, some minor variations have been described in the parameters of the precipitation mixture.

Microliter volumes of lysate:NET buffer:antiserum:*S. aureus* (10% w/v), as described with Q_{10} isolate of papaya ringspot virus translation products, were applied at ratios of 80:40:20:20. The same ratio was reported for immunoprecipitation of pepper mottle virus and papaya ringspot virus - watermelon strain (5,6), bean yellow mosaic (4), Araujia mosaic virus (20), tobacco vein mottling virus and potato virus Y (23), squash mosaic virus (21), and maize rayado fino virus (9). Fifty μl of antiserum were used in studies with pepper mottle virus and tobacco etch virus (8). The same concentration of NP-40 immunoprecipitated maize stripe mosaic virus (10) and peanut mottle virus (38) translation products. With the latter virus, a ratio of 30:15:9:15 composed the precipitation mixture. The immunoprecipitation of turnip mosaic virus translation products included a buffer of NET, 1% Triton X-100, 0.5 mg/ml ovalbumin in a ratio of 20:100:10:20; 20 μl with Protein A or 50 μl with *S. aureus* as the immunoprecipitant (35).

Tobacco vein mottling virus-RNA translation products (18,34,36) were immunoprecipitated in a buffer system of 10 mM sodium phosphate, pH 7.5, and 15 mM NaCl containing 1.5% Triton X-100 and 1.5% deoxycholate. Ratios of lysate:buffer:serum:IgGSorb (90 mg/ml) were 3:175:2:15 The buffer used in immunoprecipitation of cowpea mosaic virus translation products (12,27) was phosphate pH 7.2, 1% Triton X-100, 0.5% deoxycholate and 0.1% SDS, and the ratio of components was 5-10:100:5:25. The precipitator was *S. aureus* (10%) in buffer containing 1% bovine serum albumin (BSA). In both studies with TVMV and CPMV, the precipitate was separated from other components by centrifugation through a sucrose cushion. Similar parameters were used for immunoprecipitation of translation products of beet necrotic yellow vein virus (39) and the 190 S virus infecting *Helminthosporium victoriae* (13).

Tobacco ringspot virus RNA translation products (11) were immunoprecipitated in a buffer system of 1 M NaCl, 1% Triton X-100, 1 mM phenylmethylsulfonyfluoride and 0.01% sodium azide at reaction ratios of 40:120:8:60. The precipitator was 62.5 mg/ml Protein A sepharose CL4B with 10 mg/ml BSA. The precipitate was washed 6 times in 0.75 M NaCl with 1% Triton X-100. Similar parameters were used with pea early browning virus (26) and with Daphne virus X (14). Protein A sepharose CL4B also immunoprecipitated translation products of brome mosaic virus (24). The lysates were diluted 20-fold in IP buffer (50 mM Tris, pH 7.5, 150 mM NaCl, 2 mM EDTA, 1% NP-40) then mixed with 1:50 serum before immunoprecipitation.

A precursor involved in coat protein synthesis was identified in the immunoprecipitation analysis of translation products of tobacco ringspot virus RNA (28). Three volumes of buffer of 50 mM Tris, pH 8.0, 200 mM NaCl, 1% NP-40 and 1% BSA were mixed with 50 μl of 30% Pansorbin (*S. aureus* Protein A). Tomato bushy stunt virus RNA translation products were immunoprecipitated in NET buffer of 0.15 M NaCl, 0.5 M EDTA, 0.05 M Tris-HCl, pH 7.9, containing 0.05% NP-40 and 1% BSA. The reaction mixture was 20 μl lysate, 50 μl serum and 100 μl formalin-fixed *S. aureus* (16). Translation products of three strains of soilborne wheat mosaic virus (25) were immunoprecipitated in a mixture of 10 μl lysate, 20 μl sterile distilled water, 5 μl serum-IgG (4 mg/ml), 5 μl of 10X TBS-Triton buffer (100 mM Tris-HCl, pH 7.2, 1.5 M NaCl, 10% Triton X-100) containing 5 μl 10% deoxycholate and 5 μl BSA (10 mg/ml). The precipitator was 25 μl of 10% fixed S. aureus cells.

2. Immunoprecipitation of Viral Products from Infected Chloroplast

Immunoprecipitation techniques were applied to extracts of disrupted chloroplast from cowpea tissue infected with cymbidium ringspot virus and to translation products from the viral RNA (3). Thirty μl of extract or lysate was diluted 10-fold in buffer (10 mM sodium phosphate, pH 7.2, 0.9% NaCl, 1% Triton X-100, 0.5% deoxycholate and 0.1% SDS), mixed with 10 μl viral antiserum and with 40-60 μl 10% Protein A sepharose. The immunoprecipitate was recovered by centrifugation through a cushion of 30% sucrose containing 0.5% SDS. Coat protein (M_r 38 kd) of carnation mottle virus (15) was identified by immunoprecipitation of extracts from infected protoplasts of Chenopodium quinoa. Fifty-μl samples were mixed with 150 μl of solvent (1 M NaCl, 1% Triton X-100) 20 μl viral antiserum-IgG and 60 μl Protein A-sepharose in 25 mM Tris-HCl, pH 7.5. The precipitate was washed in 0.75 M NaCl and 1% Triton X-100. Similar studies were reported with red clover necrotic mosaic virus from infected cowpea protoplasts (32). Extracts in 50 μl of buffer (10 mM KCl, 2 mM EDTA, 5 mM 2-merceptoethanol, 50 mM Tris-HCl, pH 7.6) were mixed with antiviral IgG and precipitated with formalin-fixed S. aureus cells. Viral polyamines isolated from protoplast of Chinese cabbage infected with turnip yellow mosaic virus (2) were immunoprecipitated before HPLC analysis. The buffer contained 20 mM 2-[N-morpholinolethane sulfonic acid] (MES)/NaOH, pH 6.4, 0.1 M NaCl, 60 mM $MgCl_2$. Samples (200 μl) containing 4-10 μg virus were diluted to 1.4 ml buffer, antiviral IgG was added in excess, then the mixture was precipitated with 125 μl of 20% formalin-fixed S. aureus. Structural viral proteins of sonchus yellow net virus from infected cowpea protoplasts were identified by immunoprecipitation (37) in a buffer of TBS (20 mM Tris, pH 8.0, 0.4 M NaCl), 1 mM $MgCl_2$, 1 mM $CaCl_2$, 1.5% NP40, 0.15% deoxycholate. A pellet of 10^5 protoplasts in 10 μl H_2O was added to 100 μl buffer, mixed with 5 μl antiviral serum (1:10 in TBS) and precipitated with 10 μl 10% Protein A sepharose in TBS.

Viruses Reported in Immunoprecipitation Studies

1. Bromovirus: brome mosaic (24)
2. Comovirus: cowpea mosaic (12,27), squash mosaic (21)
3. Dianthovirus: red clover necrotic mosaic (32)
4. Maize rayado fino virus (9)
5. Nepovirus: tobacco ringspot (11,28)
6. Potexvirus: Daphne X (14)
7. Potyvirus: papaya ringspot - watermelon mosaic virus 1 and Q_{10} isolate of papaya ringspot - W (5,6,33), peanut mottle (38), pepper mottle (8), potato Y (23), tobacco etch (8), tobacco vein mottling (18,23,34,36), turnip mosaic (35), Araujia mosaic (20)
8. Rhabdovirus: sonchus yellow net (37)
9. Rice stripe virus: maize stripe (10)

10. Tobamovirus: tobacco mosaic (30,31)
11. Tobravirus: pea early browning (26)
12. Tombusvirus: carnation mottle (15), cymbidium ringspot (3), tomato yellow stunt (16)
13. Tymovirus: turnip yellow mosaic (2)
14. Wheat soilborne mosaic virus: beet necrotic yellow vein (39), soilborne wheat mosaic virus (25)
15. 196 S virus of _Helminthosporium_ _victoriae_ (13)

Literature Cited

1. Anderson, D. J., and Blobel, G. 1983. Immunoprecipitation of proteins from cell-free translations. Pages 111-120 in: Methods in Enzymology. Vol. 96. Academic Press. New York. 901 pp.
2. Balint, R., and Cohen, S. S. 1985. The incorporation of radiolabeled polyamines and methionine into turnip yellow mosaic virus in protoplasts from infected plants. Virology 144:181-193.
3. Burgyan, J., Russo, M., and Gallitelli, D. 1986. Translation of cymbidium ringspot virus RNA in cowpea protoplasts and rabbit reticulocyte lysates. J. Gen. Virol. 67:1149-1160.
4. Chang, C.-A., Hiebert, E., and Purcifull, D. E. 1988. Analysis of _in_ _vitro_ translation of bean yellow mosaic virus RNA: Inhibition of proteolytic processing by antiserum of the 49 K nuclear inclusion protein. J. Gen. Virol. 69:1117-1122.
5. deMejia, M. V. G., Hiebert, E., and Purcifull, D. E. 1985. Isolation and partial characterization of the amorphous cytoplasmic inclusions associated with infections caused by two potyviruses. Virology 142:24-33.
6. deMejia, M. V. G., Hiebert, E. Purcifull, D. E., Thornbury, D. W., and Pirone, T. P. 1985. Identification of potyviral amorphous inclusion protein as a nonstructural, virus-specific protein related to helper component. Virology 143:34-43.
7. Doughtery, W. G., and Carrington, J. C. 1988. Expression and function of potyviral gene products. Annu. Rev. Phytopathol. 26:123-143.
8. Doughtery, W. G., and Hiebert, E. 1980. Translation of potyvirus RNA in a rabbit reticulocyte lysate: Identification of nuclear inclusion proteins as products of tobacco etch virus RNA translation and cylindrical inclusion protein as a product of the potyvirus genome. Virology 104:174-182.
9. Espinoza, A. M., Ramirez, P., and Leon, P. 1988. Cell-free translation of maize rayado fino virus genomic RNA. J. Gen. Virol. 69:757-762.
10. Falk, B. W., Tsai, J. H., and Lommel, S. A. 1987. Differences in levels of detection for the maize stripe virus capsid and major non-capsid proteins in plant and insect hosts. J. Gen. Virol. 68:1801-1811.
11. Forster, R. L. S., and Morris-Krsinich, B. A. M. 1985. Synthesis and processing of the translation products of tobacco ringspot virus in rabbit reticulocyte lysates. Virology 144:516-519.
12. Franssen, H., Goldbach, R., Broekhuijsen, M., Moerman, M., and van Kammen, A. 1982. Expression of middle-component RNA of cowpea mosaic virus: In vitro generation of a precursor to both capsid proteins by a bottom-component RNA-encoded protease from infected cells. J. Virol. 41:8-17.
13. Ghabrial, S. A., Bibb, J. A., Price, K. H., Havens, W. M., and Lesnaw, J. A. 1987. The capsid polypeptides of the 190 S virus of _Helminthosporium_ _victoriae_. J. Gen. Virol. 68:1791-1800.

14. Guilford, P. J., and Forster, R. L. S. 1986. Detection of polyadenylated subgenomic RNAs in leaves infected with the potyvirus Daphne virus X. J. Gen. Virol. 66:2597-2604.
15. Harbison, S. A., Davies, J. W., and Wilson, T. M. A. 1985. Expression of high molecular weight polypeptides by carnation mottle virus RNA. J. Gen. Virol. 66:2597-2604.
16. Hayes, R. J., Brunt, A. A., and Buck, K. W. 1988. Gene mapping and expression of tomato bushy stunt virus. J. Gen. Virol. 69:3047-3057.
17. Hellmann, G. M., Thornbury, D. W., Hiebert, E., Shaw, J. G., Pirone, T. P., and Rhoads, R. E. 1983. Cell-free translation of tobacco vein mottling virus RNA II. Immunoprecipitation of products by antisera to cylindrical inclusions, nuclear inclusion, and helper component proteins. Virology 124:434-444.
18. Hellmann, G. M., Shaw, J. G., and Rhoads, R. E. 1988. In vitro analysis of tobacco vein mottling virus NI_a cistron: Evidence for a virus-encoded protease. Virology 163:554-562.
19. Hiebert, E. 1981. Immunological analyses of the in vitro translation products of plant viral RNAs. Abstr. P2/06. Proceedings of Fifth International Congress of Virology, Strasbourg, France.
20. Hiebert, E., and Charudattan, R. 1984. Characterization of Araujia mosaic virus by in vitro translation analysis. Phytopathology 74:641-646.
21. Hiebert, E., and Purcifull, D. E. 1981. Mapping of the two coat protein genes on the middle RNA component of squash mosaic virus (Comovirus group). Virology 113:630-636.
22. Hiebert, E., Purcifull, D. E., and Christie, R. G. 1984. Purification and immunological analyses of plant viral inclusion bodies. Pages 225-280 in: Methods in Virology, VIII. K. Maramorosch and H. Koprowski, eds. Academic Press, Inc., Orlando, FL. 396 pp.
23. Hiebert, E., Thornbury, D. W., and Pirone, T. P. 1984. Immunoprecipitation analysis of potyviral in vitro translation products using antisera to helper component of tobacco vein mottling virus and potato virus Y. Virology 135:1-9.
24. Horikoshi, M., Mise, K., Furusawa, I., and Shishiyama, J. 1988. Immunological analysis of brome mosaic virus replicase. J. Gen. Virol. 69:3081-3087.
25. Hsu, Y., and Brakke, M. K. 1985. Cell-free translation of soilborne wheat mosaic virus RNA. Virology 143:272-279.
26. Hughes, G., Davies, J. W., and Wood, K. R. 1986. In vitro translation of the bipartite genomic RNA of pea early browning virus. J. Gen. Virol. 67:2125-2133.
27. Jaegle, M., Wellink, J., and Goldbach, R. 1987. The genome-linked protein of cowpea mosaic virus is bound to the 5′ terminus of virus RNA by a phosphodiester linkage to serine. J. Gen. Virol. 68:627-632.
28. Jobling, S. A., and Wood, K. R. 1985. Translation of tobacco ringspot virus RNA in reticulocyte lysate: Proteolytic processing of the primary translation products. J. Gen. Virol. 66:1589-2596.
29. Kessler, S. W. 1975. Rapid isolation of antigens from cells with a Staphylococcal protein A-antibody absorbent: Parameters of the interaction of antibody-antigen complexes with protein A. J. Immunol. 15:1617-1624.
30. Oliver, M. J., Deom, C. M., De, B. K., and Beachy, R. N. 1986. In vitro transcription and translation of cloned cDNA encoding the 80 KDa protein gene of TMV. Virology 155:277-283.

31. Ooshika, I., Watanabe, Y., Meshi, T., Okada, Y., Igano, K., Inouye, K., and Yoshida, N. 1984. Identification of the 30 KDa protein of TMV by immunoprecipitation with antibodies directed against a synthetic peptide. Virology 132:71-78.
32. Osman, T. A., and Buck, K. W. 1987. Replication of red clover necrotic mosaic virus RNA in cowpea protoplasts: RNA 1 replicates independently of RNA 2. J. Gen. Virol. 68:289-296.
33. Quiot-Douine, L., Purcifull, D. E., Hiebert, E., and deMejia, M. V. G. 1986. Serological relationships and in vitro translation of an antigenically distinct strain of papaya ringspot virus. Phytopathology 76:346-351.
34. Shahabuddin, M., Shaw, J. G., and Rhoads, R. E. 1988. Mapping of the tobacco vein mottling virus VPg cistron. Virology 163:635-637.
35. Shields, S. A., and Wilson, T. M. A. 1987. Cell-free translation of turnip mosaic virus RNA. J. Gen. Virol. 68:69-180.
36. Siaw, M. F. E., Shahabuddin, M., Ballard, S., Shaw, J. G., and Rhoads, R. E. 1985. Identification of a protein covalently linked to the 5' terminus of tobacco vein mottling virus RNA. Virology 142:134-143.
37. van Beck, N. A. M., Derksen, A. C. G., and Dijkstra, J. 1986. Synthesis of sonchus yellow net virus proteins in infected cowpea protoplasts. J. Gen. Virol. 67:1702-1709.
38. Xiong, Z., Hiebert, E., and Purcifull, D. E. 1988. Characterization of the peanut mottle virus genome by in vitro translation. Phytopathology 78:1128-1134.
39. Ziegler, V., Richards, K., Guilley, H., Jonard, G., and Putz, C. 1985. Cell-free translation of beet necrotic yellow vein virus: Read through of the coat protein cistron. J. Gen. Virol. 66:2079-2087.

Section V.D.1.

IMMUNOLOGICAL TECHNIQUES FOR MOLECULAR PLANT PATHOLOGY

I. Screening of DNA Libraries

C. S. Orser and C. Beck
University of Idaho
Moscow, ID 83843

Principle

Researchers working with phytopathogenic bacteria at the molecular level are usually faced with one of two situations: either they are in possession of a cloned piece of DNA that complements a mutant deficient in producing a particular protein product, or they have a supply of antibody to a protein and they wish to isolate the piece of DNA that encodes the gene product. For direct screening of a DNA library with an antibody probe, an in vitro protein synthesis strategy can be employed (16) where individual members of the library are used to make mRNA molecules that are added to an in vitro translation system. The newly synthesized protein can be identified by Western blotting with an antibody probe or by 2-dimensional protein gel electrophoresis. However, a more direct approach to gene isolation is immunoscreening of DNA libraries constructed in expression vectors. Usually optimum expression is accomplished by combining the strength of an expression vector with an expression strain, constructed to allow the functioning of foreign DNA coding regions in *Escherichia coli*. Cloning of a particular piece of DNA into an expression vector can result in a chimeric gene fusion. The gene fusion consists of the foreign DNA with a strong bacterial or phage promotor-ribosome binding site region from, for example, the lactose or tryptophan operons or one from phage lambda. The chimeric gene fusions produce fusion proteins that contain the amino terminal end of the reporter gene product fused to the foreign gene product. Young and Davis (13,14,15) devised one such method for the isolation of eukaryotic and procaryotic genes by screening *E. coli* expression libraries with antibody probes using the bacteriophage expression vector lambda gt11.

Materials

1. Recombinant DNA library constructed in an appropriate expression vector (such as lambda gt11 or pKK223-3, available from Pharmacia).
2. Appropriate bacterial expression strain (such as *E. coli* Y1090 for lambda gt11 or JM105 for pKK223-3)
3. Antibody probe obtained by standard procedures.

NOTE: Polyclonal antibodies have the advantage that they can recognize multiple epitopes on any given protein. This can be important if only a portion of the polypeptide is expressed by *E. coli*. Alternatively, multiple monoclonal antibodies identifying different epitopes of a protein can be used.

4. Nitrocellulose filters
5. Reagents: IPTG (isopropyl-β-D-thiogalactopyranoside)
   DNase I
   Lysozyme
   NP-40 (Nonidet P-40)
   bovine serum albumin (BSA)
6. Buffers: TBS buffer (150 mM NaCl, 50 mM Tris-HCL, pH 8.1)
   TS buffer (150 mM NaCl, 50 mM Tris-HCl, pH 7.5)
7. Diaminobenzidine-HCl (DAB) (Aldrich) solution: 25 μg DAB/100 ml TS buffer, plus 25 μl of 30% H_2O_2, prepared and mixed immediately before use (5)
8. Biotin-labeled secondary antibody/avidin-peroxidase complex (Vector)
9. ^{125}I-labeled Protein A probe (ICN or Amersham)
10. Standard microbiological media components, chemicals, reagents, sterile glassware, pipets, and Petri plates

Procedures

1. Construct a DNA library of interest in an appropriate expression vector. Typically, an expression vector will contain a multiple cloning site at the 5′ end of a reporter gene, such as the lacZ gene of E. coli. This allows for the construction of β-galactosidase fusion proteins providing that the introduced DNA fragments are in the proper orientation and reading frame. The following procedure is outlined for processing of library clones constructed in a plasmid expression vector.
2. Remove anti-E. coli antibodies from antiserum by screening E. coli protein lysates. This is accomplished by transferring E. coli colonies (without foreign DNA inserts) onto nitrocellulose filters, treating as described below and incubating with antibodies to absorb the anti-E. coli antibodies from the serum (11). This procedure may need to be repeated to effectively remove all of the anti-E. coli antibodies from the serum.
3. Plate out the DNA library by diluting the cells to yield 100 to 200 colonies per antibiotic selection agar plate. However, if the immunoenzymatic method of Hsu et al (5) is to be followed for detection of antigen, up to 1000 colonies per 100 cm^2 plate can be plated. Incubate plates until colonies just become visible.
4. In a typical colony immunoassay procedure, recombinant colonies are transferred to nitrocellulose (proteins bind to nitrocellulose). Alternatively, the bacteria can be directly plated onto nitrocellulose filters overlayed on selective agar plates and grown up overnight. The filter is marked, removed, and a duplicate laid in its place for regrowth of colonies.
5. Filters are suspended in a chloroform vapor chamber for 15 min to lyse the bacterial colonies. The filter is then placed in a Petri dish or on a piece of clear plastic wrap with 10 ml of TS (50 mM Tris-HCl, pH 7.5, 150 mM NaCl) containing 5 mM $MgCl_2$, 10 μg/ml DNase I, 400 μg/ml lysozyme, and 1% BSA (w/v) and incubated at room temperature for 30 min (5). The bacterial debris remaining on the filter can be scraped off with a gloved finger and the solution replaced with 10 ml of the same solution and incubated an additional 30 min.
6. The filters are rinsed two times in TS Buffer.
7. The filters are now incubated with the diluted primary antibody at room temperature for 1 to 2 hr and washed three times in TS buffer plus 0.1% (v/v) Triton X-100.
8. Depending upon the method of detection for cross-reactive colonies or plaques, follow either procedure I for immunoscreening with ^{125}I-labeled

secondary antibody, followed by autoradiography; or procedure II for immunoenzymatic screening using biotin-labeled secondary antibody, followed by a rapid chromogenic reaction (5).

Colony-detection Procedure I

a. Treat filters with ^{125}I-labeled (>30 mCi/mg) Protein A probe in TBS plus 0.1% BSA for 2 hr, using 1 μCi of label per filter.
b. Wash filters 15 min each with three changes of TBS buffer plus 0.1% BSA plus 0.1% NP-40 (Nonidet P-40). Then wash once in TBS plus 0.1% BSA (11).
c. Dry filters and mount on X-ray film with Dupont Cronex lightening plus intensifying screen. Overnight exposure at -70 C usually produces good autoradiographic signals.

Colony-detection Procedure II

a. Treat filters with avidin-biotin complex (e.g., 1:200 dilution of biotinylated goat anti-rabbit antibody) in TS buffer for 30 min.
b. Wash filters 10 min each with three changes of TS buffer plus 0.1% (v/v) Triton X-100.
c. Add 10 ml of the avidin-peroxidase complexes (2.5 μg avidin/ml of 1:200 diluted biotin peroxidase) (5) to each filter and incubate for 15 min. Wash filters 10 min each in TS buffer plus 0.1% (v/v) Triton X-100.
d. Filters are developed by the addition of 10 ml of DAB solution for 2 to 5 min. The reaction is stopped by the addition of 10% acetic acid. The usually rare colonies making immunologically cross-reactive proteins are easily recognized using the enzyme-activated chromophores.

9. Confirmation of gene identity. The insert from the recombinant clone can be used to obtain a complete gene by southern hybridization analysis. A further confirmation may be obtained by sequencing the retrieved clone and the deduced amino acid sequence compared with the known protein sequence.

Comments

1. This procedure is designed for well-characterized proteins. The limitation of this procedure is that the protein of interest may not be a major cellular protein.
2. The immunoenzymatic method has two obvious advantages over the iodinated immunoassay method:
   a. positive clones are identifiable within minutes on the original filter; that is, there is no need for X-ray film exposure and interpretation.
   b. freedom from handling dangerous and costly radioactive materials.
3. The antibody solutions can be reused from both procedures; however, the iodinated probe is subject to decay.
4. A potential limitation of foreign DNA expression in _E. coli_ cells is protein instability. It is common for bacteria to degrade foreign polypeptides. The instability of foreign proteins can be decreased through the use of expression vectors resulting in gene fusions. Foreign proteins fused to the amino terminus of β-galactosidase prolongs the stability of hybrid proteins (12). Another solution is to use host cells defective in protein degradation (10). Again, this increases the stability of hybrid and foreign

proteins produced from expression vectors in E. coli cells.

5. If an inducible promoter is being used, it is necessary to incubate the filters on an appropriate induction medium for optimization, such as IPTG. However, cloning pUC-derived genomic or cDNA expression libraries into certain E. coli strains, such as DH1 or DH5, will circumvent the induction step because expression of the fusion proteins will be constitutive (4).

Applications

There have been several colony immunoassays reported with adequate sensitivity to distinguish antigenic determinants on fused polypeptides (1,2,3). However, the identification of cloned genes by colony immunoassays has not been common until recently. In the early immunoassays, each polypeptide molecule needed to bind to two different antibodies. Originally, the first antibody was impregnated to a solid support such as chemically activated paper. Its role was to bind the polypeptide at the position of the lysed colony. A second antibody was incorporated to again bind to the hybrid polypeptide to allow autoradiographic detection. Kemp and Cowman (6) devised a direct immunoassay for detecting E. coli colonies that contain polypetides encoded by cloned DNA segments. The use of a single antibody was the biggest advancement of their method. The specificity of their immunoassay was also improved by incorporation of monoclonal antibodies.

Another useful application of expression vectors is in the identification and isolation of unknown native proteins encoded by cloned DNAs. By flagging an uncharacterized polypeptide of interest with a portion of a well characterized polypeptide (e.g., β-galactosidase), the location of the uncharacterized polypeptide can be determined. Isolation and purification of the hybrid protein from recombinant host cells will allow antibody production. The antibody specificity will include specificity to the foreign moeity of the fusion peptide as well as to the β-galactoside moeity. The β-galactoside antibodies can be cross-absorbed and the remaining specific antibodies can be used as a probe to both locate and isolate the native protein of interest. If the DNA sequence information is available, synthetic polypeptide antigens can be generated for the same purpose (8). Of course, the advantage of the above procedure, described by Young and Davis (13) and Snyder et al (11), is that it obviates the need for a DNA sequence to produce synthetic polypeptides. A useful troubleshooting guide on DNA library immunoassay procedure is presented by Mierendorf et al (9).

Literature Cited

1. Broome, S., and Gilbert, W. 1978. Immunological screening method to detect specific translation products. Proc. Natl. Acad. Sci. USA 75:2746-2749.
2. Clarke, L., Hitzeman, R., and Carbon, J. 1979. Selection of specific clones from colony banks by screening with radioactive antibody. Pages 436-442 in: Methods in Enzymology, Vol. 68. R. Wu, ed. Academic Press, New York, N. Y. 555 pp.
3. Erlich, H., Cohen, S. N., and McDevitt, H. O. 1978. A sensitive radioimmunoassay for detecting products translated from cloned DNA fragments. Cell 13:681-689.
4. Helfman, D. M., and Hughes, S. H. 1987. Use of antibodies to screen cDNA expression libraries prepared in plasmid vectors. Pages 451-457 in: Methods in Enzymology, Vol. 152. S. L. Berger and A. R. Kimmel, eds. Academic Press, New York, N. Y. 812 pp.

5. Hsu, P. L., Hsu, S. M., and Appella, E. 1985. Immunological screening of a cDNA library using avidin-biotin-peroxidase method. Gene Anal. Techn. 2:30-36.
6. Kemp, D. J., and Cowman, A. F. 1981. Direct immunoassay for detecting Escherichia coli colonies that contain polypeptides encoded by cloned DNA segments. Proc. Natl. Acad. Sci. USA 78:4520-4524.
7. Kupper, H., Keller, W., Kurtz, C., Forss, S., Schaller, H., Franze, R., Strommaier, K., Marquardt, O., Zaslavsky, V., and Hofschneider, P. H. 1981. Cloning of cDNA of major antigen of foot and mouth disease virus and expression in E. coli. Nature 289:555-557.
8. Lerner, R. A. 1982. Tapping the immunological repertoire to produce antibodies of predetermined specificity. Nature 299:592-596.
9. Mierendorf, R. C., Percy, C., and Young, R. A. 1987. Gene isolation by screening lambda gt11 libraries with antibodies. Pages 458-468 in: Methods in Enzymology, Vol. 152. S. L. Berger and A. R. Kimmel, eds. Academic Press, New York, N.Y. 812 pp.
10. Mount, D. W. 1980. The genetics of protein degradation in bacteria. Annu. Rev. Genet. 14:279-319.
11. Snyder, M., Elledge, S., Sweetser, D., Young, R. A., and Davis, R. W. 1987. Lambda gt11: Gene isolation with antibody probes and other applications. Pages 107-128 in: Methods in Enzymology, Vol. 154. S. P. Colowick and N. O. Kaplan, eds., Academic Press, New York, N.Y. 576 pp.
12. Stanley, K. 1983. Solubilization and immune-detection of β-galactosidase hybrid proteins carrying foreign antigenetic determinants. Nucleic Acid Res. 11:4077-4081.
13. Young, R. A. and Davis, R. W. 1983. Efficient isolation of genes by using antibody probes. Proc. Natl. Acad. Sci. USA 80:1194-1198.
14. Young, R. A. and Davis, R. W. 1984. Yeats RNA polymerase II genes: Isolation with antibody probes. Science 222:778-780.
15. Young, R. A. and Davis, R. W. 1985. Immunoscreening lambda gt11 recombinant DNA expression libraries. Genet. Eng. 7:29-42.
16. Zubay, G. 1973. In vitro synthesis of protein in microbial systems. Annu. Rev. Genet. 7:267-288.

Section V.D.2.

IMMUNOLOGICAL TECHNIQUES FOR MOLECULAR PLANT PATHOLOGY

II. Polysome Immunopurification

C. S. Orser & C. Beck
University of Idaho
Moscow, ID 83843

Principle

Ribosomes operate 'bumper-to-bumper' as they translate a single mRNA. Varying lengths of nascent protein chains protrude from each ribosome, increasing in length as the ribosome progresses down the RNA strand. The entire assembly of mRNA, ribosomes, and nascent proteins is known as a polyribosome, or polysome. Most polysomes, or at least those whose mRNA is long enough to accommodate several ribosomes, can be isolated by ultracentrifugation of cell lysates. Individual polysome populations synthesizing identical proteins can be isolated by precipitation or by affinity purification with antibodies which recognize a nascent protein while it is still attached to the ribosomes. Messenger RNA from purified polysomes can be used in in vitro translation systems or to make cDNA to screen libraries of phytopathogen genomes for proteins of interest (e.g., wilt-causing glycopeptides of a bacterium, or antibacterial substances of plant tissues).

Isolation of intact polysomes for immunopurification is difficult and requires extreme precautions to prevent RNase degradation of mRNA. No single method of isolation works for every system. Anyone attempting polysome immunopurification should first consider the alternatives and determine which methods best apply to the specific needs. Many procedures are available for isolating plant polysomes from infected or non-infected plants (3,5,16,19) but not all yield polysomes with intact nascent polypeptides and would not be suitable for immunopurification.

Materials for Antibody Preparation

RNase removal. All glassware, metal spatulas, pipets, etc. must be baked at 285 C for 2-4 hr. Plasticware is treated with 0.5% diethylpyrocarbonate in 70% ethanol, rinsed in ethanol, then with suitable water. Water for processing tissues or preparation of solutions should be glass distilled or treated for 20 min with 0.5% diethylpyrocarbonate and then autoclaved. Boil dialysis tubing in 1 mM KOH, then boil in 1 mM EDTA/0.5% diethylpyrocarbonate, and rinse in suitable water (10). Gloves must be worn to prevent contamination of materials. For more information consult Maniatis et al (12), or Berger & Kimmel (1).

1. Polyclonal or monoclonal antibody against a specific protein of interest
2. 3 g Protein A-Sepharose CL-4B (Pharmacia)
3. 0.1 M phosphate buffer, pH 7.5
4. Coarse, sintered glass filter

5. Column (10 ml capacity)
6. 1 M acetic acid, pH 3
7. 0.1 M glycine-HCl buffer, pH 3
8. 1.0 M Tris base
9. Polysome buffer: 10 mM Tris-HCl, pH 7.5, 1.5 mM $MgCl_2$, 50 mM KCl
10. Dialysis tubing (optional)
11. Ultrafiltration apparatus (Amicon concentrator with UM10 membrane, or Millipore submersible CX10 negative pressure concentrator, or Amicon Centriprep 30 concentrator)

Procedures for Antibody Preparation

1. Prepare the Protein A-sepharose as instructed by Pharmacia
   a. Soak the required amount of Protein A-sepharose in 0.1 M phosphate buffer, pH 7.5, for 15 min.
   b. Wash the gel on a sintered glass filter with about 200-500 ml of this phosphate buffer.
   c. Form a slurry of the phosphate buffer and gel and pack the 10 ml column.
   d. Wash the column with 10 volumes of 1 M acetic acid to remove ribonucleases.
   e. Wash the column with 10 volumes of the phosphate buffer.
2. Mix the antibody (2-5 ml) 1:1 (v/v) with 0.1 M phosphate buffer, pH 7.5.
3. Slowly apply the antibody to the Protein A-sepharose column (2 passes may be necessary to prevent loss of antibody).
4. Wash the column with 100 ml of phosphate buffer.
5. Elute the immunoglobulin G (IgG) fraction from the column with 0.1 M glycine, pH 3, or with 1 M acetic acid, pH 3.
6. Read A_{280} to determine concentrations of fractions.
7. Pool antibody fractions and neutralize with 0.1 volume of 1.0 M Tris base (11).
8. To prepare the antibody for polysome immunopurification, reduce the concentration of Tris by dialyzing against polysome buffer or by concentrating to 1-2 ml and adding 10 volumes of polysome buffer.
9. Concentrate to 1.5-3 ml by ultrafiltration.
10. Aliquot and freeze at -70 C.

Comments

Polyclonal or monoclonal antibodies produced against a viral or bacterial phytopathogen protein must recognize the N-terminus region of the nascent protein as it comes off the ribosome for use in this procedure. The protein will probably not show the folding or post-translational modifications it would have in its active structure so it may be advisable to use a denatured form of the protein in immunization (9). Polyclonal antibodies are advantageous in this respect, because they contain antibodies specific for native and degraded forms of the protein. Polyclonal antibodies produced against pathogen proteins purified from infected plants may contain some antibodies specific for contaminating plant components. The desired antibodies can be made more specific by cross-absorbing the contaminating antibodies from serum with non-infected plant materials. Protein A-Sepharose purification of the antibody is needed to remove contaminating ribonucleases.

Polysome Isolation

This procedure, an adaptation of Davies et al. (4), employs medium or high ionic strength and high pH to ensure the isolation of intact polysomes.

Materials for Polysome Isolation

1. Grinding buffer A: 0.2 M Tris-HCl, pH 8.5, 0.2 M sucrose, 30 mM $MgCl_2$, 60 mM KCl.
2. Buffer B: 20 mM Tris-HCl, pH 8.5, 10 mM $MgCl_2$ 20 mM KCl
3. 1.5 M Sucrose in buffer B.
4. Polysome buffer C: 10 mM Tris-HCl, pH 7.5, 1.5 mM $MgCl_2$, 50 mM KCl (3 mg/ml heparin).
5. Buffer D: 20 mM Tris-HCl, pH 8.5, 10 mM $MgCl_2$, 20 mM KCl (optional)
6. 125-500 mg/ml Sucrose for gradient in buffer D (optional)
   a. 125 mg/ml Sucrose in buffer D
   b. 250 mg/ml Sucrose in buffer D
   c. 375 mg/ml Sucrose in buffer D
   d. 500 mg/ml Sucrose in buffer D

Procedures for Polysome Isolation

NOTE: All procedures performed at 0 to 4 C unless otherwise indicated.

1. Freeze 3-15 g of infected plant tissue on dry ice or with liquid nitrogen.
2. Grind the tissue using a mortar and pestle or a Waring blender (3 5-sec bursts) in 5-10 volumes of buffer A.
3. Filter the slurry through Miracloth.
4. Centrifuge the filtrate for 20 min at 29,000 g.
5. Gently layer the supernatant liquid over 4-10 ml pads of 1.5 M sucrose in buffer B and centrifuge for 90 min at 95,000 g.
6. Gently rinse and resuspend the pellet in 1 volume of polysome buffer C (3 mg/ml heparin should be added to prevent ribonuclease degradation of polysomes at low pH).
7. Measure A_{260} to determine the yield.
8. Subdivide the resuspended pellet and use in immunopurification or check the polysome population on a sucrose gradient (optional).
9. (Optional). Layer 0.5 ml aliquot on sucrose gradients in buffer C. The gradient is prepared by adding 2 ml of sucrose (500 mg/ml) to cellulose nitrate tubes followed by a 4 ml layer of sucrose (375 mg/ml), a 4 ml layer of sucrose (250 mg/ml), and 2 ml layer of sucrose (125 mg/ml). The gradient is equilibrated at 2 C for 24-48 hr prior to use. Centrifuge for 75 min at 122,000 g. Monitor and separate the fractions at A_{254} or A_{260}.

Comments

Triton-X 100 and sodium deoxycholate are not needed in this procedure, but 0.5-3.0% of each may be used after step #2 or #3 when using isolation buffers with lower pH and lower ionic strength, or in systems where membrane bound polysomes are difficult to obtain, or when chloroplast polysomes are desired. Aliquots can be frozen at step #8 with the addition of ribonuclease inhibitors such as placental ribonuclease inhibitor (2) or heparin (15). The polysome yield differs greatly from one tissue or organism to another. Palmiter (1974) determined the A_{260} units/g wet weight of various eukaryotic tissues (slime mold, silk moth, rabbit, & chicken) and obtained values between 18.2-191.0 Values in plant systems tend to be low. Some isolation procedures employ Ethylene-diaminetetraacetic acid (EDTA) or Ethylene-bis(oxyethylenenitrilo)tetraacetic acid (EGTA) chelate Ca^{2+} which is thought to activate heat-labile nucleases (8). Magnesium is inhibitory to nucleases (6) and is beneficial to polysome isolation and immunopurification.

Materials for Immunopurification of Polysomes

NOTE: Glassware used for immunopurification of polysomes should be siliconized to reduce background precipitation (18).

1. Polysome buffer: 10 mM Tris-HCl, pH 7.5, 1.5 mM $MgCl_2$, 50 mM KCl (add 3 mg/ml heparin to inhibit RNase)
2. Glycine quenched cyanogen bromide-activated Sepharose column (Pharmacia) (equilibrated with heparin-containing polysome buffer)
3. Protein A-Sepharose column equilibrated with heparin-containing polysome buffer
4. Dissociation buffer: 25 mM Tris, pH 7.5, 20 mM EDTA, 0.2 mg/ml heparin

Procedures for Immunopurification of Polysomes

This method is a modification of a procedure by Lynch (11).

1. Prepare a 10 ml column of cyanogen bromide-activated Sepharose as instructed by Pharmacia.
2. Add polysomes to antibody at a ratio of 160 A_{260} units polysomes per mg of IgG (10) or 5-15 mol IgG/mol of ribosome (1 A_{260} unit = 22.5 pM of ribosome) (2) which is equivalent to 1.5 mg antibody per 35-100 A_{260} units of ribosomes.
2. Incubate at 4 C for 1-4 hr.
3. Dilute with an equal volume in polysome buffer.
4. Pass over a column of glycine quenched cyanogen bromide-activated Sepharose (Pharmacia) to capture materials which would non-specifically bind in later steps.
5. Collect the flow-through and the first several milliliters of rinse.
6. Load onto protein-A column (reapply the flow-through). The second flow-through can be frozen for future isolations of other polysome populations.
7. Wash with 20 volumes of polysome buffer.
8. Dissociate polysomes with 2 volumes of 25 mM Tris, pH 7.5, 20 mM EDTA, 0.2 mg/ml heparin. The nascent proteins and antibodies remain bound to the column.
9. The mRNAs are then purified by ethanol precipitation or oligo (dT) Sepharose column separation for the isolation of polyadenylated mRNA.

Comments

Heparin must be removed from the isolated mRNA preparations prior to use in in vitro translation because it is inhibitory to protein synthesis (17). Most polysome immunopurification has been done with animal tissues. Polysomes synthesizing viral or plant proteins are more easily obtained than bacterial polysomes due to the high nuclease content of bacteria. Milhausen et al. (13) successfully used eukaryotic polysome immunopurification methods to isolate bacterial flagellar mRNA.

Applications

There are numerous number of applications for polysome immunopurification appropriate to phytopathology. For example, messenger RNA produced by both susceptible and resistant plants in response to viral, bacterial, or fungal pathogens is different from that of normal, unchallenged plants (7). Crude antibodies produced against the pathogen-induced plant proteins may be purified by cross-absorbing non-specific antibodies to normal plant materials. The antibodies could then be used to immunopurify polysomes, and the mRNA extracted from them may be used to study proteins induced by pathogens in resistant and susceptible plants. Once the mRNAs are isolated, an appropriate translation system may be used to produce the protein encoded by the mRNA. Purified protein produced by the translation system, and the mRNA from which it was produced, can be used in the study of the pathogen's or plant's regulation of that protein, in identification of methods to control the spread of the pathogen, and in the production of monoclonal antibodies for use in rapid identification of disease resistant and susceptible plants. Alternatively, reverse transcriptase may be used to make a cDNA copy from the mRNA isolated from the polysomes. The cDNA may serve as a hybridization probe in screening cDNA libraries of the plant's or pathogen's genome to locate mRNA sequences responsible for the protein's production. Polysome immunopurification can be used successfully in many instances, but it is best to consider all of the potential problems before attempting it.

Literature Cited

1. Berger, S. L. 1987. Section VII, Preparation and characterization of RNA. Pages 215-219 in: Methods in Enzymology, Vol. 152. S. L. Berger, and A. R. Kimmel, eds. Academic Press, Inc. 812 pp.
2. Boyer, S. H., Smith, K.D., Noyes, A. N., and Young, K. E. 1983. Adjuvants to immunological methods for mRNA purification. J. Biol. Chem. 258:2068-2071.
3. Breen, M. D., Whitehead, E. I., and Kenefick, D. G. 1971. Requirements for extraction of polyribosomes from barley tissue. Plant Physiol. 49:733-739.
4. Davies, E., Larkins, B. A., and Knight, R. H. 1972. Polysomes from peas. Plant Physiol. 50:581-584.
5. Dyer, T. A., and Scott, K. J. 1972. Decrease in chloroplast content of barley leaves infected with powdery mildew. Nature 236:237-238.
6. Gough, N. M., and Adams, J. M. 1978. Immunoprecipitation of specific polysomes using <u>Staphyloccocus aureus</u>: Purification of the immunoglobulin <u>K</u> chain messenger RNA from the mouse myeloma MPC11. Biochemistry 17:5560-5566.
7. Greenland, A., and Shaw, M. 1986. Polyribosome fractions and protein synthesis in stem rust infected wheat. Can. J. Bot. 64:1916-1927.
8. Jackson, A. O., and Larkins, B. A. 1976. Influence of ionic strength, pH, and chelation of divalent metals on isolation of polyribosomes from tobacco leaves. Plant Physiol. 57:5-10.
9. Korman, A. J., Knudsen, P. J., Kaufman, J. F., and Strominger, J. L. 1982. cDNA clones for the heavy chain of HLA-DR antigens obtained after immunopurification of polysomes by monoclonal antibody. Proc. Nat. Acad. Sci. USA 79:1844-1848.

10. Kraus, J. P. 1985. Purification of Low-abundance messenger RNAs by polysome isolation with polyclonal antibodies. Pages 407-417 in: Hybridoma Technology in the Biosciences and Medicine. T. A. Springer, ed. New York, Plenum Press. 602 pp.
11. Lynch, D. C. 1987. Use of antibodies to obtain specific polysomes. Pages 248-253 in: Methods in Enzymology, Vol. 152. S. L. Berger, and A. R. Kimmel, eds. Academic Press, Inc. 812 pp.
12. Maniatis, T., Fritsch, E. F., and Sambrook, J. 1982. Molecular Cloning: a Laboratory Manual. Cold Spring Harbor Laboratory, Cold Spring Harbor, New York. 545 pp.
13. Milhausen, M., Gill, P. R., Parker, G., and Agabian, N. 1982. Cloning of developmentally regulated flagellin genes from Caulobacter crescentus via immunoprecipitation of polyribosomes. Proc. Natl. Acad. Sci. USA 79:6847-6851.
14. Palmiter, R. D. 1974. Magnesium precipitation of ribonucleoprotein complexes: Expedient techniques for the isolation of undegraded polysomes of messenger ribonucleic acid. Biochemistry 13:3606-3615.
15. Payvar, F., and Schimke, R. T. 1979. Improvements in immunoprecipitation of specific messenger RNA: Isolation of highly purified conalbumin mRNA in high yield. Eur. J. Biochem. 101:271-282.
16. Pure, G. A., Chakravorty, A. K., and Scott, K. J. 1980. Changes in wheat leaf polysomal messenger RNA populations during the early stages of rust infection. Plant Physiol. 66:520-524.
17. Schimke, R. T., Palacios, R., Sullivan, D., Kiely, M. L., Gonzales, C., and Taylor, J. M. 1974. Immunoadsorption of ovalbumin synthesizing polysomes and partial purification of ovalbumin messenger RNA. Pages 631-648 in: Methods in Enzymology, Vol 20. K. Moldave, and L. Grossman, eds. Academic Press, New York and London. 586 pp.
18. Shapiro, S. Z., and Young, A. R. 1981. An immunochemical method for mRNA purification. J. Biol. Chem. 256:1495-1498.
19. Tomizawa, K., Komeda, Y., Sato, N., Nagarani, A., Iino, T., and Furuya, M. 1986. Isolation of cDNA for pea phytochrome using an expression vector. Plant Cell Physiol. 27:1101-1108.

GLOSSARY OF SEROLOGICAL TERMS

From: Dictionary of Immunology
W. J. Herbert & P. C. Wilkinson
Blackwell Scientific Publications
1977, 194 pp., and contributions by the editors, authors, and reviewers.

Ab 1 - first antibody produced by an animal's B lymphocytes in response to the stimulus provided by an antigen.

Ab 2 alpha - an antibody which recognizes an idiotope on Ab 1 which is extraneous to the paratope of Ab 1. Ab 1/Ab 2 alpha interactions are not usually antigen inhibitable.

Ab 2 beta - an antibody (also known as a homobody, an internal image antibody or an antigen-mimicking antibody) which interacts with the paratope of Ab 1. Ab 1/Ab 2 beta interactions are antigen inhibitable.

Ab 2 epsilon - an antibody (also known as an epibody) which interacts with an idiotope on Ab 1 and the antigen against which Ab 1 was derived.

Ab 2 gamma - an antibody which interacts with an idiotope on Ab 1 which is closely associated with the paratope of Ab 1. Ab 1/Ab 2 gamma interactions are antigen inhibitable.

Absorption - Use of antibody or antigen to remove corresponding antigen or antibody from a mixture.

Active immunity - Protection due to development of an immune response in an individual following stimulation with antigen, e.g., in a vaccine or during infection.

Active immunization - Stimulation of an individual's immune responses so as to confer protection against disease. Effected by exposure to protective antigens either during the course of infection (which may be subclinical) or by vaccination. The protection effected takes a week or more to develop, but is then long lasting and rapidly revived by a booster dose.

Adjuvant - Substance injected with antigens (usually mixed with them but sometimes given prior to or following the antigen) which non-specifically enhances or modifies the immune response to that antigen. Thus, antibody production or the reactions of well-mediated immunity are more vigorous than would be the case were the antigen injected without adjuvant. In addition, the response may be modified qualitatively, e.g., antibody of different immunoglobulin types may be stimulated.

Adsorption - Non-specific binding of molecules to solid surfaces, typically in very thin or monomolecular layers, e.g., binding of antibodies or antigens to glass, plastic, or nitrocellulose surfaces.

Affinity - The strength of the antigen-antibody interaction. The strength of bonds formed by this reversible interaction, like other reversible biomolecular equilibrium reactions, determines the rate of association between antibody (i.e., monospecific, such as a monoclonal antibody) and antigen (e.g., a hapten), versus the rate of dissociation. High-affinity antibodies have a higher rate of association with antigen and bind more antigen (e.g., 10^8 to 10^{12} mol^-) than low-affinity antibodies, which have a lower rate of associaton with antigen and bind less antigen (e.g., 10^5 to 10^7 mol^-).

Agglutination - Visible clumping of particulate antigens, e.g., red blood cells, bacteria, etc. when reacted with specific antibody, or visible clumping of inert particles to which antibodies are adsorbed, e.g., latex microspheres, when reacted with specific antigen. Clumping is caused by antigen-antibody bridging between adjacent cells or particles.

Allotype - Genetically inherited antigenic sites present on a class of antibodies in some individuals and absent from other individuals of the same species.

Anamnesis - Immune memory or secondary immune response.

Anaphylaxis - An acute immediate hypersensitivity following administration of antigen to a primed subject. The reaction is caused by release of histamine or other vasoactive substances when antigen combines with antibody on the surfaces of cells. Anaphylaxis may be generalized or localized to the site of injection.

Antibody - Immunoglobulin molecule produced by an animal's B lymphocytes in response to an antigenic stimulus and capable of binding to an antigen.

Antigen - A substance, usually protein or polysaccharide, capable of eliciting production of specific antibodies and capable of specifically binding to antigen binding sites of antibody molecules (see Immunogen). All immunogens are antigenic, but not all antigens are immunogenic (e.g., haptens).

Antigen binding capacity - A measurement of the ability of an antibody to bind antigen, based on the effects of dilution of the antibody.

Antigen binding site - = Fab; Paratope. That area of an antibody molecule that binds to antigen.

Antigenic determinant - The small site on an antigen molecule to which antibody is specifically bound, as determined by structural complementarity between antibody (Fab) and antigen molecules. An antigen molecule may comprise more than one determinant. Attached haptens add further determinants (see epitope).

Anti-idiotope antibody - an antibody which interacts with an idiotope of another antibody.

Antigen-mimicking antibody - see Ab 2 beta.

Antigen presenting cells (APC) - those cells involved in the immune response which initially interact with, engulf, break down and present an antigen on the cell's surface to T cells.

Antiserum - Serum from any animal containing antibodies to a specified antigen, e.g., anti-ovalbumin.

Autobody - an antibody with a paratope that reacts with an idiotope on the same antibody as well as with the antigen against which the antibody was generated. The self-binding characteristics of an autobody are antigen inhibitable.

Avidity - The stability of the antigen-antibody complex as governed by the affinity of the antibody for the antigen, the valency of the antibody or antigen, and the spatial arrangement of the epitope and respective paratope. An IgM antibody (i.e., ten paratopes potentially interacting with ten epitopes) is capable of forming a more stable antigen-antibody complex than an IgG antibody (two paratopes, for example, with the same epitope affinities as IgM).

B cells - Antibody producing cells found mainly in the blood, lymph nodes and spleen of the body.

Booster dose - a dose of immunogen given several days, weeks, months or years after the initial immunizing dose to stimulate continued production of antibody.

C region - = Fc. carboxyterminal portion of immunoglobulin H or L chains that is identical for a given class or subclass.

Cellular response - That part of the immune response which involves the interaction between cells and an antigen.

CH_{50} unit - The amount of serum (dilution) required to inhibit lysis of antibody-coated red blood cells by 50%, in a standard assay for hemolytic complement.

Coagulation system - System of 12 proteins in serum that results in formation of fibrin and blood clotting.

Combining site - The site on the antibody molecule that binds specifically with its corresponding antigenic determinant. Present on the Fab portion of IgG. The heavy and light chains are both involved and the combining site is formed from contiguous areas of the variable regions of both chains, involving five to fifteen amino acids.

Competitive inhibition - Process in which one population of molecules inhibits reaction of a second molecule population by competing for common binding sites on molecules of a third population, e.g., antigen A competing with antigen B for the antigen binding site of an antibody.

Complement - A system of at least 13 serum proteins that are activated by enzymatic cleavages and aggregations to produce components with biological activity.

Complementary determining region (CDR) - See hypervariable region conformational epitope - an antigenic site composed of amino acids arranged in a specific conformation.

Complement fixation - A test for antigen-antibody reactions that depends on binding (consumption) of complement by antigen-antibody complexes.

Conjugate - The product of joining two or more dissimilar molecules by covalent bonds. In immunological contexts, one is usually a protein and the other either a hapten or a label such as fluorescein, ferritin, or enzyme.

Constant region - Portions of either the heavy or light chains of immunoglobulins which exhibit conserved amino acid sequences.

Continuous or contiguous epitope - An antigenic site composed of two or more groups of peptides linked to one another.

Coombs' test - In vitro agglutination test for nonagglutinating antibodies to red blood cells, employing anti-Ig. **Direct Coombs' test**: anti-IG added to test red cells; agglutination indicates cells were coated with antibody. **Indirect Coombs' test**: serum added to red cells, cells washed, and anti-Ig added; agglutination indicates serum contains antibody to red blood cells.

Covalent bonding - Sharing of electrons between adjacent atoms in a way that reduces the total electronic energy. Classically, two shared electrons constitute one covalent bond. This is the bonding which holds atoms together in molecules, in contrast to the electrostatic attractions which act between ions, dipoles and induced dipoles. Covalent bonds, unlike these electrostatic forces, show characteristic bond lengths and bond angles which are responsible for molecular shape, and atoms so bonded cannot be separated by physicochemical techniques such as chromatography and electrophoresis.

Cross-reacting antigen - Antigen capable of combining with antibody produced in response to a different antigen. May cross-react due to sharing of determinants by the two antigens or because the antigenic determinants of each, although not identical, are closely enough related stereochemically to combine with antibody against one of them.

Antigen of identical structure in two strains of bacteria, so that antibody produced against one strain will react with the other.

Cross-reactive idiotope (CRI) - An idiotope (also known as a public idiotope or IdX) which is shared by two or more antibodies which may or may not share similar antigenic specificities.

Cryptotope - An antigenic site that is hidden, 'buried', or unavailable to an antibody until the molecule bearing the antigenic site is structurally altered.

Discontinuous or discontiguous epitopes - An antigenic site which arises either when two separate peptide chains are brought together or when a single peptide chain is folded.

Disulfide bonds - Chemical bonds between sulfhydryl-containing amino acids that bind together polypeptide chains.

Domains - Segments of polypeptide chains that are folded together and stabilized by disulfide bonds.

Double-diffusion test - Antigen-antibody precipitation reaction in agar or a similar gel in which antigen and antibody are allowed to diffuse toward one another and react at equivalence in the agar.

EA - Erythocyte coated with antibody; will be lysed if complement is added.

Electrophoresis - Separation of molecules in an electrical field.

ELISA - (enzyme-linked immunosorbent assay). An immunoassay system employing enzyme bound to antibody as the immunologic probe determining the extent of antigen-antibody reaction.

Endotoxin - Classically, a bacteriologist's term for toxins that are only released on the death of the cell, as opposed to the diffusible exotoxins that are produced by living bacteria. Today, however, the term is used almost exclusively as a synonym for the lipopolysaccharides q.v. of Gram negative bacteria.

Epibody - See Ab 2 epsilon

Epitope - An antigenic determinant of defined structure, e.g. an identified oligosaccharide, or a chemical hapten.

Epitype - A family of restricted epitopes; a set of epitopes.

Equilibrium dialysis - A technique used to measure the affinity of antibody-antigen binding, based on diffusion of unbound antigen across a dialysis membrane.

Equivalence - The ratio of antibody to antigen that gives the maximum precipitation in the quantitative precipitation reaction

Exotoxins - Diffusible toxic substances produced by microorganisms.

Fab fragment - Fragment obtained by papain hydrolysis of immunoglobulin molecules. The Fab fragment (ca. M_r 45,000) consists of one light chain linked to the N-terminal half of the contiguous heavy chain. Two Fab fragments are obtained from each 7S antibody molecule. Each contains one antigen-combining site and can combine with antigen as univalent antibody but cannot form precipitates.

$F(ab')_2$ - Fragment obtained by pepsin hydrolysis of an immunoglobulin molecule. The $F(ab')_2$ fragment consists of two Fab fragments joined by disulfide bonds.

Fc - "Crystallizable fragment": the non-antigen-binding fragment of an immunoglobulin molecule produced after digestion with papain comprising two heavy-chain fragments. When first recognized using rabbit IgG, this fragment formed crystals, but the Fc of most species does not crystallize.

Fc' - Fragment obtained by pepsin digestion of an immunoglobulin molecule. The Fc' consists of the non-antigen binding portion of IgG, comprising two heavy chain fragments.

Fd fragment - Part of the heavy chain of immunoglobulin that lies on the N-terminal side of the site of papain hydrolysis.

Flocculation - Precipitation reaction of antibody and antigen in which the precipitate appears as flakes of insoluble protein, usually seen with horse antibody-toxin reactions.

Fluorescein - Yellow dye used to label antibody for fluorescence technique.

Framework region - Sequences of amino acids in variable regions of H or L immunoglobulin chain other than hyper-variable sequences.

Freund's adjuvant - A mixture of mineral oil and lanolin that enhances immune responses when emulsified with antigen for immunization. Freund's complete adjuvant includes killed mycobacteria; Freund's incomplete adjuvant does not contain mycobacteria.

Gamma globulin = γ-globulin -Obsolete term for immunoglobulin G (IgG).

H antigens - The flagellar antigens of motile Gram-negative enterobacteria such as the Salmonella group.

H chain (heavy chain) - One pair of identical polypeptide chains, each $M_r \simeq$ 50,000, of four-chain immunoglobulin molecules. The other pair of chains are the L (light) chains.

Hapten - Substance that can combine with antibody, but cannot initiate an immune response unless it is bound to a carrier before introduction into the body. Most haptens are small molecules (M_r <1,000) and carry only one or two antigenic determinants but some macromolecules, e.g. pneumococcal polysaccharide, are haptenic. Haptenic groups can be conjugated to carriers in vitro and may then be regarded as isolated determinants capable of reacting with antibody but requiring the carrier molecule in order to become immunogenic in vivo.

Hemagglutinin - Antibody or substance that causes agglutination of red blood cells.

Hemolysis - Process of lysis of red blood cells.

Heterologous - Derived from a different species.

Heterophil antigen - An antigen found in different apparently unrelated organisms and/or tissues, usually carbohydrate.

Hidden determinant - cryptotope - Antigenic determinant which is so positioned on a molecule or cell that it is not accessible for recognition by lymphocytes or antibody, and is neither capable of binding specifically to them, nor of stimulating an immune response, unless some stereochemical change causes the hidden determinant to be revealed.

Hinge region - Amino acid segment between the first and second constant regions of the immunoglobulin heavy chain of IgG that permits bending of the molecule.

Homobody - See Ab 2 beta.

Homologous - The specific relationship between an antigen and the antibody induced by its interaction with animal immune systems.

Hybridoma - A cell or cell line formed from the fusion of a lymphocyte with a tumor cell.

Hyperimmunization - A condition resulting from immunization, usually with extensive booster doses, designed to stimulate the production of high affinity antibody, (relatively late in the immune response).

Hypervariable region - Portions of either the heavy or light chains of an immunoglobulin which exhibit extreme amino acid sequence variability and which make up part of the immunoglobulin's (antibody's) antigen combining site.

Idiotope - An antigenically distinct region associated with an immunoglobulin's paratope or hypervariable region.

IdI (private idiotope) - an idiotope exhibited by only those antibodies derived from a common B lymphocyte.

Idiotype - The total antigenic signature associated with an immunoglobulin's paratope or hypervariable region.

Idiotype network - A series of idiotype-anti-idiotype reactions that is postulated to control production of idiotype-bearing antibody molecules or cells.
IdX (public idiotope) - see cross-reactive idiotope.
Ig - Immunoglobulin.
IgA - Secretory immunoglobulin class, dimeric in secretions, monomeric in blood.
IgD - Immunoglobulin class on surface of B cells.
IgE - Immunoglobulin class that fixes to mast cells and is responsible for anaphylactic and allergic sensitivity.
IgG - The major immunoglobulin class in the serum of man. Similar immunoglobulins are found in most species from amphibians upwards, but are not present in the serum of fish. Human IgG has been intensively studied and its structure is known in some detail. Molecular weight of human IgG 150,000, $S_{20,w}$ = 7S, fixes complement, crosses human placenta.
IgM - Pentameric immunoglobulin; first class of antibody produced to most antigens during an immune response.
Immune adherence - Binding of antibody-antigen complexes or antibody-coated particles to primate erythrocytes, rabbit platelets, or white blood cells due to activation of C3 and formation of C3b.
Immunization - Administration of antigen to an animal so as to produce an immune response to that antigen. Administration either of antigen to produce active immunity, or of antibody to produce passive immunity, so as to confer protection against the harmful effects of antigenic substances or organisms.
Immunoblot (Western blot) - Reaction of labeled antibodies with proteins adsorbed to nitrocellulose paper.
Immunoelectrophoresis - A technique involving separation of proteins in a gel, using an electric field, followed by a precipitation reaction in the gel with antibodies to the separated proteins.
Immunogen - A substance that elicits a specific immune response when introduced into the tissues of an animal. This may take the form either of antibody production together with the development of cell-mediated immunity, or of specific immunological tolerance. To stimulate a response, immunogens must normally be foreign to the animal to which they are administered, of a molecular weight greater than 1,000, and of protein or polysaccharide nature. However, the definition of an antigen is arbitrary, since specific responsiveness is a property of the host tissues, not of the injected substance (see Antigen).
Immunogenic - Capable of inducing an immune response
Immunoglobulin classes - Subfamily of immunoglobulins based on large differences in H-chain amino acid sequence: IgA, IgD, IgE, IgG, IgM (isotypes).
Immunoglobulin G (IgG) - Serum globular glycoprotein with least mobility to positive electrode during electrophoresis, constituting a distinct class of antibodies.
Immunoglobulin subclasses - Subpopulations of an Ig class based on more subtle structural or antigenic differences in the H chains than are class differences (e.g., IgG1, IgG2, IgG3, IgG4).
Incomplete antibody - Antibody that binds antigen but does not precipitate or agglutinate the antigen.
Internal image antibody - see Ab 2 beta.
Isotype - The class or subclass of antibodies within a given animal species.
J chain - A polypeptide chain that joins individual four-chain immunoglobulin units to form polymeric immunoglobulins (IgA, IgM).
Kappa light chain - one of two antigenically distinct regions associated with the constant region of an immunoglobulin's light chain.

L chain (light chain) - A polypeptide chain of M_r 22,000 present in all immunoglobulin molecules in two forms, κ or λ. Each four-chain Ig molecule has either two κ or two λ light chains.

Monoclonal antibody - A homogeneous antibody population produced by a clone of antibody-forming cells.

Monomer - A single polypeptide chain. Basic components of Ig molecules are four monomers: two H chains and two L chains.

Neotope - An antigenic site that arises only after two or more molecules interact with one another.

Optimal proportions - In precipitin tests, if antibody and antigen are present in such a ratio that the maximum combination between antibody combining sites and antigenic determinants can occur, they are said to be present in optimal proportions or 'at equivalence'. At this ratio, maximum precipitation occurs.

Oudin tube - Single diffusion in agar. Antigen in solution is placed over agar containing antibody in a test tube. Antigen diffuses into agar and forms a precipitin band at equivalence.

Papain - Proteolytic enzyme from papaw plant used to hydrolyze immunoglobulin to Fab and Fc fragments.

Paratope - The antigen combining site of an immunglobulin (antibody) molecule.

Pepsin - Gastric proteolytic enzyme used to hydrolyze immunoglobulin to $F(ab')_2$ and Fc fragments.

Polyclonal - Derived from many clones.

Precipitation - In immunology, the formation of a visible complex on the addition of soluble antibody to soluble antigen. Such complexes are detected in the test tube as a sediment and in agar gels as a white line appearing where the antigen and antibody interact.

Primary response - First immune response to an immunogen not previously recognized; predominantly IgM antibody.

Private idiotope - see IdI.

Protein A - (Staphylococcal Protein A). Protein found in extracts of Staphylococcus aureus which reacts serologically with a majority of human sera. This reaction is due to binding of Protein A to the Fc fragment of human and some other mammalian IgG.

Prozone - Suboptimal precipitation or agglutination that occurs in antibody or antigen excess, depending upon which component is of variable concentration in the assay.

Public idiotope (IdX) - See cross-reactive idiotope (CRI).

Radioimmunoassay - A sensitive and verstile system of protocols similar to those of enzyme-linked immunosorbent assay (ELISA), with solid phase binding of either antibodies or antigens, but substituting radiolabeled antibodies or antigens for enzyme-IgG conjugation and recording data by means of a radiation (gamma or beta) counter.

Rocket electrophoresis - A technique to measure antigens by electrophoresing antigen into an agar layer containing antibody, resulting in a rocket-like pattern of precipitation in agar.

Sc chain - secretory component found associated with an IgA dimer.

Sequential epitope - An antigenic site composed of 5-7 sequential peptides in their unfolded random coil form.

Serotype - Subtype within a bacterial species, distinguished by serological methods, i.e. by detecting differences in the surface antigens of bacteria on addition of specific antibody.

Serum - Fluid expressed from a blood clot as it contracts after coagulation of the blood. The essential difference between plasma and serum is that the latter does not contain fibrinogen. Serum is more commonly used than is plasma in immunological procedures because there is no danger of a clot forming when other materials are added to it.

Single diffusion - A gel diffusion test in which antigen diffuses into agar containing antiserum, resulting in a ring of antigen-antibody precipitation.

Specificity - A term defining selective reactivity between substances, e.g. of an antigen with its corresponding antibody or primed lymphocyte.

Titer - In serological reactions, a relative measure of the amount of antibody in an antiserum per unit volume of original serum. The antibody is serially diluted and antigen is added. Serum titer is indicated as the reciprocal of the highest serum dilution producing a discernible antigen-antibody reaction.

Tolerance - Active state of unresponsiveness to a given immunogen; immune responses to other immunogens are normal.

Valence (of antibody) - The number of epitopes with which one antibody can combine. The valence of most antibodies is 2. IgM may have a valence up to 10; IgA has different valences depending on degree of polymerization.

SOURCES OF SUPPLY

Mention of a trademark, proprietary product, or vendor does not constitute a guarantee or warranty of the product by APS, authors, or associated institutions or agencies, and does not imply approval to the exclusion of other products or vendors that may be suitable.

Abbott Laboratories, 14th and Sheridan Rd., Chicago, IL 60064
Advanced Magnetics Inc., 61 Momey St., Cambridge, MA 02138
Agdia, Inc., Michawaka, IN 48544
Aldrich Chemical Co, Inc., 940 W. St. Paul Ave., Milwaukee, WI 53233
Amicon Division, W. R. Grace & Co.-Conn., 24 Cherry Hill Drive, Danvers, MA 01923
American Type Culture Collection, Rockville, MD 20852
Amersham, Arlington Heights, IL 60005
Baker, J. P. Baker Chemical Co., 222 Red School Lane, Phillipsburg, NJ 08865
Baltimore Biological Laboratories (BBL), Division of B-D Laboratories, Inc., 2201 Aisquith St., Baltimore, MD 21218
Bellco Glass, Vineland, NJ 08360
Bio-Rad Laboratories, Richmond, CA 98801
Biotechnology Incorporated, Piscataway, NJ 08854
Biotek Co., Winooski, VT 05404
Biotrace RB, Gelman Sciences, 600 South Wagner Rd., Ann Arbor, MI 48106
Boehringer-Mannheim Biochemicals, Indianapolis, IN 46250
BRL, Gibco/BRL Laboratories, 3175 Staley Rd., Grand Island, NY 14072
Brinkman Instruments, Westbury, NY 11590
Calbiochem-Behring, La Jolla, CA 92037
Cel-Line Associates, Inc., P.O. Box 82, Newfield, NJ
Clay-Adams, (Becton Dickinson Primary Care Diagnostics), One-T Becton Dr., Franklin Lakes, NJ 07417
Colab Laboratories, Inc., Chicago Heights, IL 60411
Collaborative Research, Bedford, MA 01730
Corning Science Products, P.O. Box 5000, Corning, NY 14830
Costar, 16152 Covello St., Van Nuys, CA 91405
Costar, 205 Broadway, Cambridge, MA 02139
Difco Laboratories, Inc., P.O. Box 1058A, Detroit, MI 48232
Dynatech Laboratories, Inc., Chantilly, VA 20201
Falcon, (Becton Dickinson Labware), 2 Bridgewater Lane, Lincoln Park, NJ 07035
Firma MEKU, Am Weingarten 144, D 3015, Wennigsen, West Germany
Fisher Chemical Co., Chicago, IL 60143
Flow Laboratories, Inc., McLean, VA 22107
Fluka AG Chemiche Fabrik, Buchs, Switzerland
FMC Corp., Marine Colloids Division, Rockland, ME 04841
Fuji Photo Optical Co., Ltd., 10-T High Point Drive, Wayne, NJ 07470
Gelman Sciences, 600 South Wagner Rd., Ann Arbor, MI 48106
Gibco/BRL Laboratories, 3175 Staley Rd., Grand Island, NY 14072
Hana Biologics, 626 Bancroft Way, Berkeley, CA 94710
Harris Co., North Billerica, MA 01862
Helena Laboratories, P.O. Box 752, Beaumont, TX 77704
Hyclone Laboratories, Inc., Logan, UT 84321
Hoefer Scientific Instruments, San Francisco, CA 94107
ICN Biomedicals, Inc., 3300 Hyland Ave., Costa Mesa, CA 92626
Immobilon, Millipore, Bedford, MA 01730

Institute for Medical Research, Camden, NJ 08103
Instruments for Research and Industry, Inc., Cheltenham, PA 19012
International Potato Center, Peru
Jackson Immunoresearch Laboratory, West Grove, PA, 19390
Janke and Kunkel K. G., Staufen, West Germany
Janssen Life Sciences Products, Piscataway, NJ 08854
Kodak, Rochester, NY 14650
Lab Systems, Inc., 6200 Oakton St., Morton Grove, IL 90053
Lab-Tek, Miles Scientific, 30 W 475 N. Aurora, Naperville, IL 60540
Lerner Laboratories, New Haven, CT 06513
Litton Bionetics (see NetPath, Inc.)
Merck & Co., Inc., P.O. Box 2000, Rahway, NJ 07065
Miles Laboratories, Inc., P.O. Box 2000, Elkhart, IN 46515
Miles Scientific, 30 W 475 N. Aurora, Naperville, IL 60540
Millipore, Bedford, MA 01730
Molecular Devices, Palo Alto, CA 94304
NetPath, Inc., One Malcolm Ave., Peterboro, NJ 07608 (a subsidiary of Corning Glassworks)
NIGMS (National Institute of General Medical Sciences) Cell Repository, Bethesda, MD
Nunc Inter Med, Naperville, IL 60566
Nytran, Schleicher and Schull Inc., Keene, NH 03431
Oxoid Ltd., Wade Rd., Basingstoke, RG24 OPW, Hants, England
Pharmacia, (LKB Biotechnology Inc.), 800 Centennial Ave., Piscataway, NJ 08854
Pierce, P.O. Box 117, Rockford, IL 61105
Polysciences, Inc., 400 Valley Rd., Warrington, PA 18976-2590
Polyscientific Research & Development Corp., 70 Cleveland Ave., Bay Shore, NY 11706
Precision Plastic Ball Co., Chicago, IL 60641
Queue Systems, Inc., P.O. Box 5366, North Branch, NJ 08876
Research Products International Corp., 410 North Business Center Dr., Mount Prospect, IL 60056
Revco, 275 Aiken Rd., Asheville, NC 28804
Ribi Immunochemical, Inc., P.O. Box 1409, Hamilton, MT 59840
Schleicher and Schull Inc., Keene, NH 03431
Scotch Brite, 3M, 3M Center, St. Paul, MN 55144-1000
Scripps Laboratories, Inc., San Diego, CA 92131
Seradyn, Inc., P.O. Box 1210, Indianapolis, IN 46206
Sigma Chemical Co., P.O. Box 14508, St. Louis, MO 63178
Sorvall, Du Pont Co., Biotechnology Systems, BRML-Chandler Mill, Wilmington, DE 19898
Titer Tech, Flow Laboratories, 936 Hyde Park Blvd., Inglewood, CA 90302
Turner Associates, Palo Alto, CA 94304
U. S. Biochemicals Corp., P.O. Box 22400, Cleveland, OH 44122
Vector Laboratories, 1429 Rollins Rd., Burlingame, CA 94010
V-Max, Molecular Devices, 4700 Bohannon Dr., Menlo Park, CA 94025
Zeta Probe/Zetabind, Commercial Intertech Corp., 1777 Logan Ave., Youngstown, OH 44501
Zymed, Inc. South San Francisco, CA 94080

INDEX, GENERAL

INDEX, CHEMICALS